Springer Undergraduate Mathematics Series

T0222672

For further volumes:
http://www.springer.com/series/3423

Hartmut Logemann · Eugene P. Ryan

Ordinary Differential Equations

Analysis, Qualitative Theory and Control

 Springer

Hartmut Logemann
Department of Mathematical Sciences
University of Bath
Bath
UK

Eugene P. Ryan
Department of Mathematical Sciences
University of Bath
Bath
UK

ISSN 1615-2085 ISSN 2197-4144 (electronic)
ISBN 978-1-4471-6397-8 ISBN 978-1-4471-6398-5 (eBook)
DOI 10.1007/978-1-4471-6398-5
Springer London Heidelberg New York Dordrecht

Library of Congress Control Number: 2014933526

Mathematics Subject Classification: 34A12, 34A30, 34A34, 34C25, 34D05, 34D20, 34D23, 34H05, 34H15, 93C15, 93D05, 93D10, 93D15, 93D20

Printed on acid-free paper

Springer is part of Springer Science+Business Media (www.springer.com)

Preface

This text is based on various courses taught, over many years, by the authors at the University of Bath. The intention is a rigorous – and essentially self-contained – treatment of initial-value problems for ordinary differential equations. The material is presented at a technical level accessible by final year undergraduate students of mathematics and appropriate also for students in the early stages of postgraduate study, both in mathematics and mathematically-oriented engineering. Only a basic grounding in linear algebra (e.g. finite-dimensional vector spaces, norms, inner products, linear transformations and matrices, Jordan form) and analysis (e.g. uniform continuity, uniform convergence, compactness in a finite-dimensional setting, elementary differential and integral calculus) is assumed: the typical UK undergraduate attains this level of mathematical maturity by the end of his/her second year of study in mathematics. In an appendix, these basics are assembled to provide the mathematical framework underpinning the book. In the main body of the text, diverse results are presented pertaining to existence and uniqueness of solutions of initial-value problems, continuous dependence on initial data, flows, qualitative behaviour of solutions, limit sets, stability theory, invariance principles, introductory control theory, stabilization by feedback. The latter aspects, namely the coverage of control theoretic concepts, is a distinguishing feature. This thread runs from essentially classical linear control theory, through developments in absolute stability of feedback systems, and terminates with an introductory account of more recent notions of feedback stabilizability and input-to-state stability. The book has no pretensions to comprehensiveness. On the one hand, the permeating thread of control reflects a bias towards synthesis: the bringing of stable behaviour to potentially or inherently unstable processes through appropriate choice of inputs. On the other hand, the book does not contain material relating to the theory of bifurcations or chaos (these topics are treated in numerous other texts on ordinary differential equations).

Finally, we would like to thank Mr Elvijs Sarkans for his valuable comments on an earlier draft of the book.

Hartmut Logemann & Eugene P. Ryan
Bath
November 2013

Notation

$\mathbb{N}$	the natural numbers $\{1, 2, 3, \ldots\}$
$\mathbb{N}_0$	the non-negative integers $\mathbb{N} \cup \{0\}$
$\mathbb{Z}$	the integers
$\mathbb{Q}$	the field of rational numbers
$\mathbb{R}$	the field of real numbers
$\mathbb{C}$	the field of complex numbers
$\mathbb{R}_+$	the non-negative reals $[0, \infty)$
$\mathbb{C}_+$	the open right half complex plane $\{z \in \mathbb{C} : \operatorname{Re} z > 0\}$
$\mathbb{C}_-$	the open left half complex plane $\{z \in \mathbb{C} : \operatorname{Re} z < 0\}$
$\mathbb{F}$	either $\mathbb{R}$ or $\mathbb{C}$
$\mathbb{F}^N$	the vector space of ordered N-tuples of numbers from $\mathbb{F}$
$\mathbb{F}^{P \times N}$	the vector space of $P \times N$ matrices with entries from $\mathbb{F}$
$GL(N, \mathbb{F})$	the group of invertible $\mathbb{F}^{N \times N}$ matrices (general linear group)
$*$	(in superscript) Hermitian/conjugate transpose of a matrix
$\oplus$	direct sum (of subspaces of $\mathbb{F}^N$)
$S^\perp$	orthogonal complement of a subspace S of $\mathbb{F}^N$
im	image (of an element of $\mathbb{F}^{N \times P}$): $\operatorname{im} M = \{Mx \in \mathbb{F}^N : x \in \mathbb{F}^P\}$
ker	kernel (of an element of $\mathbb{F}^{N \times P}$): $\ker M = \{x \in \mathbb{F}^P : Mx = 0\}$
rk	matrix rank
tr	trace of a square matrix: the sum of the diagonal entries
det	determinant (of a square matrix)
adj	adjugate (of a square matrix)
σ	spectrum (set of eigenvalues of a square matrix): $\sigma(M) = \{\lambda \in \mathbb{C} : \lambda \text{ is an eigenvalue of } M\}$
span	span of a set of vectors
$\|\cdot\|$	generic symbol for a norm
$\langle \cdot, \cdot \rangle$	inner product (on $\mathbb{F}^N$)

$\mathbb{B}(x,r)$	the open ball of radius $r > 0$ centred at $x \in X$, X a metric space
∂S	boundary of a set S in a metric space
$\mathrm{cl}(S)$, $\overline{S}$	closure of a set S in a metric space
dist	distance from a point to a set or distance between two sets: for $z \in \mathbb{R}^N$ and non-empty sets $X, Y \subset \mathbb{R}^N$, $\mathrm{dist}(z, X) := \inf\{\|z - x\| : x \in X\}$ $\mathrm{dist}(X, Y) := \inf\{\|x - y\| : x \in X,\ y \in Y\}$
dom	domain of a map
$C(K,Y)$	vector space of continuous functions $K \subset X \to Y$, (X, Y normed spaces)
$\|\cdot\|_\infty$	the supremum norm on $C(K,Y)$ with K compact: $\|f\|_\infty := \sup_{x \in K} \|f(x)\|$, $\|\cdot\|$ a norm on Y
$\mathcal{K}$	the class of continuous and strictly increasing functions $a: \mathbb{R}_+ \to \mathbb{R}_+$ with $a(0) = 0$.
$\mathcal{K}_\infty$	the class of unbounded $\mathcal{K}$ functions
$\mathcal{KL}$	the class of functions $b: \mathbb{R}_+ \times \mathbb{R}_+ \to \mathbb{R}_+$ such that, for all $t \in \mathbb{R}_+$, $b(\cdot, t) \in \mathcal{K}$ and, for all $s \in \mathbb{R}_+$, $b(s, \cdot)$ is decreasing with $b(s, t) \to 0$ as $t \to \infty$
$PC(I,Y)$	vector space of piecewise continuous functions $I \to Y = \mathbb{F}^{P \times Q}$, $I \subset \mathbb{R}$ an interval
$PC^1(I,Y)$	vector space of piecewise continuously differentiable functions
$\star$	convolution (of functions)
$\circ$	composition (of functions)
∂_i	i-th partial derivative: for a function $X \subset \mathbb{R}^N \to \mathbb{R}$, $(x_1, \dots, x_N) = x \mapsto f(x)$, $(\partial_i f)(x)$ denotes the derivative at x of f with respect to the i-th component x_i of its argument x
∇	gradient of a function $X \subset \mathbb{R}^N \to \mathbb{R}$, $(x_1, \dots, x_n) = x \mapsto f(x)$: $(\nabla f)(x) = (\partial_1 f, \dots, \partial_N f)(x)$
D	differentiation in the Fréchet sense. For a function $f: X \subset \mathbb{R}^N \to \mathbb{R}^M$, $(Df)(x) := \big((\partial_j f_i)(x)\big)$ is the $M \times N$ matrix of partial derivatives at x of components f_i of f with respect to components x_j of its argument
$C^1(X, \mathbb{R}^M)$	vector space of continuously differentiable functions defined on $X \subset \mathbb{R}^N$ with values in $\mathbb{R}^M$.
$f^{-1}(Y)$	pre-image of a set $Y \subset \mathbb{R}^M$ under the map $f: X \subset \mathbb{R}^N \to \mathbb{R}^M$, that is, the set $\{x \in X : f(x) \in Y\}$
$f^{-1}(y)$	pre-image of a point $y \in \mathbb{R}^M$ under the map $f: X \subset \mathbb{R}^N \to \mathbb{R}^M$, that is, the set $\{x \in X : f(x) = y\} = f^{-1}(\{y\})$
$I(\tau, \xi)$	maximal interval of existence of the solution of the non-autonomous initial-value problem $\dot{x}(t) = f(t, x(t))$, $x(\tau) = \xi$ (assuming uniqueness)

I_ξ maximal interval of existence of the solution of the autonomous initial-value problem $\dot{x}(t) = f(x(t))$, $x(0) = \xi$ (assuming uniqueness)

$\mathbb{F}(s)$ the field of rational functions in s with coefficients in $\mathbb{F}$

$\mathcal{L}$ Laplace transform

$[\![a,b]\!]$ line segment joining two points $a, b \in \mathbb{R}^2$:
$[\![a,b]\!] := \{(1 - \mu)a + \mu b \colon 0 \leq \mu \leq 1\}$

$\square$ indicates end of proof

$\triangle$ indicates end of example

Throughout, by an *interval* $J \subset \mathbb{R}$ we mean a non-degenerate interval, that is, an interval with endpoints $a \geq -\infty$ and $b \leq \infty$ satisfying $a < b$. An interval may be open or closed, neither open nor closed, bounded or unbounded.

We denote, by $\mathbb{F}^N$, the set of all ordered N-tuples x with components $x_1, \ldots, x_N$ in $\mathbb{F}$, An element $x \in \mathbb{F}^N$ can be viewed as a column ($N \times 1$ matrix), that is,

$$x = \begin{pmatrix} x_1 \\ \vdots \\ x_N \end{pmatrix},$$

or, alternatively, $x \in \mathbb{F}^N$ can be viewed as a row ($1 \times N$ matrix), that is,

$$x = \begin{pmatrix} x_1, \ldots, x_n \end{pmatrix}.$$

Throughout, we exploit the notational flexibility afforded by the two equivalent representations of elements of $\mathbb{F}^N$: in some situations, we adopt the column form and, in other situations, we opt for the row alternative. For example, in linear algebraic contexts, the column form is appropriate in matrix manipulation: for a $P \times N$ matrix $M \in \mathbb{F}^{P \times N}$ and $x \in \mathbb{F}^N$, both x and $Mx \in \mathbb{F}^P$ should be interpreted in column form. On the other hand, if f is a (nonlinear) map $\mathbb{F}^N \to \mathbb{F}^P$ and we wish to express $f(x) \in \mathbb{F}^P$, $x \in \mathbb{F}^N$, in explicit componentwise form, then we adopt

$$f(x) = \begin{pmatrix} f_1(x_1, \ldots, x_N), \ldots, f_P(x_1, \ldots, x_n) \end{pmatrix}$$

as the preferred alternative to its typographically cumbersome column form. Our view is that the benefits to typography and layout available through selective use of the equivalent representations of elements of $\mathbb{F}^N$ outweigh any potential for confusion.

Contents

1
Introduction

The theory of differential equations impinges on many branches of pure and applied mathematics. Moreover, many diverse areas in science (e.g. biology, physics), economics and engineering give rise to problems which, when mathematically formulated, correspond to a study of differential equations. In essence, differential equations play a central role in understanding the evolution in time of many natural processes. This book develops a rigorous framework for the study of initial-value problems for systems of ordinary differential equations. Fundamental questions of existence, uniqueness and maximal extension of solutions are addressed. Dependence of solutions on initial data is studied. Qualitative behaviour (and *stability*, in particular) of solutions is investigated. These issues are analytical in nature: given a system of differential equations, the task is to analyse the nature and properties of its solutions (if solutions exist). *Analysis* of differential equations has a long history, dating back to the remarkable era of Isaac Newton (1643-1727), Gottfried von Leibniz (1646-1716) and the Bernoulli brothers (Jacob (1654-1705) and Johann (1667-1748)). A second viewpoint on differential equations, and with its origins in the seminal 1868 paper[1] "On Governors" by James Clerk Maxwell (1831-1879), is that of *synthesis*. From this viewpoint, a system of differential equations is not regarded as immutable but, instead, is deemed to be an object which can be altered, through choice of extraneous inputs, in such a way that its solutions exhibit prescribed properties. Frequently, such inputs are generated by a *feedback* process (a governor in Maxwell's paper) in which the requisite inputs are generated

---

[1] J.C. Maxwell. On Governors, *Proceedings of the Royal Society of London*, Vol. 16 (1867-1868), pp. 270-283.

H. Logemann and E. P. Ryan, *Ordinary Differential Equations*,
Springer Undergraduate Mathematics Series,
DOI: 10.1007/978-1-4471-6398-5_1, © Springer-Verlag London 2014

by some suitable operations on available outputs from, or observations on, the system: in such circumstances, the terminology *feedback synthesis* applies. The synthesis viewpoint forms the basis of *control theory*. The book also encompasses an introduction to control theory and addresses fundamental questions of controllability (an examination of the extent to which solutions may be influenced through choice of input), observability (an examination of the extent to which the internal "state" of a system can be deduced from available outputs or observations) and stabilizability & stabilization (an examination of the extent to which stable behaviour of solutions can be synthesized through feedback).

1.1 Examples

Our goal is the study of systems of ordinary differential equations of the form

$$\dot{x}(t) = f(t, x(t)).$$

Here $f: J \times G \to \mathbb{R}^N$ is a suitably regular function, $J \subset \mathbb{R}$ is an interval and G is a non-empty open subset of $\mathbb{R}^N$. If f is independent of t (that is, $f : G \to \mathbb{R}^N$), then the above differential equation is said to be *autonomous*. Systems of differential equations arise naturally in diverse areas of science and engineering. We start by looking at some simple illustrative examples.

1.1.1 An example from circuit theory

The study of electrical circuits is a source of many important differential equations. Consider, for example, the circuit shown in Figure 1.1 consisting of a parallel connection of a capacitor (with capacitance C), an inductor (with inductance L), and a nonlinear resistor.

Let $J = \mathbb{R}$. At any time $t \in J$, the current $\imath_R(t)$ through the resistor is related to the voltage $v_R(t)$ across the resistor by a nonlinear function g, that is,

$$\imath_R(t) = g(v_R(t)).$$

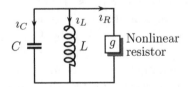

Figure 1.1

For example, if g is given by

$$g(\zeta) = -\zeta + \zeta^3, \quad \forall \zeta \in \mathbb{R}, \tag{1.1}$$

then this corresponds to a particular component known as a twin-tunnel diode.

The voltage v_L across the inductor is related to the current $\imath_L$ through the inductor by Faraday's law

$$v_L(t) = L\frac{d\imath_L}{dt}(t), \quad L \text{ constant}.$$

The voltage v_C across the capacitor and the corresponding current $\imath_C$ satisfy the relation

$$C\frac{dv_C}{dt}(t) = \imath_C(t), \quad C \text{ constant}.$$

Kirchoff's current and voltage laws give

$$\imath_R(t) + \imath_L(t) + \imath_C(t) = 0 \quad \forall t \in J$$

and

$$v_R(t) = v_L(t) = v_C(t) \quad \forall t \in J.$$

Eliminating the variables $\imath_C$, $\imath_R$, v_L and v_R from the above relations, yields the system of two differential equations

$$L\frac{d\imath_L}{dt}(t) = v_C(t), \qquad C\frac{dv_C}{dt}(t) = -\imath_L(t) - g(v_C(t)).$$

Defining $x_1(t) := L\imath_L(t)$ and $x_2(t) := v_C(t)$, we obtain

$$\dot{x}_1(t) = x_2(t), \qquad \dot{x}_2(t) = -\mu_1 x_1(t) - \mu_2 g(x_2(t)),$$

where $\mu_1 := 1/(CL)$ and $\mu_2 := 1/C$. Setting

$$f(z) = f(z_1, z_2) := \big(z_2, -\mu_1 z_1 - \mu_2 g(z_2)\big)$$

for all $z = (z_1, z_2) \in G := \mathbb{R}^2$ defines a function $f \colon G \to \mathbb{R}^2$ and, on writing $x(t) = (x_1(t), x_2(t))$, the above pair of differential equations can be written as the autonomous system $\dot{x}(t) = f(x(t))$.

Now assume, for simplicity, that $\mu_1 = \mu_2 = 1$ and consider again the case of a twin-tunnel diode described by the characteristic (1.1), in which case f is given by

$$f(z) = f(z_1, z_2) = \big(z_2, -z_1 + z_2 - z_2^3\big).$$

Note that $f(z) = 0$ if, and only if, $z = 0$. Let $\mathcal{A}$ be the "annular" region in the plane, as in Figure 1.2, wherein the inner boundary is the circle of unit radius centred at $(0,0)$ and the outer boundary is a polygon with vertices as shown. A straightforward calculation reveals that there is no point (z_1, z_2) of either the inner or the outer boundary at which the vector $f(z_1, z_2)$ is directed to the exterior of the annulus (equivalently, at each point z of each boundary, the vector $f(z)$ is either tangential or directed inwards). An immediate consequence of this observation is the following fact: if $x \colon J \to G$ is a solution on the (that is, a continuously differentiable function with $\dot{x}(t) = f(x(t))$ for all $t \in J$) with $x(0) \in \mathcal{A}$, then $x(t) \in \mathcal{A}$ for all $t \geq 0$.

The set $\mathcal{A}$ is said to be *positively invariant*: solutions starting in the set are trapped within the set in forwards time. We now know that the set $\mathcal{A}$ is positively invariant and is such that $f(z) \neq 0$ for all $z \in \mathcal{A}$. These two properties are sufficient to ensure (via the Poincaré-Bendixson theorem – to be stated and proved in Section 4.6) that the system has at least one periodic solution (that is, a solution $x\colon J \to G$ with the property that, for some $T > 0$, $x(t) = x(t+T)$ for all $t \in J$), the *orbit* of which (that is, the set $\{x(t)\colon t \in \mathbb{R}\} = \{x(t)\colon t \in [0,T)\}$) is contained in $\mathcal{A}$. The exis-

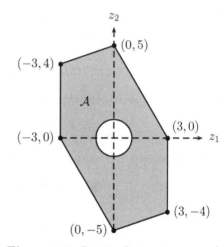

Figure 1.2 Positively invariant set $\mathcal{A}$

tence of a periodic orbit is reflected in the terminology "nonlinear oscillator" commonly used in the context of this circuit.

1.1.2 The nonlinear pendulum

One end P of a (weightless) rigid rod of length l is attached to a pivot, and a point mass m is attached to the other end (see Figure 1.3). The system moves in

Figure 1.3

a vertical plane under the influence of a uniform gravitational field (with gravitation constant g) directed vertically downward. We assume that the frictional force resisting the motion is proportional (with a coefficient of friction $c \geq 0$) to the speed of the point mass and acts perpendicular to the rod. If $\theta(t)$ denotes the angle between the vertical and the rod (measured in the counterclockwise sense) at time $t \in J := \mathbb{R}$, then an application of Newton's second law yields the equation of motion $ml\ddot{\theta}(t) = -mg\sin\theta(t) - cl\dot{\theta}(t)$, from which we obtain the following second-order differential equation

$$\ddot{\theta}(t) + a\,\dot{\theta}(t) + b\sin\theta(t) = 0\,,$$

where $a := c/m$ and $b := g/l$. Writing $(x_1(t), x_2(t)) = (\theta(t), \dot{\theta}(t))$ we obtain

$$\dot{x}_1(t) = x_2(t), \quad \dot{x}_2(t) = -ax_2(t) - b\sin x_1(t).$$

Setting $G := \mathbb{R}^2$, $x(t) = (x_1(t), x_2(t))$ and introducing the function $f\colon G \to \mathbb{R}^2$ given by $f(z) = f(z_1, z_2) := (z_2, -az_2 - b\sin z_1)$, the above equation may be expressed in the form of an autonomous system $\dot{x}(t) = f(x(t))$.

1.1.3 The controlled inverted pendulum

With reference to Figure 1.4, assume that the pendulum is frictionless and is inverted with its pivot P placed at the centre of a trolley (of mass $M > 0$ and subject to a horizontal control force u) which may move, without friction on a horizontal plane. If $\theta(t)$ denotes the angle between the vertical and the rod (measured in the clockwise sense) at time $t \in J := \mathbb{R}$ and $\xi(t)$ denotes the horizontal displacement of the trolley (measured from some fixed reference point) at time t, then horizontal displacement of the pendulum at time t is given by $\xi(t) + l\sin\theta(t)$. An application of Newton's second law yields the equations of motion in the form of two coupled second-order differential equations:

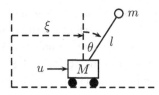

Figure 1.4 Inverted pendulum

$$(M + m)\ddot{\xi}(t) + ml\left(\ddot{\theta}(t)\cos\theta(t) - \dot{\theta}^2(t)\sin\theta(t)\right) = u(t),$$

$$\ddot{\xi}(t)\cos\theta(t) + l\ddot{\theta}(t) - g\sin\theta(t) = 0.$$

A straightforward calculation then gives

$$\ddot{\theta}(t) = \frac{\left((M + m)g - ml\dot{\theta}^2(t)\cos\theta(t)\right)\sin\theta(t) - u(t)\cos\theta(t)}{l\left(M + m - m\cos^2\theta(t)\right)}$$

$$\ddot{\xi}(t) = \frac{m\sin\theta(t)\left(l\dot{\theta}^2(t) - g\cos\theta(t)\right) + u(t)}{M + m - m\cos^2\theta(t)}$$

Set $G := \mathbb{R}^4$. With each forcing function u, we associate a function $f_u\colon J \times G \to \mathbb{R}^4$ given by

$$f_u(t, z) = f_u(t, z_1, z_2, z_3, z_4) := \left(z_2, f_2(z_1, z_2, u(t)), z_4, f_4(z_1, z_2, u(t))\right),$$

with functions $f_2, f_4 \colon \mathbb{R} \times \mathbb{R} \times \mathbb{R} \to \mathbb{R}$ defined by

$$\left. \begin{aligned} f_2(z_1, z_2, v) &:= \frac{\left((M + m)g - mlz_2^2\cos z_1\right)\sin z_1 - v\cos z_1}{l\left(M + m - m\cos^2 z_1\right)}, \\ f_4(z_1, z_2, v) &:= \frac{m\sin z_1\left(lz_2^2 - g\cos z_1\right) + v}{M + m - m\cos^2 z_1}. \end{aligned} \right\} \tag{1.2}$$

Writing $x(t) := (\theta(t), \dot{\theta}(t), \xi(t), \dot{\xi}(t))$, the system may be expressed in the form

$$\dot{x}(t) = f_u(t, x(t)). \tag{1.3}$$

Clearly each distinct choice of forcing function gives rise to a different function f_u. A natural question arises: can one find a function u such that solutions of the associated system (1.3) exhibit prescribed behaviour. For example, given any initial configuration of the system

$$x(0) = (\theta(0), \dot{\theta}(0), \xi(0), \dot{\xi}(0)),$$

does there exist a function u (a control) such that the pendulum asymptotically approaches the vertically upright position whilst the trolley approaches the 0 rest position, expressed symbolically as $(\theta(t), \dot{\theta}(t), \xi(t), \dot{\xi}(t)) = x(t) \to 0$ as $t \to \infty$? This *control problem* may be naively visualized as that of balancing a broom vertically upright on the palm of the hand.

If we consider only "small" deviations of the pendulum from the vertically upright rest position $((\theta(t), \dot{\theta}(t), \xi(t), \dot{\xi}(t)) = 0$ for all $t)$, and approximate the nonlinear terms in (1.2), for "small" values of the arguments z_1, z_2 and v as follows

$$\sin z_1 \approx z_1, \quad \cos z_1 \approx 1, \quad z_2^2 \sin z_1 \approx 0, \quad z_2^2 \cos z_1 \approx 0,$$

then we arrive at the *linearized model* given by

$$\left. \begin{array}{l} \dot{x}(t) = Ax(t) + Bu(t), \\[2mm] A := \begin{pmatrix} 0 & 1 & 0 & 0 \\ (M+m)g/Ml & 0 & 0 & 0 \\ 0 & 0 & 0 & 1 \\ -mg/M & 0 & 0 & 0 \end{pmatrix}, \quad B := \begin{pmatrix} 0 \\ -1/Ml \\ 0 \\ 1/M \end{pmatrix}. \end{array} \right\} \tag{1.4}$$

This inhomogeneous linear system of differential equations "approximately" governs the behaviour of the inverted pendulum "near" the vertically upright rest position. Systems of linear differential equations will be investigated in detail in Chapters 2 and 3.

1.1.4 Satellite dynamics

One example, to which we will return to in Chapter 3, is that of a satellite of mass m (considered as a point mass) in an inverse square law force field. The motion is governed by a pair of second order equations in the radial distance r from the origin O and the angle θ, as shown in Figure 1.5. If we assume that the satellite has the capability of thrusting in the radial direction with a thrust v_1

and in the "tangential" direction with thrust v_2, then the equations of motion are given by

$$\ddot{r}(t) = r(t)\dot{\theta}^2(t) - \frac{k}{r^2(t)} + \frac{v_1(t)}{m}, \quad \ddot{\theta}(t) = -\frac{2\dot{\theta}(t)\dot{r}(t)}{r(t)} + \frac{v_2(t)}{mr(t)},$$

where $k > 0$ is a constant. For details relating to the derivation of these equations, we refer to [15, Section A.4].

If $v_1(t) = v_2(t) = 0$ for all $t \geq 0$, then, for each pair of real numbers σ and ω satisfying $\sigma^3\omega^2 = k$, these equations admit the solution (r, θ) given by

$$r(t) = \sigma, \quad \theta(t) = \omega t. \tag{1.5}$$

This shows that circular orbits are possible. The positivity of k implies that $\sigma > 0$ and $\omega \neq 0$.

In the following we will reformulate the above system of second-order equations in first-order form. To this end, let σ and ω be real numbers such that $\sigma^3\omega^2 = k$, let x_1, x_2, x_3 and x_4 be given by

$$x_1(t) = r(t) - \sigma, \quad x_2(t) = \dot{r}(t), \quad x_3 = \theta(t) - \omega t, \quad x_4 = \dot{\theta}(t) - \omega; \quad \forall t \geq 0$$

and define

$$u_1(t) = \frac{v_1(t)}{m}, \quad u_2(t) = \frac{v_2(t)}{m}; \quad \forall t \geq 0.$$

Then

$$\left.\begin{aligned}
\dot{x}_1(t) &= x_2(t), \\
\dot{x}_2(t) &= (x_1(t) + \sigma)(x_4(t) + \omega)^2 - \frac{\sigma^3\omega^2}{(x_1(t) + \sigma)^2} + u_1(t), \\
\dot{x}_3(t) &= x_4(t), \\
\dot{x}_4(t) &= -\frac{2x_2(t)(x_4(t) + \omega)}{x_1(t) + \sigma} + \frac{u_2(t)}{x_1(t) + \sigma}.
\end{aligned}\right\} \tag{1.6}$$

Note that if $u_1(t) = u_2(t) = 0$ for all $t \geq 0$, then 0 is an equilibrium of the above first-order system, that is, $(x_1, x_2, x_3, x_4) = 0$ is a solution. Obviously, this solution corresponds to the circular orbit (1.5) of the original system.

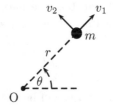

Figure 1.5

Since $r(t) > 0$ for all t, we see that x_1 is required to take its values in $(-\sigma, \infty)$. With a given $\mathbb{R}^2$-valued forcing (or input) function $u = (u_1, u_2)$ and defining $G := (-\sigma, \infty) \times \mathbb{R}^3$, we associate a function $f_u \colon \mathbb{R} \times G \to \mathbb{R}^4$ given by

$$f_u(t, z) = f_u(t, z_1, z_2, z_3, z_4) := \left(z_2, \; f_2(z_1, z_4, u_1(t)), \; z_4, \; f_4(z_1, z_2, z_4, u_2(t))\right),$$

where the functions $f_2 \colon (-\sigma, \infty) \times \mathbb{R}^2 \to \mathbb{R}$ and $f_4 \colon G \to \mathbb{R}$ are defined by

$$\left.\begin{aligned}
f_2(z_1, z_4, w_1) &:= (z_1 + \sigma)(z_4 + w)^2 - \frac{\sigma^3 w^2}{(z_1 + \sigma)^2} + w_1, \\
f_4(z_1, z_2, z_4, w_2) &:= -\frac{2 z_2 (z_4 + w)}{z_1 + \sigma} + \frac{w_2}{z_1 + \sigma}.
\end{aligned}\right\} \tag{1.7}$$

Writing $x := (x_1, x_2, x_3, x_4)$, system (1.6) may be expressed in the form

$$\dot{x}(t) = f_u(t, x(t)). \tag{1.8}$$

Clearly each distinct choice of $u = (u_1, u_2)$ gives rise to a different function f_u.

If we consider only "small" deviations from the circular orbit (1.5) and approximate the nonlinear terms on the right-hand sides of (1.7) for "small" values of the relevant arguments as follows

$$\left.\begin{aligned}
(z_1 + \sigma)(z_4 + w)^2 &\approx w^2 z_1 + 2w\sigma z_4 + \sigma w^2, \quad \frac{1}{(z_1 + \sigma)^2} \approx \frac{1}{\sigma^2} - \frac{2 z_1}{\sigma^3}, \\
\frac{z_2 (z_4 + w)}{z_1 + \sigma} &\approx \frac{w z_2}{\sigma}, \quad \frac{w_2}{z_1 + \sigma} \approx \frac{w_2}{\sigma},
\end{aligned}\right\} \tag{1.9}$$

then we arrive at the linearized model

$$\left.\begin{aligned}
\dot{x}(t) &= Ax(t) + Bu(t), \\
A &:= \begin{pmatrix} 0 & 1 & 0 & 0 \\ 3w^2 & 0 & 0 & 2w\sigma \\ 0 & 0 & 0 & 1 \\ 0 & -2w/\sigma & 0 & 0 \end{pmatrix}, \quad B := \begin{pmatrix} 0 & 0 \\ 1 & 0 \\ 0 & 0 \\ 0 & 1/\sigma \end{pmatrix}.
\end{aligned}\right\} \tag{1.10}$$

This inhomogeneous linear system of differential equations "approximately" governs the behaviour of the satellite "near" the circular orbit. Assuming that r and θ can both be measured, we associate with the linearized system an $\mathbb{R}^2$-valued *output* or *observation* y defined by

$$y(t) := Cx(t), \quad \text{where} \quad C = \begin{pmatrix} 1 & 0 & 0 & 0 \\ 0 & 0 & 1 & 0 \end{pmatrix} \tag{1.11}$$

In Chapter 3 it will be shown that the controlled and observed linear system given by (1.10) and (1.11) has the following *controllability* and *observability* properties.

(a) For every initial state $x(0) = x^0 \in \mathbb{R}^4$, every target state $x^1 \in \mathbb{R}^4$ and every time $T > 0$, there exists a control function $u \colon [0, T] \to \mathbb{R}^2$ such that $x(T) = x^1$.

(b) For every time $T > 0$, the state $x(T)$ can be determined from the knowledge of the input and output signals u and y on the interval $[0, T]$.

1.1.5 Population dynamics

The predator-prey model of Lotka-Volterra. Consider two interacting species, a predator species and a prey species. The size of predator population at time t is denoted by $p(t)$, and that of the prey by $q(t)$. In the Lotka-Volterra[2] system of differential equations

$$\dot{p}(t) = p(t)\big(-a + bq(t)\big), \quad \dot{q}(t) = q(t)\big(c - dp(t)\big),$$

which describes the predator-prey interaction, a, b, c, d are positive constants. The prey population is assumed to have ample resources and so, in the absence of predators (that is, setting $p(t) = 0$ for all t)

$$\text{growth rate} = \text{birth rate} - \text{mortality rate} = c > 0\,,$$

and the prey population increases (exponentially) in size in accordance with the linear differential equation $\dot{q}(t) = cq(t)$. In the presence of predators, and assuming that the rate of attrition is proportional (constant of proportionality $d > 0$) to the size of the predator population, the prey growth rate reduces from c to $c - dp(t)$, which may become negative. The situation is reversed for the predator population: in the absence of prey $(q(t) = 0$ for all $t)$, the predator population decreases (exponentially) in accordance with the equation $\dot{p}(t) = -ap(t)$ (without an adequate food supply, the predator mortality rate exceeds its birth rate); in the presence of prey, predation on the prey population enlarges the growth rate to $-a + bq(t)$, which, for a sufficiently large prey population, is positive.

On setting $x(t) = (p(t), q(t))$, $G := (0, \infty) \times (0, \infty)$ and defining $f\colon G \to \mathbb{R}^2$ by $f(z_1, z_2) = \big(z_1(-a + bz_2), z_2(c - dz_1)\big)$, the above system of differential equations can be written in the form of the autonomous system $\dot{x}(t) = f(x(t))$. Note that $(z_1, z_2) = (c/d, a/b)$ is the only point in G such that $f(z_1, z_2) = 0$. Associated with this point is the constant (equilibrium) solution $(p(t), q(t)) \equiv (c/d, a/b)$. All other solutions are non-constant and we will show in Section 4.6 that they are periodic.

Cooperating populations: a two-caste colony. Consider a biological system of two cooperating populations. Let $w(t)$ and $q(t)$ represent the populations of *workers* and *reproductives* (queens and males) at time t in a two-caste colony. The role of the workers is to forage and to rear the offspring of the reproductives. The latter aspect has an implicit mechanism for caste determination: offspring can become either reproductives or sterile workers. With reference to Figure 1.6, let $r(t)$ denote the resources of the colony at

---

[2] Alfred James Lotka (1880-1940), naturalized US American.
   Vito Volterra (1860-1940), Italian.

time t, that is, the productive capacity of the colony which is to be divided between the production of workers and reproductives. Resources are generated/sustained/replenished by workers and so it is assumed that $r(t)$ is proportional to the population $w(t)$. At time t, $u(t) \in [0,1]$ denotes the fraction of resources dedicated to enlarging the worker population: the remaining fraction $1 - u(t)$ is dedicated to enlarging the population of reproductives $q(t)$. The evolution of

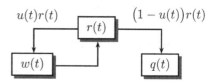

Figure 1.6 Two-caste colony

the populations is modelled by the following pair of ordinary differential equations

$$\dot{w}(t) = (au(t) - \mu)w(t), \quad \dot{q}(t) = -\nu q(t) + b(1 - u(t))w(t),$$

with the constraint $u(t) \in [0,1]$. Here, μ, $\nu > 0$ are mortality parameters and a, $b > 0$ are "resource to population" conversion constants. It is assumed that $a > \mu - \nu > 0$. Clearly, the evolution of the populations is influenced by the choice of function u, which may be chosen by the colony. With a given function $u \colon \mathbb{R} \to [0,1]$, we associate $f_u \colon \mathbb{R} \times G \to \mathbb{R}^2$, with $G := (0, \infty) \times (0, \infty)$, given by

$$f_u(t, z) = f_u(t, z_1, z_2) := \big((au(t) - \mu)z_1, \, -\nu z_2 + b(1 - u(t))z_1\big).$$

Writing $x(t) = (w(t), q(t))$, the above equations may be expressed in the form $\dot{x}(t) = f_u(t, x(t))$.

Assuming that the colony operates on a cyclical basis, with a season $[0,1]$ of unit duration and that a measure of colony "fitness" is the size of population of reproductives at the end of the season, then the principle of "survival of the fittest" would suggest that the colony attempts to allocate its resources during the season so as to maximize $q(1)$ (in which case, it is best equipped to propagate into the next season). This can be formulated as the following *optimal control problem*: maximize $q(1)$ over functions $u \colon [0,1] \to [0,1]$. Whilst the theory of *optimal* control is outside the scope of the book, this example does serve to highlight the synthesis aspect of the study of differential equations: determine the function u so as to achieve prescribed behaviour. Suffice it to say here that the following is the optimal strategy

$$u(t) = \begin{cases} 1, & 0 \le t < t_s \\ 0, & t_s \le t \le 1 \end{cases}, \quad t_s = 1 - \frac{1}{\mu - \nu}\ln\left(\frac{a}{a - \mu + \nu}\right).$$

In words, the optimal strategy is to devote initially all resources to enlarging the worker population, then, at time t_s, switch all resources to enlarging the reproductive population. Note that this strategy is independent of the initial populations $w(0)$ and $q(0)$ (each assumed positive). It is interesting to note that the qualitative features of this strategy have been observed in certain colonies of wasps[3]. For the illustrative parameter values $a = 2.5$, $b = 4$, $\mu = 3$ and $\nu = 2$, the evolution of the populations, with initial values $(w(0), q(0)) = (2, 1)$, is depicted below.

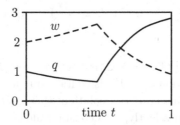

Figure 1.7 Optimal evolution of populations

Exercise 1.1

Satellite Dynamics
Consider the satellite example described in Section 1.1.4.

(a) Derive the affine linear approximations (see (1.9)) of the nonlinear terms on the right-hand sides of (1.7).

(b) Let $\varepsilon > 0$ and define $\sigma_\varepsilon := \left(k/(\omega+\varepsilon)^2\right)^{1/3}$. Verify that (x_1, x_2, x_3, x_4) defined by

$$x_1(t) = \sigma_\varepsilon - \sigma, \quad x_2(t) = 0, \quad x_3(t) = \varepsilon t, \quad x_4(t) = \varepsilon; \quad \forall t \geq 0$$

is a solution of the nonlinear system (1.6) with $u_1 = u_2 = 0$. (Note that, in terms of r and θ, this solution corresponds to the circular orbit given by $r(t) = \sigma_\varepsilon$ and $\theta(t) = (\omega + \varepsilon)t$.)

(c) Show that the equilibrium 0 of the uncontrolled (that is, $u_1 = u_2 = 0$) nonlinear system (1.6) is "unstable" in the sense that for every $\delta > 0$ there exists an initial vector $\xi = (\xi_1, \xi_2, \xi_3, \xi_4) \in G\backslash\{0\}$ with $\|\xi\| < \delta$ and an unbounded solution $x = (x_1, x_2, x_3, x_4)$ of the uncontrolled nonlinear system (1.6) with $x(0) = \xi$ and $\|x(t)\| \to \infty$ as $t \to \infty$.

Exercise 1.2

First integrals
Here, we consider the autonomous system

$$\dot{x}(t) = f(x(t)), \tag{1.12}$$

---

[3] See Section 2.6 in G.F. Oster and E.O. Wilson, *Caste and Colony in the Social Insects*, Princeton University Press, Princeton, New Jersey, 1978.

where $f\colon G \to \mathbb{R}^N$ is continuous and $G \subset \mathbb{R}^N$ is a non-empty open set. A non-constant continuously differentiable function $E\colon G \to \mathbb{R}$ is called a *first integral* for (1.12) if $\langle(\nabla E)(z), f(z)\rangle = 0$ for all $z \in G$, where ∇E denotes the gradient of E and $\langle\cdot,\cdot\rangle$ denotes the Euclidean inner product on $\mathbb{R}^N$. A set of the form

$$E^{-1}(\gamma) := \{z \in G\colon E(z) = \gamma\}, \quad \text{where } \gamma \in \mathbb{R},$$

is called a *level set* of E.

(a) Assume that $E\colon G \to \mathbb{R}$ is a first integral for (1.12). Show that the image of every solution of (1.12) is contained in a level set of E, that is, if $x\colon I \to G$ (I an interval) is a solution, then $\{x(t) : t \in I\} \subset E^{-1}(\gamma)$ for some $\gamma \in \mathbb{R}$ (equivalently, E is constant along solutions of (1.12)).

(b) Let $f\colon \mathbb{R}^2 \to \mathbb{R}^2$ be given by $f(z) = f(z_1, z_2) = \big(z_2, g(z_1)\big)$, where $g\colon \mathbb{R} \to \mathbb{R}$ is continuous. Define $E\colon \mathbb{R}^2 \to \mathbb{R}$ by

$$E(z) = E(z_1, z_2) := -\int_0^{z_1} g(s)\mathrm{d}s + z_2^2/2.$$

Show that E is a first integral for (1.12).

(c) Consider the nonlinear pendulum of Section 1.1.2 in the absence of friction $a = 0$. Determine a first integral.

(d) Consider the Lotka-Volterra predator-prey equations of Section 1.1.5:

$$\dot{p} = p(-a + bq), \quad \dot{q} = q(c - dp), \quad a, b, c, d \text{ positive constants},$$

on the open quadrant $G = (0, \infty) \times (0, \infty)$. Show that

$$E\colon G \to \mathbb{R}, \ (x, y) \mapsto dx - c\ln x + by - a\ln y$$

is a first integral.

(e) Motivate the inclusion of the non-constancy condition in the above concept of first integral.

1.2 Initial-value problems

Let $f\colon J \times G \to \mathbb{R}^N$, where $J \subset \mathbb{R}$ is an interval and G is a non-empty open subset of $\mathbb{R}^N$. Consider the ordinary differential equation

$$\dot{x}(t) = f(t, x(t)), \tag{1.13}$$

which, when subject to the condition

$$x(\tau) = \xi, \quad (\tau, \xi) \in J \times G, \tag{1.14}$$

is referred to as an *initial-value problem*. In order to make sense of the notion of a solution of the initial-value problem (1.13)-(1.14), it is necessary to impose some regularity on the function f. We distinguish two cases which will be treated separately:

- the function f is (jointly) continuous;

- the function f is locally Lipschitz with respect to its second argument (a notion to be made precise in due course) and, for each continuous function $y \colon J \to G$, the function $t \mapsto f(t, y(t))$ is piecewise continuous.

1.2.1 Continuous righthand side

In Chapter 4, the following basic result will be established: if f is continuous, then (1.13)-(1.14) has at least one solution, that is, a continuously differentiable function $x \colon I \to G$, on some interval $I \subset J$ containing τ, with $x(\tau) = \xi$ and satisfying the differential equation (1.13) for all $t \in I$. Moreover, it will be shown that every solution is either *maximal* (in the sense of being defined on a maximal interval of existence) or can be extended to a maximal solution, that is, a solution that has no proper extension which is also a solution (a proper extension of $x \colon I \to G$ is a function $\tilde{x} \colon \tilde{I} \to G$, where $\tilde{I} \subset J$ is an interval such that $I \subset \tilde{I}$ and $I \neq \tilde{I}$, and $\tilde{x}(t) = x(t)$ for all $t \in I$).

Example 1.1

Let $J = \mathbb{R} = G$ and consider the scalar initial-value problem

$$\dot{x}(t) = t x^{1/3}(t), \quad x(0) = 0.$$

By inspection, we see that the zero function $x = 0$ on $\mathbb{R}$ is a (maximal) solution. Moreover, it is readily verified that, for each $c \geq 0$, the continuously differentiable function

$$\mathbb{R} \to \mathbb{R}, \quad t \mapsto x(t) := \begin{cases} \left((t-c)/\sqrt{3}\right)^3, & t \geq c \\ 0, & t < c \end{cases}$$

is also a (maximal) solution, and so there exist uncountably many maximal solutions. $\triangle$

The above example serves to illustrate the fact that, whilst continuity of f ensures the existence of a maximal solution of the initial-value problem (1.13)-(1.14), one cannot conclude that the solution is unique. A natural question then arises: under what additional condition on f is the existence of a unique maximal solution assured? An answer to this question is provided in Chapter 4 in the form of a *Lipschitz*[4] *condition*. A function $f: J \times G \to \mathbb{R}^N$ is said to be *locally Lipschitz* with respect to its second argument if, for each $(t, z) \in J \times G$, there exist neighbourhoods $T \subset J$ and $Z \subset G$ of t and z, respectively, and a number $L \geq 0$ (in general depending on T and Z) such that

$$\|f(s, z_1) - f(s, z_2)\| \leq L\|z_1 - z_2\| \quad \forall s \in T, \ \forall z_1, z_2 \in Z.$$

Equipped with this concept, the following fact is proved in Chapter 4: if f is continuous and locally Lipschitz in its second argument, then, for each $\tau \in J$ and $\xi \in G$, the initial-value problem (1.13)-(1.14) has a unique maximal solution. Observe that the function $f : (t, z) \mapsto tz^{1/3}$ adopted in the above example (with uncountably many maximal solutions) fails to be locally Lipschitz with respect to its second argument (consider the point $(t, z) = (0, 0)$).

Example 1.2

Let $J = \mathbb{R} = G$ and consider the scalar initial-value problem

$$\dot{x}(t) = tx^3(t), \quad x(0) = 1.$$

The function $f : (t, z) \mapsto tz^3$ is locally Lipschitz with respect to its second argument. To see this, we argue as follows. Let $(t, z) \in \mathbb{R} \times \mathbb{R}$ be arbitrary, $T := [t - 1, t + 1]$, $Z := [z - 1, z + 1]$ and define $L := 3(|t| + 1)(|z| + 1)^2$. Then

$$|sz_1^3 - sz_2^3| = |s||z_1^2 + z_1 z_2 + z_2^2||z_1 - z_2| \leq L|z_1 - z_2| \ \forall s \in T, \ \forall z_1, z_2 \in Z.$$

Therefore, the initial-value problem has a unique maximal solution. By separation of variables and integration (see Exercise 1.3(b) below), this solution is found to be

$$(-1, 1) \to \mathbb{R}, \quad t \mapsto x(t) := \frac{1}{\sqrt{1 - t^2}}.$$

This is an example of an initial-value problem with a finite maximal interval of existence: the solution "blows up" as t approaches either end of its interval of existence $(-1, 1)$.   △

---

[4] Rudolf Otto Lipschitz (1832-1903), German.

Exercise 1.3

Separation of variables

Let $G, J \subset \mathbb{R}$ be intervals with G open, let $h \colon J \to \mathbb{R}$ and let $k \colon G \to \mathbb{R}$ be continuous. Define $f \colon J \times G \to \mathbb{R}$ by setting $f(t, z) := h(t)k(z)$. Consider the initial-value problem

$$\dot{x}(t) = f(t, x(t)) = h(t)k(x(t)), \quad x(\tau) = \xi, \quad (\tau, \xi) \in J \times G. \quad (1.15)$$

By a solution of (1.15) we mean a continuously differentiable function $x \colon I \to G$ with $x(\tau) = \xi$ such that $\dot{x}(t) = h(t)k(x(t))$ for all $t \in I$, where $I \subset J$ is an interval containing τ. Heuristically, we can "solve" this problem by separation of variables and integration:

$$\int_\xi^{x(t)} \frac{ds}{k(s)} = \int_\tau^t h(s)ds.$$

The purpose of this exercise is to identify conditions under which the above heuristic procedure is valid.

In parts (a)–(d) below, assume that $k(\xi) \neq 0$, let $U \subset G$ be an open interval with $\xi \in U$ and such that $k(z) \neq 0$ for all $z \in U$, and define

$$K(z) := \int_\xi^z \frac{ds}{k(s)} \ \forall z \in U \quad \text{and} \quad H(t) := \int_\tau^t h(s)ds \ \forall t \in J.$$

(a) Show that the function $K \colon U \to K(U)$ is invertible.

(b) Show that there exists an open interval $I \subset J$ with $\tau \in I$ and such that $H(I) \subset K(U)$.

(c) With I as in part (b), deduce that the function $x \colon I \to G$ given by

$$x(t) := K^{-1}(H(t)) \ \forall t \in I$$

is a solution of (1.15). Furthermore, show that this solution is the only solution on I.

(d) Reconsider the initial-value problem in Example 1.2. Apply the results in (a)-(c) to derive the unique maximal solution

$$x \colon (-1, 1) \to \mathbb{R}, \quad t \mapsto x(t) = \frac{1}{\sqrt{1 - t^2}} \, .$$

In parts (e) and (f) below, we consider the case wherein $k(\xi) = 0$.

(e) Now, consider the case in which $k(\xi) = 0$. By inspection, we see that, for every interval $I \subset J$ containing τ, the constant function $x \colon I \to G$, $t \mapsto \xi$, is a solution of (1.15). There may be other solutions. The purpose of this part of the exercise is to identify conditions under which there are no other solutions.

Assume that there exists $\delta > 0$ such that $[\xi - \delta, \xi + \delta] \subset G$ and $k(z) \neq 0$ for all $z \in [\xi - \delta, \xi) \cup (\xi, \xi + \delta]$. Furthermore, assume that the improper integrals

$$\int_{\xi - \delta}^{\xi} \frac{\mathrm{d}s}{k(s)} \quad \text{and} \quad \int_{\xi}^{\xi + \delta} \frac{\mathrm{d}s}{k(s)}$$

are divergent, that is,

$$\left| \int_{\xi - \delta}^{z} \frac{\mathrm{d}s}{k(s)} \right| \to \infty \ \text{ as } z \uparrow \xi \quad \text{and} \quad \left| \int_{z}^{\xi + \delta} \frac{\mathrm{d}s}{k(s)} \right| \to \infty \ \text{ as } z \downarrow \xi. \quad (1.16)$$

Prove that, if $x \colon I \to G$ is a solution of (1.15), then $x(t) = \xi$ for all $t \in I$. (*Hint.* Argue by contradiction.)

(f) Reconsider the initial-value problem in Example 1.1, wherein $k \colon z \mapsto z^{1/3}$, $(\tau, \xi) = (0, 0)$ and so $k(\xi) = 0$. We have seen that this initial-value problem has uncountably many solutions on $J = \mathbb{R}$ (including the constant solution $t \mapsto x(t) = \xi = 0$). Explain why this is not at variance with the result in (d) above.

1.2.2 Righthand side with discontinuous time dependence

The assumption of continuity of f, and continuity with respect to its first argument in particular, is difficult to justify in many situations. For example, in the biological example of a two-caste colony in Section 1.1.4, we met a perfectly reasonable model in which the righthand side of the underlying differential equation is *discontinuous* in its t-dependence (recall that the optimal control $t \mapsto u(t)$ has a discontinuity at $t = t_s$). What can we say about existence of solutions in such cases: indeed, how do we even define the concept of solution? Unavoidably, we have to contend with the possibility of "solutions" of (1.13)-(1.14) which fail to be continuously differentiable. In order to arrive at a sensible notion of solution, consider the integrated version of (1.13)-(1.14):

$$x(t) = \xi + \int_{\tau}^{t} f(s, x(s)) \mathrm{d}s. \quad (1.17)$$

We now deem a solution of (1.13)-(1.14) to be a continuous function $x \colon I \to G$, where $I \subset J$ is an interval containing τ, such that (1.17) holds for all $t \in I$. For this definition to have substance, the integral on the righthand side must make sense. The integral does indeed make sense (as a Lebesgue[5]

---

[5] Henri Lebesgue (1875-1941), French.

integral) if, for continuous x, the integrand $s \mapsto f(s, x(s))$ is a locally Lebesgue integrable function. A prerequisite for proceeding down this particular avenue is familiarity with the theory of *Lebesgue integration*. However, as we cannot reasonably assume that the undergraduate readership of the book has this familiarity, we choose instead to operate within the framework of Riemann[6] integration.

In particular, we work with the notion of piecewise continuity. Let $I \subset \mathbb{R}$ be an interval. We deem a function $g \colon I \to \mathbb{R}^N$ to be *piecewise continuous* if the following hold: for every $a, b \in I$ with $a < b$, the interval $[a, b]$ admits a finite partition $a = t_1 < t_2 < \cdots < t_{n-1} < t_n = b$ such that g (i) is continuous on every subinterval (t_i, t_{i+1}), $i = 1, \ldots, n-1$, (ii) has right limit at t_1, (iii) has left limit at t_n, and (iv) has both left and right limits at every t_i, $i = 2, \ldots, n-1$. Let $g \colon I \to \mathbb{R}^N$ be piecewise continuous, $c \in I$ and $\tau \in I$ be a point of continuity of g. The indefinite integral $\Gamma \colon I \to \mathbb{R}^N$, $t \mapsto \int_c^t g(s)\mathrm{d}s$ is differentiable at τ, with derivative $\Gamma'(\tau) = g(\tau)$. This is the fundamental theorem of calculus in the context of piecewise continuous functions (see Theorem A.30 in the Appendix).

Assumption A. $f \colon J \times G \to \mathbb{R}^N$ satisfies the following:

1. f is locally Lipschitz with respect to its second argument;

2. for every continuous function $y \colon J \to G$, the function $t \mapsto f(t, y(t))$ is piecewise continuous.

Then, as we shall prove in Chapter 4, for each $(\tau, \xi) \in J \times G$, the initial-value problem (1.13)-(1.14) has unique maximal solution (in the sense of solving (1.17)) $x \colon I \subset J \to G$. The above assumption on f ensures that $g \colon s \mapsto f(s, x(s))$ is piecewise continuous and so the fundamental theorem of calculus applies to conclude that x is differentiable at all points $t \in I$ of continuity of g and so, for every subinterval $[a, b] \subset I$, we have

$$\dot{x}(t) = g(t) = f(t, x(t)) \; \forall\, t \in [a, b]\backslash E,$$

where E is the finite set of points $t \in [a, b]$ at which g fails to be continuous.

Example 1.3

It is not difficult to verify that, for every piecewise continuous u, the function f_u in each of examples in Section 1.1 (viz. the inverted pendulum, satellite dynamics and the two-caste colony) satisfies assumption **A** and so the associated initial-value problem has a unique maximal solution in each case. △

---

[6] Bernhard Riemann (1826-1866), German.

1.2.3 Linear systems

A highly-structured subclass of initial-value problems is that of linear systems of the form

$$\dot{x}(t) = A(t)x(t) + b(t), \quad x(\tau) = \xi \in \mathbb{R}^N,$$

where $A\colon J \to \mathbb{R}^{N \times N}$ (the space of real $N \times N$ matrices) and $b\colon J \to \mathbb{R}^N$ are piecewise continuous functions, and J is an interval with $\tau \in J$. It is clear that the function $f\colon (t, z) \mapsto A(t)z + b(t)$ satisfies assumption **A** and so, anticipating the theory of Chapter 4, the linear initial-value problem has a unique solution defined on J, that is, a continuous function $x\colon J \to \mathbb{R}^N$ with the properties: $x(\tau) = \xi$ and, for every compact interval $K \subset J$, $\dot{x}(t) = A(t)x(t) + b(t)$ for all $t \in K \backslash E$, where E is the finite set of points $t \in K$ at which either A or b (possibly both) fails to be continuous. The linear structure of the problem renders it amenable to a more detailed analysis: in particular, we can establish not only existence and uniqueness but can provide an explicit formula for the unique solution of the initial-value problem. This we do in the next chapter.

Exercise 1.4

Let J be an interval, let $a\colon J \to \mathbb{R}$ be piecewise continuous and consider the initial-value problem

$$\dot{x}(t) = a(t)x(t), \quad x(\tau) = \xi, \tag{1.18}$$

for fixed (but arbitrary) $\tau \in J$ and $\xi \in \mathbb{R}$.

(a) Consider the specific case wherein $\xi = 0$. Clearly, the zero function, given by $x(t) = 0$ for all $t \in J$, is a solution. Prove that this is the only solution on J.

(b) Now consider the general case of arbitrary $\xi \in \mathbb{R}$. Verify that

$$x\colon J \to \mathbb{R}, \quad t \mapsto \exp\left(\int_\tau^t a(s)\mathrm{d}s\right) \xi$$

is a solution and, using the result in (a), deduce that this is the unique maximal solution of the initial value problem (1.18).

(c) Assume that $J = \mathbb{R}$. Prove that, if $\xi \neq 0$, then the unique maximal solution $x\colon \mathbb{R} \to \mathbb{R}$ of the initial-value problem (1.18) has the property $x(t) \to 0$ as $t \to \infty$ if, and only if,

$$\int_\tau^t a(s)\mathrm{d}s \to -\infty \quad \text{as} \ \ t \to \infty.$$

(d) Assume that $J = \mathbb{R}$ and that a is periodic of period $T > 0$ (that is, $a(t + T) = a(t)$ for all $t \in \mathbb{R}$). Prove that, if $\xi \neq 0$, then the unique maximal solution $x\colon \mathbb{R} \to \mathbb{R}$ of (1.18) satisfies $x(t) \to 0$ as $t \to \infty$ if, and only if,

$$\int_0^T a(s)\mathrm{d}s < 0.$$

Exercise 1.5

Let J be an interval, let $a, b\colon J \to \mathbb{R}$ be piecewise continuous and consider the initial-value problem

$$\dot{x}(t) = a(t)x(t) + b(t), \quad x(\tau) = \xi, \tag{1.19}$$

for fixed (but arbitrary) $\tau \in J$ and $\xi \in \mathbb{R}$.

Multiply both sides of the differential equation in (1.19) by the function $\mu\colon J \to \mathbb{R}$ (called an *integrating factor*) given by

$$\mu(t) := \exp\left(-\int_\tau^t a(s)\mathrm{d}s\right)$$

and integrate to obtain a solution $x\colon J \to \mathbb{R}$ of the initial-value problem (1.19). Deduce that x is the unique maximal solution of (1.19).

Exercise 1.6

Consider again the model of cooperating populations described in Section 1.1.4. Given that the optimal control is a piecewise constant function $[0,1] \to [0,1]$, with one discontinuity at $t_s \in (0,1)$, of the form

$$t \mapsto u(t) = \begin{cases} 1, & t \in [0, t_s) \\ 0, & t \in [t_s, 1], \end{cases}$$

prove that $t_s = 1 - \ln\left(a/(a - \mu + \nu)\right)/(\mu - \nu)$.

1.3 Related texts

In our treatment of initial-value problems for ordinary differential equations with emphasis on analysis, qualitative behaviour, stability and control, we have endeavoured to make the text as self-contained as possible. Nevertheless, the reader may wish to consult other works. The literature on ordinary differential equations is vast: contributions that have a similar flavour to ours include [2], [14] and [21]. We remark, however, that [2] is pitched at a more advanced level than that adopted here. The text [4] contains a blend of the theory of ordinary differential equations with applications in a wide variety of specific problems. Other contributions to the textbook literature on ordinary differential equationa include [1], [10] and [18]. The associated aspects of control and stabilization also have an extensive bibliography, from which we suggest the following texts as appropriate sources: [3], [8], [19] and [20].

2

Linear differential equations

Systems of linear differential equations form the focus of our first line of investigation. In particular, we will develop a *theory of existence and uniqueness* of solutions of homogeneous initial-value problems of the form $\dot{x}(t) = A(t)x(t)$, $x(\tau) = \xi$, under the assumption that A is piecewise continuous. The special case of constant A forms an important sub-class for which, as we shall see, the solution x of the initial-value problem is given in terms of the matrix exponential function by $x(t) = \exp(A(t - \tau))\xi$ for all $t \in \mathbb{R}$. Then, we extend the existence and uniqueness theory to inhomogeneous initial-value problems of the form $\dot{x}(t) = A(t)x(t) + b(t)$, $x(\tau) = \xi$, where b is a piecewise continuous extraneous input or forcing function. In certain circumstances, the function b is open to choice, and may be chosen so as to ensure that the unique solution of the initial-value problem has some desirable properties: questions relating to the extent to which solutions may be influenced through the choice of input form the basis of *linear control theory* - fundamentals of which form the focus of Chapter 3.

For a periodic function A (that is, a function A with the property that, for some $p > 0$, $A(t + p) = A(t)$ for all $t \in \mathbb{R}$), it is intuitively reasonable to surmise the existence of periodic solutions of the homogeneous differential equation $\dot{x}(t) = A(t)x(t)$: we investigate this and related issues pertaining to such periodic differential equations, within the framework of what is traditionally referred to as *Floquet theory*[1].

In this chapter, we make free use of the material presented in Appendices

---

[1] Gaston Floquet (1847-1920), French.

H. Logemann and E. P. Ryan, *Ordinary Differential Equations*, Springer Undergraduate Mathematics Series, DOI: 10.1007/978-1-4471-6398-5_2, © Springer-Verlag London 2014

A.1-A.3, including generalized eigenspaces, matrix norms, the concepts of piece-wise continuous and piecewise continuously differentiable functions, and the triangle inequality for integrals.

2.1 Homogeneous linear systems

Whilst we are primarily interested in linear differential equations over the real field $\mathbb{R}$, the ensuing analysis applies equally to differential equations over the complex field $\mathbb{C}$. On occasions, it will prove notationally and analytically convenient to consider the complex case. For this reason, we develop the theory in the context of a field $\mathbb{F}$ which is either $\mathbb{R}$ or $\mathbb{C}$ (precisely which of these being largely immaterial).

Let J be an interval and let $A \colon J \to \mathbb{F}^{N \times N}$ be a piecewise continuous function (see Appendix A.3) from J to the space $\mathbb{F}^{N \times N}$ of $N \times N$ matrices with entries in $\mathbb{F}$ and equipped with the norm induced by the 2-norm on $\mathbb{F}^N$:

$$\|L\| := \sup_{z \neq 0} \frac{\|Lz\|}{\|z\|}$$

(see Appendix A.2).

First, we will consider the issue of existence and uniqueness of solutions of the linear homogeneous initial-value problem

$$\dot{x}(t) = A(t)x(t), \quad x(\tau) = \xi \tag{2.1}$$

for initial data $(\tau, \xi) \in J \times \mathbb{F}^N$. Since A is not continuous, but only piecewise continuous, it would be unreasonable to expect that there exists a continuously differentiable function $x \colon J \to \mathbb{F}^N$ satisfying the initial-value problem (2.1).

Exercise 2.1

Let $N = 1$, $J = [-1, 1]$ and $\tau = 0$. Provide an example of a piecewise continuous function $A \colon J \to \mathbb{R}$ and $\xi \in \mathbb{R}$ with the property that there does not exist a continuously differentiable function $x \colon J \to \mathbb{R}$ such that $x(0) = \xi$ and $\dot{x}(t) = A(t)x(t)$ for all $t \in [-1, 1]$

By a *solution* of (2.1) we mean a continuous function $x \colon J_x \to \mathbb{F}^N$ satisfying

$$x(t) = \xi + \int_\tau^t A(\sigma)x(\sigma) \, \mathrm{d}\sigma \quad \forall \, t \in J_x,$$

where $J_x \subset J$ is an interval such that $\tau \in J_x$. Note that, by Theorems A.30 and A.31 (generalized fundamental theorems of calculus), $x \colon J_x \to \mathbb{F}^N$ is a solution of (2.1) if, and only if, x is piecewise continuously differentiable (Appendix A.3), with $x(\tau) = \xi$ and

$$\dot{x}(t) = A(t)x(t) \ \ \forall t \in J_x \backslash E,$$

where E is the set of points in J at which A fails to be continuous. Since A is piecewise continuous, the set E is "small" in the sense, that, for all $t_1, t_2 \in J$ with $t_1 < t_2$, the intersection $E \cap [t_1, t_2]$ has at most finitely many elements. Note that not every point in E is necessarily a point of discontinuity of a solution of (2.1) (for example, if $\xi = 0$, then the zero function is a solution).

Exercise 2.2

Provide an example of discontinuous A and $\xi \neq 0$ with the property that there exists a solution $x \colon J \to \mathbb{F}^N$ of (2.1) and a point $\sigma \in E$ such that x is continuously differentiable in an open interval containing σ.

If A is continuous on J, then every solution $x : J_x \to \mathbb{F}^N$ is continuously differentiable and (2.1) is satisfied for all $t \in J_x$.

In certain contexts, the initial condition in (2.1) is not relevant, in which case we say that a continuous function $x \colon J_x \to \mathbb{F}^N$, where $J_x \subset J$ is an interval, is a solution of the differential equation $\dot{x}(t) = A(t)x(t)$ if there exists $\tau \in J_x$ such that

$$x(t) = x(\tau) + \int_\tau^t A(\sigma)x(\sigma)\,\mathrm{d}\sigma \ \ \forall t \in J_x. \tag{2.2}$$

Note that, by Theorems A.30 and A.31, $x \colon J_x \to \mathbb{F}^N$ is a solution of the differential equation in this sense if, and only if, x is piecewise continuously differentiable and the differential equation $\dot{x}(t) = A(t)x(t)$ is satisfied for every $t \in J_x$ which is not a point of discontinuity of A. The next exercise asserts that, if (2.2) holds for some $\tau \in J_x$, then (2.2) holds for all $\tau \in J_x$.

Exercise 2.3

Let $x \colon J_x \to \mathbb{F}^N$ be a solution of the differential equation $\dot{x}(t) = A(t)x(t)$. Show that

$$x(t_2) - x(t_1) = \int_{t_1}^{t_2} A(\sigma)x(\sigma)\,\mathrm{d}\sigma \ \ \forall t_1, t_2 \in J_x.$$

Our goals are to show that, for each $(\tau, \xi) \in J \times \mathbb{F}^N$, (2.1) admits precisely one solution defined on J and to characterize that solution explicitly in terms of A, τ and ξ. In particular, we will establish the existence of a map $\Phi \colon J \times J \to \mathbb{F}^{N \times N}$ – referred to as the *transition matrix function* – such that $J \to \mathbb{F}^N, t \mapsto \Phi(t, \tau)\xi$ is the unique solution on J of (2.1).

2.1.1 Transition matrix function

To make progress, a number of preliminary technicalities are required.

Lemma 2.1

Define the sequence (M_n) of continuous matrix-valued functions $M_n \colon J \times J \to \mathbb{F}^{N \times N}$ by the recursion:

$$M_1(t, s) := I, \quad M_{n+1}(t, s) := I + \int_s^t A(\sigma) M_n(\sigma, s) \, \mathrm{d}\sigma \ \ \forall (t, s) \in J \times J, \ \forall n \in \mathbb{N}.$$

For each closed and bounded interval $[a, b] \subset J$, the sequence (M_n) is uniformly convergent on $[a, b] \times [a, b]$.

Proof

First note that

$$M_{n+1}(t, s) - M_n(t, s) = \int_s^t A(\sigma_1) \int_s^{\sigma_1} A(\sigma_2) \cdots \int_s^{\sigma_{n-1}} A(\sigma_n) \, \mathrm{d}\sigma_n \cdots \mathrm{d}\sigma_2 \, \mathrm{d}\sigma_1$$

$$\forall (t, s) \in J \times J, \ \forall n \in \mathbb{N} \quad (2.3)$$

and

$$\int_s^t \int_s^{\sigma_1} \cdots \int_s^{\sigma_{n-1}} \mathrm{d}\sigma_n \cdots \mathrm{d}\sigma_2 \mathrm{d}\sigma_1 = \frac{(t - s)^n}{n!} \ \ \forall (t, s) \in J \times J, \ \forall n \in \mathbb{N}, \quad (2.4)$$

as can be easily verified (see Exercise 2.4). Let $a, b \in J$, with $a < b$, be arbitrary and write $X := [a, b] \times [a, b]$. Since A is piecewise continuous, there exists $K > 0$ such that

$$\|A(t)\| \leq K \ \ \forall t \in [a, b],$$

which, in conjunction with (2.3), (2.4) and the triangle inequality for integrals (see Proposition A.28), yields

$$\|M_{n+1}(t, s) - M_n(t, s)\| \leq K^n \left| \int_s^t \int_s^{\sigma_1} \cdots \int_s^{\sigma_{n-1}} \mathrm{d}\sigma_n \cdots \mathrm{d}\sigma_2 \mathrm{d}\sigma_1 \right|$$

$$= \frac{K^n |t - s|^n}{n!} \leq \frac{K^n (b - a)^n}{n!} \ \ \forall (t, s) \in X, \ \forall n \in \mathbb{N}.$$

Define the real sequence (m_n) by

$$m_1 := 1, \quad m_{n+1} := \frac{K^n (b - a)^n}{n!} \ \ \forall n \in \mathbb{N},$$

and note that the series $\sum_{n=1}^{\infty} m_n$ is convergent, with limit $\exp(K(b-a))$. Let (f_n) be the sequence of functions $f_n \in C(X, \mathbb{F}^{N \times N})$ given by

$$f_1(t,s) := M_1(t,s) = I, \quad \forall\, (t,s) \in X$$
$$f_{n+1}(t,s) := M_{n+1}(t,s) - M_n(t,s) \quad \forall\, (t,s) \in X, \quad \forall\, n \in \mathbb{N}.$$

Then,

$$\|f_n(t,s)\| \leq m_n \quad \forall\, (t,s) \in X, \ \forall\, n \in \mathbb{N}.$$

By the Weierstrass[2] criterion (Corollary A.23), the series $\sum_{n=1}^{\infty} f_n$ is uniformly convergent. Equivalently, the sequence (S_n) of its partial sums $S_n := \sum_{k=1}^{n} f_k$ is uniformly convergent on X. Noting that $S_n(t,s) = M_n(t,s)$ for all $(t,s) \in X$, we may conclude that the sequence (M_n) is uniformly convergent on $X = [a,b] \times [a,b]$. $\qquad\square$

Exercise 2.4

Prove that (2.3) and (2.4) hold.

In view of Lemma 2.1 and since $[a,b] \subset J$ is arbitrary, we may define a function $\Phi \colon J \times J \to \mathbb{F}^{N \times N}$ by setting

$$\Phi(t,s) := \lim_{n \to \infty} M_n(t,s) \quad \forall\,(t,s) \in J \times J. \tag{2.5}$$

Since each M_n is continuous and, by Lemma 2.1, the sequence (M_n) converges uniformly on $X = [a,b] \times [a,b]$ for all $a,b \in J$ with $a < b$, it follows that Φ is continuous (see Proposition A.22). Moreover, for $n \geq 2$,

$$M_n(t,s) = M_1(t,s) + \sum_{k=1}^{n-1} \big(M_{k+1}(t,s) - M_k(t,s)\big) \quad \forall\,(t,s) \in J \times J,$$

and thus, invoking (2.3), we have

$$\Phi(t,s) = I + \int_s^t A(\sigma_1)\mathrm{d}\sigma_1 + \int_s^t A(\sigma_1) \int_s^{\sigma_1} A(\sigma_2)\mathrm{d}\sigma_2 \,\mathrm{d}\sigma_1$$
$$+ \int_s^t A(\sigma_1) \int_s^{\sigma_1} A(\sigma_2) \int_s^{\sigma_2} A(\sigma_3)\mathrm{d}\sigma_3\,\mathrm{d}\sigma_2\,\mathrm{d}\sigma_1 + \cdots$$
$$\forall\,(t,s) \in J \times J. \tag{2.6}$$

Note that $\Phi(t,t) = I$ for all $t \in J$. The function Φ is referred to as the *transition matrix function*; A is said to be its *generator* or, alternatively, we say that Φ *is generated by* A. The series representation of Φ given in (2.6) is the *Peano-Baker*[3] *series*. It converges to Φ uniformly on $[a,b] \times [a,b]$ for every interval $[a,b] \subset J$.

---

[2] Karl Theodor Wilhelm Weierstrass (1815-1897), German.
[3] Giuseppe Peano (1858-1932), Italian; Henry Frederick Baker (1866-1956), British.

Example 2.2

Let $\mathbb{F} = \mathbb{R}$ and let $A: \mathbb{R} \to \mathbb{R}^{2 \times 2}$ be given by $A(t) = \begin{pmatrix} 0 & 2t \\ 0 & 0 \end{pmatrix}$.

Noting that $A(t)A(s) = 0$ for all $t, s \in \mathbb{R}$, we see that the Peano-Baker series terminates after two terms to give

$$\Phi(t, s) = I + \int_s^t A(\sigma) \, d\sigma = \begin{pmatrix} 1 & t^2 - s^2 \\ 0 & 1 \end{pmatrix} \quad \forall\, (t, s) \in \mathbb{R} \times \mathbb{R}.$$

$\triangle$

If $J = \mathbb{R}$ and A is constant, then the Peano-Baker series gives

$$\Phi(t, \tau) = I + (t - \tau)A + \frac{(t - \tau)^2 A^2}{2!} + \cdots = \sum_{k=0}^{\infty} \frac{(t - \tau)^k}{k!} A^k$$

$$= \exp(A(t - \tau)) \quad \forall\, t, \tau \in \mathbb{R} \tag{2.7}$$

and so we identify Φ with the matrix exponential function: in particular,

$$\Phi(t, 0) = \sum_{k=0}^{\infty} \frac{t^k A^k}{k!} = \exp(At) \quad \forall\, t \in \mathbb{R},$$

$$\Phi(t, \tau) = \Phi(t - \tau, 0) \quad \forall\, t, \tau \in \mathbb{R}.$$

For further details on the matrix exponential, see Proposition A.27.

Exercise 2.5

Assume that $A: \mathbb{R} \to \mathbb{F}^{N \times N}$ is such that, for all $t, s \in \mathbb{R}$, the matrices $A(t)$ and $A(s)$ commute. Show that the transition matrix function Φ is given by

$$\Phi(t, \tau) = \exp\left(\int_\tau^t A(\sigma) \, d\sigma \right).$$

We proceed to establish basic properties of the transition matrix function.

Corollary 2.3

The transition matrix function Φ satisfies

$$\Phi(t, s) = I + \int_s^t A(\sigma)\Phi(\sigma, s) \, d\sigma \quad \forall\, (t, s) \in J \times J. \tag{2.8}$$

Moreover, for each $s \in J$, the function $t \mapsto \Phi(t, s)$ is piecewise continuously differentiable with derivative

$$\partial_1 \Phi(t, s) = A(t)\Phi(t, s) \quad \forall\, t \in J \backslash E$$

where $E \subset J$ is the set of points at which A fails to be continuous.

Proof

The identity (2.8) follows from (2.5), the defining equation for Φ, in conjunction with Lemma 2.1 and Theorem A.32. The remaining claims are an immediate consequence of (2.8) and Theorem A.30. $\qquad\square$

The next result (the so-called Gronwall[4] lemma) is a basic tool in differential and integral equations. It will not only be used in this chapter, but it will also be invoked, in Chapter 4, in the context of nonlinear differential equations.

Lemma 2.4 (Gronwall's lemma)

Let $I \subset \mathbb{R}$ be an interval, let $\tau \in I$, and let $g, h : I \to [0, \infty)$ be continuous. If, for some constant $c \geq 0$,

$$g(t) \leq c + \left| \int_\tau^t h(\sigma) g(\sigma) \, d\sigma \right| \quad \forall\, t \in I, \tag{2.9}$$

then

$$g(t) \leq c \exp\left(\left| \int_\tau^t h(\sigma) \, d\sigma \right| \right) \quad \forall\, t \in I. \tag{2.10}$$

Note that whilst (2.9) (the hypothesis in Lemma 2.4) is an inequality in g (involving c and h), the inequality (2.10) (the conclusion in Lemma 2.4) provides a bound for g in terms of c and h.

Proof of Lemma 2.4

Define $G, H : I \to [0, \infty)$ by setting

$$G(t) := c + \left| \int_\tau^t h(\sigma) g(\sigma) \, d\sigma \right| \quad \text{and} \quad H(t) = \left| \int_\tau^t h(\sigma) \, d\sigma \right| \quad \forall\, t \in I.$$

By hypothesis, $0 \leq g(t) \leq G(t)$ for all $t \in I$. Let $t \in I$ be arbitrary. We consider two cases: $t \geq \tau$ and $t < \tau$.
Case 1. Assume that $t \geq \tau$. The inequality in (2.10) evidently holds for $t = \tau$. Hence, without loss of generality we may assume that $t > \tau$. Then

$$G(s) = c + \int_\tau^s h(\sigma) g(\sigma) \, d\sigma \quad \text{and} \quad H(s) = \int_\tau^s h(\sigma) \, d\sigma \quad \forall\, s \in [\tau, t].$$

Differentiation yields

$$G'(s) = h(s) g(s) \leq h(s) G(s) = H'(s) G(s) \quad \forall\, s \in [\tau, t].$$

---

[4] Thomas Hakon Grönwall (1877-1932), Swedish.

Therefore,

$$\big(G(s)\exp\big(-H(s)\big)\big)' = (G'(s) - H'(s)G(s))\exp\big(-H(s)\big) \le 0 \;\; \forall\, s \in [\tau, t]$$

which, on integration, gives

$$G(t)\exp\big(-H(t)\big) \le G(\tau) = c.$$

Hence, we arrive at the requisite inequality

$$g(t) \le G(t) \le c\exp\big(H(t)\big) = c\exp\left(\left|\int_\tau^t h(s)\mathrm{d}s\right|\right).$$

Case 2. Assume that $t < \tau$. In this case,

$$G(s) = c + \int_s^\tau h(\sigma)g(\sigma)\,\mathrm{d}\sigma \;\; \text{and} \;\; H(s) = \int_s^\tau h(\sigma)\,\mathrm{d}\sigma \;\; \forall\, s \in [t, \tau],$$

and differentiation yields

$$G'(s) = -h(s)g(s) \ge -h(s)G(s) = H'(s)G(s) \;\; \forall\, \sigma \in [t, \tau].$$

An argument analogous to that used in Case 1 gives the desired inequality. □

Exercise 2.6

In the above proof, complete Case 2 by providing an argument similar to that of Case 1.

We are now in a position to state and prove the existence and uniqueness result which asserts that the initial-value problem (2.1) has precisely one solution defined on J.

Theorem 2.5

Let $(\tau, \xi) \in J \times \mathbb{F}^N$. The function

$$x\colon J \to \mathbb{F}^N, \;\; t \mapsto x(t) := \Phi(t, \tau)\xi. \tag{2.11}$$

is a solution of the initial-value problem (2.1). Moreover, if $y\colon J_y \to \mathbb{F}^N$ is also a solution of (2.1), then $y(t) = x(t)$ for all $t \in J_y$.

Proof

Let $(\tau, \xi) \in J \times \mathbb{F}^N$ be arbitrary. It is immediate that the function x given by (2.11) is a solution of (2.1), since, by Corollary 2.3,

$$x(t) = \Phi(t, \tau)\xi = \xi + \int_\tau^t A(\sigma)\Phi(\sigma, \tau)\xi\,\mathrm{d}\sigma = \xi + \int_\tau^t A(\sigma)x(\sigma)\,\mathrm{d}\sigma \;\; \forall\, t \in J.$$

Let $y \colon J_y \to \mathbb{F}^N$ be another solution of (2.1). Then

$$e(t) := x(t) - y(t) = \int_\tau^t A(\sigma)\big(x(\sigma) - y(\sigma)\big)\mathrm{d}\sigma = \int_\tau^t A(\sigma)e(\sigma)\,\mathrm{d}\sigma \quad \forall t \in J_y.$$

Invoking the triangle inequality for integrals (Proposition A.28), we conclude

$$\|e(t)\| \leq \left| \int_\tau^t \|A(\sigma)\|\|e(\sigma)\|\,\mathrm{d}\sigma \right| \quad \forall t \in J_y.$$

By Gronwall's lemma (Lemma 2.4), it follows that $e(t) = 0$ for all $t \in J_y$, showing that $y(t) = x(t)$ for all $t \in J_y$. □

Further properties of the transition matrix function readily follow.

Corollary 2.6

For all $t, \sigma, \tau \in J$,

$$\Phi(\tau, \tau) = I, \quad \Phi(t, \tau) = \Phi(t, \sigma)\Phi(\sigma, \tau) \quad \text{and} \quad \Phi^{-1}(t, \tau) = \Phi(\tau, t).$$

Proof

Let $\sigma, \tau \in J$ and $\xi \in \mathbb{F}^N$ be arbitrary. The first identity follows immediately from (2.5), the defining equation for Φ, and the definition of M_n (see Lemma 2.1). To prove the second identity, set $\zeta := \Phi(\sigma, \tau)\xi$ and define the functions $y, z \colon J \to \mathbb{F}^N$ by $y(t) := \Phi(t, \tau)\xi$ and $z(t) = \Phi(t, \sigma)\zeta$. By Theorem 2.5, y is the unique solution of the initial-value problem $\dot{x}(t) = A(t)x(t)$, $x(\tau) = \xi$, and z is the unique solution of the initial-value problem

$$\dot{x}(t) = A(t)x(t), \quad x(\sigma) = \zeta. \tag{2.12}$$

Noting that $y(\sigma) = \Phi(\sigma, \tau)\xi = \zeta$, we see that y also solves the initial-value problem (2.12). Hence, by Theorem 2.5, $y(t) = z(t)$ for all $t \in J$, and thus, in particular,

$$\Phi(t, \sigma)\Phi(\sigma, \tau)\xi = \Phi(t, \sigma)\zeta = z(t) = y(t) = \Phi(t, \tau)\xi$$

Since $\xi \in \mathbb{F}^N$ is arbitrary, we have $\Phi(t, \sigma)\Phi(\sigma, \tau) = \Phi(t, \tau)$. Finally, as an immediate consequence of this identity, we have

$$\Phi(\tau, t)\Phi(t, \tau) = \Phi(\tau, \tau) = I,$$

and so $\Phi(t, \tau)$ is invertible with inverse $\Phi^{-1}(t, \tau) = \Phi(\tau, t)$. □

Exercise 2.7

Let Φ be the transition matrix function generated by $A\colon J \to \mathbb{F}^{N \times N}$. Define $\tilde{A}$ by $\tilde{A}(t) = -A^*(t)$ for all $t \in J$. Prove that the transition matrix function $\tilde{\Phi}$ generated by $\tilde{A}$ is given by

$$\tilde{\Phi}(t,s) = \Phi^*(s,t) \quad \forall\, (t,s) \in J \times J.$$

Here M^* denotes the Hermitian transposition of a matrix M (see also Appendix A.1). (*Hint.* Prove that, if $x\colon J \to \mathbb{F}^N$ is a solution of $\dot{x}(t) = A(t)x(t)$ and $y\colon J \to \mathbb{F}^N$ is a solution of $\dot{y}(t) = -A^*(t)y(t)$, then, for some scalar c, we have $\langle x(t), y(t) \rangle = c$ for all $t \in J$.)

2.1.2 Solution space

Let $\mathcal{S}_{\text{hom}}$ denote the set of all solutions $x\colon J \to \mathbb{F}^N$ of the homogeneous differential equation $\dot{x}(t) = A(t)x(t)$, that is, the set of functions $x\colon J \to \mathbb{F}^N$ that solve the initial-value problem (2.1) for some $(\tau, \xi) \in J \times \mathbb{F}^N$. It is easy to show that the set $\mathcal{S}_{\text{hom}}$ forms a vector space, a subspace of $C(J, \mathbb{F}^N)$, the so-called *solution space* of the homogeneous differential equation. If $y_1, \ldots, y_N \in \mathcal{S}_{\text{hom}}$, then $w(t) := \det(y_1(t), \ldots, y_N(t))$ is called the *Wronskian*[5] associated with the solutions $y_1, \ldots, y_N$. Next, we establish some some properties of the solution space and the Wronskian. Recall that the *trace* of a square matrix $M = (m_{ij}) \in \mathbb{F}^{N \times N}$ is defined by $\operatorname{tr} M := \sum_{j=1}^{N} m_{jj}$, the sum if its diagonal elements.

Proposition 2.7

(1) Let $b_1, \ldots, b_N$ be a basis of $\mathbb{F}^N$ and let $\tau \in J$. Then the functions $y_j\colon J \to \mathbb{F}^N$ defined by $y_j(t) := \Phi(t, \tau)b_j$, $j = 1, 2, \ldots, N$, form a basis of the solution space $\mathcal{S}_{\text{hom}}$. In particular, $\mathcal{S}_{\text{hom}}$ is N-dimensional and, for every solution $x\colon J \to \mathbb{F}^N$, there exist scalars $\gamma_1, \ldots, \gamma_N$ such that $x(t) = \sum_{j=1}^{N} \gamma_j y_j(t)$ for all $t \in J$.

(2) Let $y_1, \ldots, y_N$ be in $\mathcal{S}_{\text{hom}}$ and let w be the associated Wronskian. Then

$$w(t) = w(\tau) \det \Phi(t, \tau) \quad \forall\, (t, \tau) \in J \times J, \tag{2.13}$$

and moreover, $\dot{w}(t) = (\operatorname{tr} A(t))w(t)$ for all $t \in J$ which are not points of discontinuity of A, and so

$$w(t) = w(\tau) \exp\left(\int_{\tau}^{t} \operatorname{tr} A(s)\, \mathrm{d}s \right) \quad \forall\, t \in J. \tag{2.14}$$

---

[5] Josef-Maria Hoëné de Wronski (1778-1853), Polish.

In particular, if $w(\tau) = 0$ for some $\tau \in J$, then $w(t) = 0$ for all $t \in J$, or, equivalently, if $w(\tau) \neq 0$ for some $\tau \in J$, then $w(t) \neq 0$ for all $t \in J$.

(3) Elements $y_1, \ldots, y_n$ of $\mathcal{S}_{\text{hom}}$, where $n \leq N$, are linearly independent (as elements in the vector space $C(J, \mathbb{F}^N)$) if, and only if, for every $t \in J$, the vectors $y_1(t), \ldots, y_n(t)$ are linearly independent (as elements of $\mathbb{F}^N$).

Proof

(1) Theorem 2.5 ensures that $y_1, \ldots, y_N$ are solutions and so are in $\mathcal{S}_{\text{hom}}$. Moreover, these solutions are linearly independent (in the vector space $C(J, \mathbb{F}^N)$). Indeed, if, for $\alpha_1, \ldots, \alpha_N \in \mathbb{F}$, we have $\sum_{j=1}^{N} \alpha_j y_j(t) = 0$ for all $t \in J$, then $\sum_{j=1}^{N} \alpha_j y_j(\tau) = \sum_{j=1}^{N} \alpha_j b_j = 0$, and so, by linear independence of $b_1, \ldots, b_N$ (in $\mathbb{F}^N$), it follows that $\alpha_1 = \ldots = \alpha_N = 0$. Next, we show that $y_1, \ldots, y_N$ form a basis of $\mathcal{S}_{\text{hom}}$. Let x be an arbitrary element of $\mathcal{S}_{\text{hom}}$. Then, by Theorem 2.5, $x(t) = \Phi(t, \tau) x(\tau)$ for all $t \in J$. Since $b_1, \ldots, b_N$ form a basis of $\mathbb{F}^N$, there exist scalars $\gamma_1, \ldots, \gamma_N$ such that $x(\tau) = \sum_{j=1}^{N} \gamma_j b_j$. Consequently, $x(t) = \sum_{j=1}^{N} \gamma_j \Phi(t, \tau) b_j = \sum_{j=1}^{N} \gamma_j y_j(t)$ for all $t \in J$, showing that $y_1, \ldots, y_N$ span $\mathcal{S}_{\text{hom}}$.

(2) Let $\tau \in J$ be fixed, but arbitrary. Since $y_j \in \mathcal{S}_{\text{hom}}$ for $j = 1, \ldots, N$, it follows from Theorem 2.5 that $y_j(t) = \Phi(t, \tau) y_j(\tau)$ for all $t \in J$ and all $j = 1, \ldots, N$. Hence, for all $t \in J$,

$$w(t) = \det \Phi(t, \tau) \det(y_1(\tau), \ldots, y_N(\tau)) = w(\tau) \det \Phi(t, \tau),$$

establishing (2.13). Moreover, writing $\Phi(t, \tau) = (\varphi_1(t, \tau), \ldots, \varphi_N(t, \tau))$, where $\varphi_j(t, \tau)$ denotes the j-th column of $\Phi(t, \tau)$, it follows, from the definition of the determinant (see (A.8) in Appendix A.1) and the product rule for differentiation, that, for all $t \in J$ which are not points of discontinuity of A,

$$\big(\partial_1 \det \Phi\big)(t, \tau) = \sum_{j=1}^{N} \det \big(\varphi_1(t, \tau), \ldots, \varphi_{j-1}(t, \tau), \partial_1 \varphi_j(t, \tau), \ldots, \varphi_N(t, \tau)\big),$$

where ∂_1 denotes the derivative with respect to the first argument. In the following we assume that τ is not a point of discontinuity of A. Then, since $\Phi(\tau, \tau) = I$, the above identity yields for $t = \tau$,

$$\big(\partial_1 \det \Phi\big)(\tau, \tau) = \sum_{j=1}^{N} \det \big(e_1, \ldots, e_{j-1}, A(\tau) e_j, e_{j+1}, \ldots, e_N\big),$$

where $e_1, \ldots, e_N$ denotes the canonical basis of $\mathbb{F}^N$. Denoting the entries of

$A(t)$ by $a_{ij}(t)$ it follows that

$$\left(\partial_1 \det \Phi\right)(\tau,\tau) = \sum_{j=1}^{N} a_{jj}(\tau) = \operatorname{tr} A(\tau).$$

Therefore, differentiation of (2.13) with respect to t at $t = \tau$ yields

$$\dot{w}(\tau) = w(\tau)\operatorname{tr} A(\tau). \tag{2.15}$$

The argument leading to (2.15) applies to any $\tau \in J$ which is not a point of discontinuity of A and therefore $\dot{w}(t) = (\operatorname{tr} A(t))w(t)$ for every $t \in J$ which is not a point of discontinuity of A. Furthermore, (2.14) now follows from Exercise 2.5 and Theorem 2.5.

(3) Let $y_1, \ldots, y_n$ be in $\mathcal{S}_{\text{hom}}$ and $\alpha_1, \ldots, \alpha_n \in \mathbb{F}$.
Sufficiency. Assume that $y_1(t), \ldots, y_n(t)$ are linearly independent vectors in $\mathbb{F}^N$ for all $t \in J$. It immediately follows that

$$\alpha_1 y_1 + \cdots + \alpha_n y_n = 0 \implies \alpha_k = 0, \ k = 1, \ldots, n$$

and so $y_1, \ldots, y_n$ are linearly independent in $\mathcal{S}_{\text{hom}}$.
Necessity. Let $y_1, \ldots, y_n$ be linearly independent in $\mathcal{S}_{\text{hom}}$. Let $\tau \in J$ be arbitrary. Assume that $\alpha_1 y_1(\tau) + \cdots + \alpha_n y_n(\tau) = 0$. Then, $y := \alpha_1 y_1 + \cdots + \alpha_n y_n$ solves the initial-value problem $\dot{x}(t) = A(t)x(t)$, $x(\tau) = 0$, which we know has unique solution 0. Therefore, $y = 0$ and so, by linear independence of the functions $y_1, \ldots, y_n$, we have $\alpha_k = 0$, $k = 1, \ldots, n$. This establishes linear independence of $y_1(\tau), \ldots, y_n(\tau)$ and, as $\tau \in J$ is arbitrary, the result follows. $\square$

Statement (3) of Proposition (2.7) says that linear independence of $y_1, \ldots, y_n \in \mathcal{S}_{\text{hom}}$ as functions is equivalent to linear independence of $y_1(t), \ldots, y_n(t)$ (as vectors in $\mathbb{F}^N$) for every $t \in J$. The following exercise shows that if $y_1, \ldots, y_n \in C(J, \mathbb{F}^N)$ are not required to be solutions of $\dot{x}(t) = A(t)x(t)$, then this equivalence does not hold.

Exercise 2.8

Show, by counterexample, that linear independence of $y_1, \ldots, y_n \in C(J, \mathbb{F}^N)$ does not imply linear independence of $y_1(t), \ldots, y_n(t) \in \mathbb{F}^N$ for all $t \in J$.

A *fundamental system* for the homogeneous differential equation $\dot{x}(t) = A(t)x(t)$ is a set of N linearly independent solutions, or, equivalently, a basis of $\mathcal{S}_{\text{hom}}$. If $\{\psi_1, \ldots, \psi_N\}$ is a fundamental system, then the matrix-valued function $\Psi \colon J \to \mathbb{F}^{N \times N}$ defined by

$$\Psi(t) := \left(\psi_1(t), \ldots, \psi_n(t)\right) \quad \forall t \in J$$

is said to be a *fundamental matrix* for the differential equation $\dot{x}(t) = A(t)x(t)$.

Proposition 2.8

Let Ψ be a fundamental matrix. Then $\Psi(t)$ is invertible for every $t \in J$ and

$$\Phi(t, \tau) = \Psi(t)\Psi^{-1}(\tau) \quad \forall t, \tau \in J.$$

Proof

By part (3) of Proposition 2.7 we may infer that $\Psi(t)$ is invertible for all $t \in J$. Let $\xi \in \mathbb{F}^N$ be arbitrary and define $x \colon J \to \mathbb{F}^N$ by setting $x(t) := \Psi(t)\Psi^{-1}(\tau)\xi$. Obviously, x is a linear combination of the columns of Ψ and consequently, x is a solution. Moreover, $x(\tau) = \xi$, so that x solves the initial-value problem (2.1). By Theorem 2.5, the function $t \mapsto \Phi(t, \tau)\xi$ is the unique solution of (2.1). Hence, $x(t) = \Phi(t, \tau)\xi$ for all $t \in J$, showing that

$$\Phi(t, \tau)\xi = \Psi(t)\Psi^{-1}(\tau)\xi \quad \forall t \in J.$$

Since ξ was arbitrary, we obtain that $\Phi(t, \tau) = \Psi(t)\Psi^{-1}(\tau)$ for all $t \in J$. $\square$

Exercise 2.9

Let $\mathbb{F} = \mathbb{R}$ and let $A \colon \mathbb{R} \to \mathbb{R}^{2 \times 2}$ be given by

$$A(t) := \begin{pmatrix} 0 & 1 \\ 0 & 2t \end{pmatrix}.$$

Find two linearly independent solutions of $\dot{x}(t) = A(t)x(t)$ and hence determine the transition matrix function Φ.

2.1.3 Autonomous systems

Let us now turn attention to the case of constant A, that is, we consider the *autonomous* homogeneous initial-value problem, with $J = \mathbb{R}$,

$$\dot{x}(t) = Ax(t), \quad x(\tau) = \xi \in \mathbb{F}^N, \quad \text{where } A \in \mathbb{F}^{N \times N}. \tag{2.16}$$

Recall that, for $M \in \mathbb{F}^{N \times N}$, $\exp(M) := \sum_{k=0}^{\infty}(1/k!)M^k$ and that, by (2.7),

$$\Phi(t, \tau) = \exp(A(t - \tau)) \quad \forall t, \tau \in \mathbb{R}.$$

In particular,

$$\Phi(t, 0) = \exp(At) \quad \forall t \in \mathbb{R} \quad \text{and} \quad \Phi(t, \tau) = \Phi(t - \tau, 0) \quad \forall t, \tau \in \mathbb{R}. \tag{2.17}$$

The following is an immediate corollary of Theorem 2.5.

Corollary 2.9

The function $\mathbb{R} \to \mathbb{F}^N$, $t \mapsto \exp(A(t-\tau))\xi$ is the unique solution of the autonomous homogeneous initial-value problem (2.16).

In view of (2.17), we see that the unique solution (on $\mathbb{R}$) $t \mapsto \Phi(t,\tau)\xi = \Phi(t-\tau,0)\xi$ of the initial-value problem (2.16) is simply a translation of the solution $t \mapsto \Phi(t,0)\xi$ (on $\mathbb{R}$) of the initial-value problem $\dot{x}(t) = Ax(t)$, $x(0) = \xi$. Consequently, we may assume without loss of generality that $\tau = 0$ in (2.16).

We briefly digress to record following some important properties of the matrix exponential function.

Lemma 2.10

Let $P, Q \in \mathbb{F}^{N \times N}$.

(1) If P is diagonal, that is, $P = \text{diag}(p_1, \ldots, p_n)$, then

$$\exp(P) = \text{diag}\big(\exp(p_1), \ldots, \exp(p_N)\big).$$

(2) $\exp(P^*) = (\exp(P))^*$.

(3) For all $t \in \mathbb{R}$,

$$\frac{\mathrm{d}}{\mathrm{d}t} \exp(Pt) = P \exp(Pt) = \exp(Pt)P.$$

(4) If $PQ = QP$, then $\exp(P)Q = Q\exp(P)$ and

$$\exp(P+Q) = \exp(P)\exp(Q). \tag{2.18}$$

(5) $\exp(-P)\exp(P) = \exp(P)\exp(-P) = I$, that is, $\exp(P)$ is invertible with inverse $\exp(-P)$.

Proof

The proofs of parts (1)-(3) are straightforward (see Exercise 2.10). To prove part (4), assume that P and Q commute, that is, $PQ = QP$. Then,

$$\exp(P)Q = \sum_{k=0}^{\infty} \frac{1}{k!} P^k Q = \sum_{k=0}^{\infty} \frac{1}{k!} Q P^k = Q \sum_{k=0}^{\infty} \frac{1}{k!} P^k = Q\exp(P).$$

Let $z \in \mathbb{F}^N$ and define $y: \mathbb{R} \to \mathbb{C}^N$ by setting $y(t) = \exp(Pt)\exp(Qt)z$. Using part (3) and the product rule, differentiation of y leads to

$$\dot{y}(t) = P\exp(Pt)\exp(Qt)z + \exp(Pt)Q\exp(Qt)z = (P+Q)y(t) \quad \forall t \in \mathbb{R},$$

where we have used the fact that $\exp(Pt)Q = Q\exp(Pt)$. Moreover, $y(0) = z$. The unique solution of the initial-value problem $\dot{x} = (P+Q)x$, $x(0) = z$, is the function $t \mapsto \exp((P+Q)t)z$, and so

$$\exp((P+Q)t)z = y(t) = \exp(Pt)\exp(Qt)z \quad \forall\, t \in \mathbb{R}.$$

As $z \in \mathbb{F}^N$ is arbitrary, we have $\exp((P+Q)t) = \exp(Pt)\exp(Qt)$ for all $t \in \mathbb{R}$.

Finally, statement (5) is an immediate consequence of statement (4) (on setting $Q = -P$). This completes the proof. $\qquad\square$

Exercise 2.10

Prove assertions (1)-(3) of Lemma 2.10.

Exercise 2.11

Let $P, Q \in \mathbb{F}^{N\times N}$. Show that, if P and Q do not commute, then (2.18) does not hold in general.

In the following, we will show how, in principle, N linearly independent solutions (or, equivalently, a fundamental matrix) of the autonomous differential equation (2.16), over the complex field $\mathbb{F} = \mathbb{C}$, can be computed. In this context, a pivotal role is played by the concepts of generalized eigenspaces and algebraic/geometric multiplicities of eigenvalues, the definitions of which (together with key results) can be found in Appendix A.1. For $A \in \mathbb{C}^{N\times N}$ it is convenient to define

$$\sigma(A) := \{\lambda \in \mathbb{C} : \lambda \text{ is an eigenvalue of } A\}.$$

The set $\sigma(A)$ (the set of all eigenvalues of A) is called the *spectrum* of $\sigma(A)$.

Theorem 2.11

Let $A \in \mathbb{C}^{N\times N}$. For $\lambda \in \sigma(A)$, let $m(\lambda)$ denote the algebraic multiplicity of λ, denote its associated generalized eigenspace by $E(\lambda) := \ker(A - \lambda I)^{m(\lambda)}$, and, for $z \in \mathbb{C}^N$, define $x_z \colon \mathbb{R} \to \mathbb{C}^N$ by $x_z(t) := \exp(At)z$.

(1) For $\lambda \in \sigma(A)$ and $z \in E(\lambda)$,

$$x_z(t) = e^{\lambda t} \sum_{k=0}^{m(\lambda)-1} \frac{t^k}{k!}(A - \lambda I)^k z \quad \forall\, t \in \mathbb{R}. \tag{2.19}$$

(2) Let $B(\lambda)$ be a basis of $E(\lambda)$ and write

$$\mathcal{B} := \cup_{\lambda \in \sigma(A)} B(\lambda),$$

The set of functions $\{x_z \colon z \in \mathcal{B}\}$ is a basis of the solution space of (2.16) (with $\mathbb{F} = \mathbb{C}$).

Proof

Let $\lambda \in \sigma(A)$ and $z \in E(\lambda)$. Then, $(A - \lambda I)^k z = 0$ for all $k \geq m(\lambda)$ and so

$$x_z(t) = \exp(At)z = e^{\lambda t}\big(\exp(A - \lambda I)t\big)z = e^{\lambda t} \sum_{k=0}^{m(\lambda)-1} \frac{t^k}{k!}(A - \lambda I)^k z \quad \forall\, t \in \mathbb{R},$$

establishing statement (1).

By the generalized eigenspace decomposition theorem (see Theorem A.8), $\mathcal{B}$ is a basis of $\mathbb{C}^N$ and so, by Proposition 2.7, $\{x_z : z \in \mathcal{B}\}$ is a basis of the solution space of (2.16) (with $\mathbb{F} = \mathbb{C}$), proving statement (2). $\qquad\square$

Theorem 2.11 shows that, by computing the eigenvalues of A and computing $m(\lambda)$ linearly independent generalized eigenvectors associated with λ for each $\lambda \in \sigma(A)$, N linearly independent solutions of (2.16) (with $\mathbb{F} = \mathbb{C}$) can be obtained by using formula (2.19).

The next result, a consequence of Theorem 2.11, says, roughly speaking, that the growth of $\|\exp(At)\|$ as $t \to \infty$ is determined by the spectrum of A.

Theorem 2.12

Let $A \in \mathbb{C}^{N \times N}$, set $\mu_A := \max\{\mathrm{Re}\,\lambda : \lambda \in \sigma(A)\}$ and

$$\Gamma_A := \{\gamma \in \mathbb{R} : \text{there exists } M_\gamma \geq 1 \text{ such that } \|\exp(At)\| \leq M_\gamma e^{\gamma t} \;\forall\, t \geq 0\}.$$

(1) $(\mu_A, \infty) \subset \Gamma_A$ and $\inf \Gamma_A = \mu_A$.

(2) $\mu_A \in \Gamma_A$ if, and only if, every $\lambda \in \sigma(A)$ satisfying $\mathrm{Re}\,\lambda = \mu_A$ is semisimple.

(3) Let $\gamma \in \mathbb{R}$. If, for all $\xi \in \mathbb{C}^N$, $\lim_{t \to \infty} \exp((A - \gamma I)t)\xi = 0$, then $\mu_A < \gamma$.

Before we prove Theorem 2.12, we state an immediate corollary.

Corollary 2.13

Let $A \in \mathbb{C}^{N \times N}$ and define μ_A as in Theorem 2.12.

(1) $\mu_A < 0$ if, and only if, $\|\exp(At)\|$ decays exponentially fast as $t \to \infty$.

(2) $\mu_A < 0$ if, and only if, $\lim_{t \to \infty} \exp(At)\xi = 0$ for every $\xi \in \mathbb{C}^N$.

(3) If $\mu_A = 0$, then $\sup_{t \geq 0} \|\exp(At)\| < \infty$ if, and only if, all purely imaginary eigenvalues of A are semisimple.

Proof of Theorem 2.12

Let $\lambda \in \sigma(A)$ and let $z \in \mathbb{C}^N$ be an associated generalized eigenvector. Then, by statement (1) of Theorem 2.11,

$$\exp(At)z = e^{\lambda t} \sum_{k=0}^{m(\lambda)-1} \frac{t^k}{k!}(A - \lambda I)^k z \quad \forall t \in \mathbb{R}, \qquad (2.20)$$

where $m(\lambda)$ denotes the algebraic multiplicity of λ. Let $z_1, \ldots, z_N$ be a basis of $\mathbb{C}^N$ consisting of generalized eigenvectors of A (such a basis exists by Theorem A.8) and define the invertible matrix $Z := (z_1, \ldots, z_n) \in \mathbb{C}^{N \times N}$.

(1) Let $\gamma \in (\mu_A, \infty)$ be arbitrary. We will show that $\gamma \in \Gamma_A$ (and so $(\mu_A, \infty) \subset \Gamma_A$). Noting that $\gamma > \operatorname{Re}\lambda$ for all $\lambda \in \sigma(A)$ and invoking (2.20), we may infer the existence of $L \geq 1$ such that

$$\|\exp(At)z_i\| \leq L e^{\gamma t}\|z_i\| \quad \forall t \geq 0, \ i = 1, \ldots, N. \qquad (2.21)$$

Let $\xi \in \mathbb{C}^N$ be arbitrary and write $\eta := Z^{-1}\xi$. Then $\xi = \sum_{i=1}^{N} \eta_i z_i$, where η_i, $i = 1, \ldots, N$, are the components of η and so

$$\|\exp(At)\xi\| \leq \sum_{i=1}^{N} |\eta_i|\|\exp(At)z_i\| \leq \|Z^{-1}\|\|\xi\| \sum_{i=1}^{N} \|\exp(At)z_i\|,$$

which, in conjunction with (2.21) and writing $M_\gamma := L\|Z^{-1}\| \sum_{i=1}^{N} \|z_i\|$, gives

$$\|\exp(At)\xi\| \leq M_\gamma e^{\gamma t}\|\xi\| \quad \forall \xi \in \mathbb{C}^N, \forall t \geq 0.$$

Since ξ is arbitrary, it follows that $\|\exp(At)\| \leq M_\gamma e^{\gamma t}$ for all $t \geq 0$. Therefore, $\gamma \in \Gamma_A$, showing that $(\mu_A, \infty) \subset \Gamma_A$. As an immediate consequence of the latter inclusion, we obtain $\inf \Gamma_A \leq \mu_A$. On the other hand, by (2.20), $\inf \Gamma_A \geq \mu_A$. Therefore, $\mu_A = \inf \Gamma_A$, completing the proof of statement (1).

(2) We proceed to prove statement (2). We will use the fact that an eigenvalue λ of A is semisimple if, and only if, the generalized eigenspace $E(\lambda)$ coincides with the eigenspace $\ker(A - \lambda I)$ (see Proposition A.10 in Appendix A.1). If all $\lambda \in \sigma(A)$ satisfying $\operatorname{Re}\lambda = \mu_A$ are semisimple, then, invoking (2.20), it is clear that for every generalized eigenvector z of A, there exists $L_z \geq 1$ such that $\|\exp(At)z\| \leq L_z e^{\mu_A t}\|z\|$ for all $t \geq 0$. By an argument identical to that used in the proof of the inclusion $(\mu_A, \infty) \subset \Gamma_A$, it follows that there exists $M \geq 1$ such that $\|\exp(At)\| \leq M e^{\mu_A t}$ for all $t \geq 0$, implying that $\mu_A \in \Gamma_A$. Conversely, assume that $\mu_A \in \Gamma_A$. Let $\lambda \in \sigma(A)$ be such that $\operatorname{Re}\lambda = \mu_A$ and let $z \in E(\lambda)$. Then, by (2.20), for all $t \in \mathbb{R}$,

$$\left\| \sum_{k=0}^{m(\lambda)-1} \frac{t^k}{k!}(A - \lambda I)^k z \right\| = \|e^{-\lambda t}\exp(At)\| = e^{-\mu_A t}\|\exp(At)\|$$

By hypothesis, $\sup_{t \geq 0} e^{-\mu_A t} \| \exp(At) \| < \infty$, and hence, $(A - \lambda I)z = 0$. This holds for every $z \in \bar{E}(\lambda)$ and consequently, λ is semisimple.

(3) Finally, to prove statement (3), let $\lambda \in \sigma(A)$ and let $v \in \mathbb{C}^N$ be an associated eigenvector. By hypothesis,

$$e^{(\lambda - \gamma)t} v = e^{-\gamma t} \exp(At)v = \exp((A - \gamma I)t)v \to 0 \ \text{ as } t \to \infty$$

and so $\operatorname{Re} \lambda < \gamma$. Since $\lambda \in \sigma(A)$ was arbitrary, we conclude that $\mu_A < \gamma$. $\quad\square$

Next, we turn attention to the special case of (2.16) over the real field $\mathbb{F} = \mathbb{R}$. In particular, we consider the initial-value problem

$$\dot{x}(t) = Ax(t), \quad x(0) = \xi \in \mathbb{R}^N, \quad A \in \mathbb{R}^{N \times N} \tag{2.22}$$

and will show how to compute N linearly independent *real* solutions. As a prelude, we set the following.

Exercise 2.12

Let $V \subset \mathbb{C}^N$ be a subspace that is closed under complex conjugation (that is, if $v \in V$, then $\bar{v} \in V$). Show that V has a real basis.

If A is a *real* $N \times N$ matrix and $\lambda \in \sigma(A)$ is a *real* eigenvalue of algebraic multiplicity $m(\lambda)$, then the associated generalized eigenspace $\ker(A - \lambda I)^{m(\lambda)}$ is closed under complex conjugation and so, by Exercise 2.12, has a *real basis*. This fact is used implicitly in the following theorem. Furthermore, for $z \in \mathbb{C}^N$, the real and imaginary parts of z, denoted by $\operatorname{Re} z$ and $\operatorname{Im} z$, respectively, should be interpreted in the natural componentwise manner.

Theorem 2.14

Let $A \in \mathbb{R}^{N \times N}$. For $\lambda \in \sigma(A)$, let $m(\lambda)$ denote the algebraic multiplicity of λ, denote its associated generalized eigenspace by $E(\lambda) := \ker(A - \lambda I)^{m(\lambda)}$, and let $B(\lambda)$ be a basis thereof, chosen to be a real basis whenever λ is real. For all $z \in \mathbb{C}^N$, define real solutions $x_z, y_z \colon \mathbb{R} \to \mathbb{R}^N$ of (2.22) by $x_z(t) := \exp(At)\operatorname{Re} z$ and $y_z(t) := \exp(At)\operatorname{Im} z$.

(1) Let B_0 (respectively, B_+) denote the union of all $B(\lambda)$ with $\lambda \in \sigma(A)$ and $\operatorname{Im} \lambda = 0$ (respectively, $\operatorname{Im} \lambda > 0$). The set of functions $\mathbb{R} \to \mathbb{R}^N$ given by

$$\{x_z \colon z \in B_0 \cup B_+\} \cup \{y_z \colon z \in B_+\},$$

forms a basis of the solution space of (2.22).

(2) If λ is a real eigenvalue of A, then, for every $z \in E(\lambda)$, the function x_z can be expressed in the form

$$x_z(t) = e^{\lambda t} \sum_{k=0}^{m(\lambda)-1} \frac{t^k}{k!}(A - \lambda I)^k \operatorname{Re} z . \tag{2.23}$$

(3) If $\lambda = \alpha + i\beta$, with $\beta \neq 0$, is an eigenvalue of A, then, for every $z \in E(\lambda)$, the functions x_z and y_z can be expressed as follows

$$x_z(t) = e^{\alpha t} \sum_{k=0}^{m(\lambda)-1} \frac{t^k}{k!} \big[\cos(\beta t)\operatorname{Re}\big((A - \lambda I)^k z\big) - \sin(\beta t)\operatorname{Im}\big((A - \lambda I)^k z\big) \big] .$$
$$\tag{2.24}$$

and

$$y_z(t) = e^{\alpha t} \sum_{k=0}^{m(\lambda)-1} \frac{t^k}{k!} \big[\cos(\beta t)\operatorname{Re}\big((A - \lambda I)^k z\big) + \sin(\beta t)\operatorname{Im}\big((A - \lambda I)^k z\big) \big] .$$
$$\tag{2.25}$$

Theorem 2.14 shows that, by computing the eigenvalues of A and computing $m(\lambda)$ linearly independent generalized eigenvectors associated with λ for each $\lambda \in \sigma(A)$, N linearly independent real solutions of (2.22) can be obtained by using formulas (2.23)-(2.25).

Proof of Theorem 2.14

Let $\lambda \in \sigma(A)$ and $z \in E(\lambda)$. By Theorem 2.11,

$$\exp(At)z = e^{\lambda t}\big(\exp(A - \lambda I)t\big)z = e^{\lambda t} \sum_{k=0}^{m(\lambda)-1} \frac{t^k}{k!}(A - \lambda I)^k z \quad \forall t \in \mathbb{R}.$$

Therefore, for all $t \in \mathbb{R}$,

$$x_z(t) := \exp(At)\operatorname{Re} z = \operatorname{Re}\big(\exp(At)z\big) = \operatorname{Re}\left(e^{\lambda t} \sum_{k=0}^{m(\lambda)-1} \frac{t^k}{k!}(A - \lambda I)^k z \right).$$

Statement (2) follows immediately.

Now assume $\lambda = \alpha + i\beta$, with $\beta \neq 0$. Then,

$$x_z(t) = e^{\alpha t}\operatorname{Re}\left(e^{i\beta t} \sum_{k=0}^{m(\lambda)-1} \frac{t^k}{k!}(A - \lambda I)^k z \right) \quad \forall t \in \mathbb{R},$$

from which (2.24) follows. Since $y_z(t) = \exp(At)\operatorname{Im} z = \operatorname{Im}(\exp(At)z)$ for all $t \in \mathbb{R}$, an analogous calculation yields (2.25). This establishes statement(3).

It remains to prove statement (1). To this end, observe that B_0 is either empty or is a set of real vectors in $\mathbb{R}^N$. Noting that complex eigenvalues of A occur in conjugate pairs, it is readily seen that, if $B(\lambda) = \{v_1, \ldots, v_p\}$ is a basis of $E(\lambda)$, then $\{\bar{v}_1, \ldots, \bar{v}_p\}$ is a basis of $E(\bar{\lambda})$. Writing $B_- := \{\bar{v} : v \in B_+\}$, it follows, by the generalized eigenspace decomposition theorem (see Theorem A.8), that $B_0 \cup B_+ \cup B_-$ is a basis of $\mathbb{C}^N$. If B_+ is non-empty, then writing $B_+ = \{v_1, \ldots, v_q\}$, we have

$$\operatorname{span}(B_+ \cup B_-) = \operatorname{span}\{v_1, \ldots, v_q, \bar{v}_1, \ldots, \bar{v}_q\} = \operatorname{span} B_1,$$

where

$$B_1 := \{\operatorname{Re} v_1, \ldots, \operatorname{Re} v_q, \operatorname{Im} v_1, \ldots, \operatorname{Im} v_q\}.$$

If $B_+ = \emptyset$, then $B_1 := \emptyset$. We may now conclude that $\mathcal{B} = B_0 \cup B_1$ is a real basis of $\mathbb{C}^N$. Moreover,

$$\mathcal{B} = \{x_z(0) : z \in B_0 \cup B_+\} \cup \{y_z(0) : z \in B_+\}$$

showing that the N functions $\mathbb{R} \to \mathbb{R}^N$ in the set $\{x_z : z \in B_0 \cup B_+\} \cup \{y_z : z \in B_+\}$ are linearly independent solutions of (2.22). This completes the proof. $\square$

2.2 Inhomogeneous linear systems

In the following, let $A : J \to \mathbb{F}^{N \times N}$ and $b : J \to \mathbb{F}^N$ be piecewise continuous and let Φ be the transition matrix function generated by A. We will consider the issue of existence and uniqueness of solutions of the linear inhomogeneous initial-value problem

$$\dot{x}(t) = A(t)x(t) + b(t), \quad x(\tau) = \xi, \quad (\tau, \xi) \in J \times \mathbb{F}^N. \tag{2.26}$$

A *solution* of (2.26) is a continuous function $x : J_x \to \mathbb{F}^N$ satisfying

$$x(t) = \xi + \int_\tau^t \big(A(\sigma)x(\sigma) + b(\sigma)\big)d\sigma \quad \forall t \in J_x.$$

where $J_x \subset J$ is an interval such that $\tau \in J_x$. By Theorems A.30 and A.31, $x : J_x \to \mathbb{R}^N$ is a solution of (2.26) if, and only if, x is piecewise continuously differentiable, $x(\tau) = \xi$ and

$$\dot{x}(t) = A(t)x(t) + b(t) \quad \forall t \in J_x \backslash E,$$

where E is the set of points in J at which A or b fail to be continuous. Piecewise continuity of A and b implies that the set E is "small" in the sense that, for all $t_1, t_2 \in J$ with $t_1 < t_2$, the intersection $E \cap [t_1, t_2]$ has at most finitely many elements. If A and b are continuous on J, then x is continuously differentiable and the differential equation in (2.26) holds for all $t \in J$.

Theorem 2.15

Let $(\tau, \xi) \in J \times \mathbb{F}^N$. The function

$$x: J \to \mathbb{F}^N, \quad t \mapsto \Phi(t, \tau)\xi + \int_\tau^t \Phi(t, \sigma)b(\sigma)\, d\sigma. \qquad (2.27)$$

is a solution of the initial-value problem (2.26). Moreover, if $y: J_y \to \mathbb{F}^N$ is another solution of (2.26), then $y(t) = x(t)$ for all $t \in J_y$.

Proof

Let $(\tau, \xi) \in J \times \mathbb{F}^N$ be arbitrary. We first show that x, given by (2.27), is a solution. Invoking Corollary 2.3, we have

$$x(t) = \left(I + \int_\tau^t A(\sigma)\Phi(\sigma, \tau) d\sigma \right)\xi$$
$$+ \int_\tau^t \left(I + \int_\sigma^t A(\eta)\Phi(\eta, \sigma)d\eta \right) b(\sigma)\, d\sigma \quad \forall\, t \in J.$$

Changing the order of integration and then relabelling the variables of integration, we find

$$\int_\tau^t \int_\sigma^t A(\eta)\Phi(\eta, \sigma)b(\sigma)\, d\eta d\sigma = \int_\tau^t \int_\tau^\eta A(\eta)\Phi(\eta, \sigma)b(\sigma)\, d\sigma d\eta$$
$$= \int_\tau^t A(\sigma) \int_\tau^\sigma \Phi(\sigma, \eta)b(\eta)\, d\eta d\sigma\,.$$

Therefore,

$$x(t) = \xi + \int_\tau^t \left(A(\sigma) \left(\Phi(\sigma, \tau)\xi + \int_\tau^\sigma \Phi(\sigma, \eta)b(\eta)\, d\eta \right) + b(\sigma) \right) d\sigma$$
$$= \xi + \int_\tau^t \left(A(\sigma)x(\sigma) + b(\sigma) \right) d\sigma \quad \forall\, t \in J$$

and so x is a solution of (2.26).

Finally, let $y\colon J_y \to \mathbb{R}^N$ be another solution of (2.26). Then

$$e(t) := x(t) - y(t) = \int_\tau^t A(\sigma)\big(x(\sigma) - y(\sigma)\big)\mathrm{d}\sigma = \int_\tau^t A(\sigma)e(\sigma)\,\mathrm{d}\sigma \quad \forall t \in J_y.$$

Therefore, e solves the initial-value problem $\dot{e}(t) = A(t)e(t)$, $e(\tau) = 0$, and so, by Theorem 2.5, e must be the zero function. Hence, $y(t) = x(t)$ for all $t \in J_y$.   □

The formula (2.27) for the (unique) solution of the inhomogeneous initial-value problem (2.26) is frequently referred to as *the variation of parameters formula*.

In certain contexts, the initial condition in (2.26) is not relevant, in which case we say that a continuous function $x\colon J_x \to \mathbb{F}^N$, where $J_x \subset J$ is an interval, is a solution of the differential equation $\dot{x}(t) = A(t)x(t) + b(t)$ if there exists $\tau \in J_x$ such that

$$x(t) = x(\tau) + \int_\tau^t \big(A(\sigma)x(\sigma) + b(\sigma)\big)\mathrm{d}\sigma \quad \forall t \in J_x. \tag{2.28}$$

Note that, by Theorems A.30 and A.31, $x\colon J_x \to \mathbb{F}^N$ is a solution of the differential equation in this sense if, and only if, x is piecewise continuously differentiable and the differential equation $\dot{x}(t) = A(t)x(t) + b(t)$ is satisfied for every $t \in J_x$ which is not a point of discontinuity of A or b. The next exercise asserts that, if (2.28) holds for some $\tau \in J_x$, then (2.28) holds for all $\tau \in J_x$.

Exercise 2.13

Let $x\colon J_x \to \mathbb{F}^N$ be a solution of the differential equation $\dot{x}(t) = A(t)x(t) + b(t)$. Show that

$$x(t_2) - x(t_1) = \int_{t_1}^{t_2} \big(A(\sigma)x(\sigma) + b(\sigma)\big)\mathrm{d}\sigma \quad \forall t_1, t_2 \in J_x.$$

Let $\mathcal{S}_{\text{ih}}$ denote the set of all solutions $x\colon J \to \mathbb{F}^N$ of the inhomogeneous differential equation $\dot{x}(t) = A(t)x(t) + b(t)$. The following result contains information on the structure of $\mathcal{S}_{\text{ih}}$.

Corollary 2.16

Let $y \in \mathcal{S}_{\text{ih}}$. Then

$$\mathcal{S}_{\text{ih}} = y + \mathcal{S}_{\text{hom}} = \{y + x : x \in \mathcal{S}_{\text{hom}}\},$$

where $\mathcal{S}_{\text{hom}}$ is the solution space of the homogeneous system $\dot{x}(t) = A(t)x(t)$.

Exercise 2.14

Prove Corollary 2.16.

Corollary 2.16 says that S_{ih} is an affine linear space: the sum of an arbitrary solution y of the inhomogeneous problem (sometimes also called a *particular solution*) and the (linear) solution space of the associated homogeneous problem.

Finally, we consider the inhomogeneous initial-value problem with constant A, namely

$$\dot{x}(t) = Ax(t) + b(t), \quad x(\tau) = \xi \in \mathbb{F}^N, \tag{2.29}$$

where $A \in \mathbb{F}^{N \times N}$, $b \colon J \to \mathbb{F}^N$ is piecewise continuous and $\tau \in J$. By Theorem 2.15, we may immediately conclude the following.

Corollary 2.17

The function

$$x \colon J \to \mathbb{F}^N, \, t \mapsto \exp(A(t - \tau))\xi + \int_\tau^t \exp(A(t - \sigma))b(\sigma) \, d\sigma$$

is a solution of the inhomogeneous initial-value problem (2.29). Moreover, if $y \colon J_y \to \mathbb{F}^N$ is also a solution of (2.29), then $y(t) = x(t)$ for all $t \in J_y$.

2.3 Systems with periodic coefficients: Floquet theory

Periodic phenomena feature prominently in the sciences and engineering: rotation of the Earth around its axis, heart beat, alternating electric current, to mention just a few examples. Correspondingly, the study of systems with periodic coefficients is a classical theme in differential equations. Here, we turn attention to linear homogeneous systems with $J = \mathbb{R}$ and a piecewise continuous periodic function $A \colon \mathbb{R} \to \mathbb{F}^{N \times N}$ with period $p > 0$:

$$\dot{x}(t) = A(t)x(t), \quad A(t + p) = A(t) \; \forall t \in \mathbb{R}. \tag{2.30}$$

It is natural to ask if there exist periodic solutions of the homogeneous system (2.30). By a periodic solution, we mean a solution x with the property that, for some $q > 0$, $x(t) = x(t + q)$ for all $t \in \mathbb{R}$. Observe that a constant solution qualifies as a periodic solution and, since $x = 0$ is a solution of (2.30), one might argue that there always exists a periodic solution. Disregarding the zero or trivial solution, our primary concern is the existence or otherwise of *non-zero* periodic solutions and, more generally, the qualitative behaviour of solutions of (2.30).

The following example illustrates the fact that non-zero periodic solutions of (2.30) need not necessarily exist.

Example 2.18

Consider the scalar initial-value problem with $\mathbb{F} = \mathbb{R}$

$$\dot{x}(t) = (1 + \sin t)x(t), \quad x(0) = \xi.$$

Here, $A\colon t \mapsto 1 + \sin t$ is periodic with period $p = 2\pi$. The unique solution of the initial-value problem is $x\colon t \mapsto \xi e^{(1+t-\cos t)}$ which fails to be periodic for all $\xi \neq 0$. $\qquad\qquad \triangle$

We briefly digress to state a result - the *spectral mapping theorem* - which will play a key role in our investigations.

Theorem 2.19 (Spectral mapping theorem)

Let $a_n \in \mathbb{C}$, $n \in \mathbb{N}_0 := \mathbb{N} \cup \{0\}$, assume that the series $\sum_{n=0}^{\infty} a_n z^n =: f(z)$ converges for all $z \in \mathbb{C}$ and let $M \in \mathbb{C}^{N \times N}$. Then the series $f(M) := \sum_{n=0}^{\infty} a_n M^n$ converges in $\mathbb{C}^{N \times N}$ and $f(M)$ has the following properties.

(1) $\sigma(f(M)) = \{f(\lambda)\colon \lambda \in \sigma(M)\}$.

(2) If f is injective on $\sigma(M)$, then, for each $\lambda \in \sigma(M)$, the algebraic multiplicities of $f(\lambda) \in \sigma(f(M))$ and λ coincide.

(3) If f is injective on $\sigma(M)$ and $f'(\lambda) \neq 0$ whenever $\lambda \in \sigma(M)$ is not semisimple, then, for each $\lambda \in \sigma(M)$, the $f(\lambda)$-eigenspace $\ker(f(M) - f(\lambda)I)$ coincides with the λ-eigenspace $\ker(M - \lambda I)$ (and so, *a fortiori*, the geometric multiplicities of $f(\lambda)$ and λ coincide).

In order to avoid disrupting the presentation of our main concern, namely, the investigation of qualitative features of solutions of (2.30), we relegate the proof of the spectral mapping theorem to the end of the current chapter (see Section 2.4) and embark on our first task of identifying conditions under which (2.30) has a periodic solution.

Let Φ be the transition matrix function generated by A (a p-periodic function $\mathbb{R} \to \mathbb{F}^{N \times N}$). Let $\xi \in \mathbb{F}^N$, $\tau \in \mathbb{R}$ and set $y(t) := \Phi(t + p, \tau + p)\xi$ for all

$t \in \mathbb{R}$. Then, by Corollary 2.3,

$$y(t) - \xi = \int_{\tau+p}^{t+p} A(\sigma)\Phi(\sigma, \tau+p)\xi \, d\sigma$$

whence

$$y(t) - \xi = \int_{\tau}^{t} A(\sigma + p)\Phi(\sigma + p, \tau + p)\xi \, d\sigma = \int_{\tau}^{t} A(\sigma)y(\sigma) \, d\sigma \quad \forall t \in \mathbb{R}$$

and so y is the unique solution of the initial-value problem $\dot{x}(t) = A(t)x(t)$, $x(\tau) = \xi$. Therefore, $\Phi(t + p, \tau + p)\xi = \Phi(t, \tau)\xi$ for all $t \in \mathbb{R}$. Since $\tau \in \mathbb{R}$ and $\xi \in \mathbb{F}^N$ are arbitrary, we may deduce the following property of Φ:

$$\Phi(t + p, \tau + p) = \Phi(t, \tau) \quad \forall (t, \tau) \in \mathbb{R} \times \mathbb{R}. \tag{2.31}$$

Therefore, for all $(t, \tau) \in \mathbb{R} \times \mathbb{R}$,

$$\begin{aligned}
\Phi(t + p, \tau) &= \Phi(t + p, \tau + p)\Phi(\tau + p, \tau) = \Phi(t, \tau)\Phi(\tau, \tau - p) \\
&= \Phi(t, \tau)\Phi(\tau, 0)\Phi(0, \tau - p) = \Phi(t, 0)\Phi(p, \tau) \\
&= \Phi(t, 0)\Phi(p, 0)\Phi(0, \tau).
\end{aligned}$$

We may now infer by induction that, for all $n \in \mathbb{N}$,

$$\Phi(t + np, \tau) = \Phi(t, 0)\Phi^n(p, 0)\Phi(0, \tau) \quad \forall (t, \tau) \in \mathbb{R} \times \mathbb{R}. \tag{2.32}$$

Exercise 2.15

Prove, by induction, that (2.32) holds for all $n \in \mathbb{N}$.

The following result gives a necessary and sufficient condition for the existence of a non-zero periodic solution of period np, where $n \in \mathbb{N}$.

Proposition 2.20

Let $n \in \mathbb{N}$. System (2.30) has a non-zero periodic solution x of period np if, and only if, $\Phi(p, 0)$ has an eigenvalue λ such that $\lambda^n = 1$.

Proof

To prove sufficiency, assume that λ is an eigenvalue of $\Phi(p, 0)$ and $\lambda^n = 1$. Let $v \in \mathbb{C}^N$ be an associated eigenvector. Then $v \neq 0$ and $\Phi^n(p, 0)v = \lambda^n v = v$. The unique solution $x \colon \mathbb{R} \to \mathbb{F}^N$ of the initial-value problem

$$\dot{x}(t) = A(t)x(t), \quad x(0) = v,$$

is given by $x(t) = \Phi(t,0)v$. Invoking (2.32), with $\tau = 0$, gives

$$x(t + np) = \Phi(t + np, 0)v = \Phi(t,0)\Phi^n(p,0)v = \Phi(t,0)v = x(t) \ \forall t \in \mathbb{R},$$

and so x is a non-zero periodic solution of period np.

We proceed to prove necessity. To this end, assume that x is a non-zero periodic solution of (2.30), with period np. Then $v := x(0) \neq 0$ (because the zero function is the unique solution of the initial-value problem $\dot{y}(t) = A(t)y(t)$, $y(0) = 0$). Invoking (2.32), with $\tau = 0$, we have

$$\Phi(t,0)v = x(t) = x(t + np) = \Phi(t + np, 0)v = \Phi(t,0)\Phi^n(p,0)v,$$

and thus, $\Phi(t,0)\big(I - \Phi^n(p,0)\big)v = 0$. Consequently $\big(I - \Phi^n(p,0)\big)v = 0$ and so 1 is an eigenvalue of $\Phi^n(p,0)$. By Theorem 2.19 (with $f(z) = z^n$),

$$\sigma(\Phi^n(p,0)) = \big\{\lambda^n \colon \lambda \in \sigma(\Phi(p,0))\big\}.$$

Therefore, $\Phi(p,0)$ has an eigenvalue λ with the property that $\lambda^n = 1$. $\square$

Example 2.21

For $\mathbb{F} = \mathbb{R}$ and $N = 3$, consider (2.30) with $A \colon \mathbb{R} \to \mathbb{R}^{3\times 3}$ (period $p = 2\pi$) given by

$$A(t) := \begin{pmatrix} 0 & 1 & \sin t \\ 0 & 0 & 1 \\ 0 & 0 & 0 \end{pmatrix}.$$

In this case, the Peano-Baker series terminates and the state transition function Φ is given by

$$\Phi(t,\tau) = I + \int_\tau^t A(s_1)ds_1 + \int_\tau^t A(s_1) \int_\tau^{s_1} A(s_2)ds_2 ds_1$$

$$= \begin{pmatrix} 1 & t-\tau & \cos\tau - \cos t + (t-\tau)^2/2 \\ 0 & 1 & t-\tau \\ 0 & 0 & 1 \end{pmatrix}.$$

Therefore,

$$\Phi(p,0) = \Phi(2\pi, 0) = \begin{pmatrix} 1 & 2\pi & 2\pi^2 \\ 0 & 1 & 2\pi \\ 0 & 0 & 1 \end{pmatrix}$$

which evidently has eigenvalue $\lambda = 1$. By Proposition 2.20, it follows that (2.30) has a non-zero periodic solution of period 2π. Inspection of the form of Φ reveals that (2.30) can have no non-constant periodic solutions. Indeed, for every $\xi = (\xi_1, \xi_2, \xi_3)^* \in \mathbb{R}^3$ and every $\tau \in \mathbb{R}$, the function x defined by $x(t) := \Phi(t,\tau)\xi$ is unbounded (and hence not periodic) if $(\xi_2, \xi_3) \neq (0,0)$ and is constant if $(\xi_2, \xi_3) = (0,0)$. We therefore conclude that all non-zero periodic solutions are constant and are of the form $x(t) = (\xi_1, 0, 0)^*$ for all $t \in \mathbb{R}$. $\triangle$

Exercise 2.16

Let $n \in \mathbb{N}$. Assume that $\Phi(p,0)$ has an eigenvalue λ such that $\lambda^n = 1$ and that the function $\mathbb{C} \to \mathbb{C}, z \mapsto z^n$ is injective on $\sigma(\Phi(p,0))$ (the latter condition holds trivially for $n = 1$).

(a) Show that $x : \mathbb{R} \to \mathbb{F}^N$ is a np-periodic solution of (2.30) if, and only if, $x(0) \in \ker(\Phi(p,0) - \lambda I)$. (*Hint.* Inspect the proof of Proposition 2.20. Make use of Theorem 2.19.)

(b) Let $\mathcal{S}_{np}$ denote the set of all np-periodic solutions of (2.30). Show that $\mathcal{S}_{np}$ is a vector space and that $\dim \mathcal{S}_{np} = \dim \ker(\Phi(p,0) - \lambda I)$.

Exercise 2.17

Let $n \in \mathbb{N}$ and $\mu \in \mathbb{C}$. Show that system (2.30) has a non-zero solution $x : \mathbb{R} \to \mathbb{C}^N$ with the property

$$x(t + np) = \mu x(t) \quad \forall t \in \mathbb{R}$$

if, and only if, $\Phi(p,0)$ has an eigenvalue λ such that $\lambda^n = \mu$.

(*Hint.* Note that the claim is a generalization of Proposition 2.20 (which corresponds to the special case of $\mu = 1$). Inspect the the proof of Proposition 2.20 and modify it in a suitable way.)

Next we present a variant of Proposition 2.20 which provides sufficient conditions for (2.30) to have a non-constant periodic solution.

Proposition 2.22

Let $n \in \mathbb{N}$ with $n \geq 2$. If the function $\mathbb{C} \to \mathbb{C}, z \mapsto z^n$ is injective on $\sigma(\Phi(p,0))$ and if $\Phi(p,0)$ has an eigenvalue λ such that $\lambda^n = 1$ and $\lambda^k \neq 1, k = 1,\ldots,n-1$, then, for each non-zero $\xi \in \ker(\Phi(p,0) - \lambda I)$, the solution of (2.30), with initial data $x(0) = \xi$, is non-constant and periodic.

Example 2.21 shows that the above proposition does not hold in the case of $n = 1$.

Proof of Proposition 2.22

Assume that $n \in \mathbb{N}$, $n \geq 2$, λ is an eigenvalue of $\Phi(p,0)$ with $\lambda^n = 1$ and $\lambda^k \neq 1$, $1 \leq k \leq n-1$. By hypothesis, the function $f : z \mapsto z^n$ is injective on $\sigma(\Phi(p,0))$. Moreover, since $\Phi(p,0)$ is invertible, $0 \notin \sigma(\Phi(p,0))$ and so $f'(\lambda) \neq 0$ for all $\lambda \in \sigma(\Phi(p,0))$. Therefore, by the spectral mapping theorem (Theorem 2.19), $\ker(\Phi^n(p,0) - I) = \ker(\Phi(p,0) - \lambda I)$. Let $\xi \in \ker(\Phi^n(p,0) - I) = \ker(\Phi(p,0) - $

λI). With initial data $x(0) = \xi$, (2.30) has unique solution $x \colon \mathbb{R} \to \mathbb{F}^N$ given by $x(t) = \Phi(t, 0)\xi$. Invoking (2.32), we obtain

$$x(t+np) - x(t) = \big(\Phi(t+np, 0) - \Phi(t, 0)\big)\xi = \Phi(t, 0)\big(\Phi^n(p, 0) - I\big)\xi = 0 \ \forall\, t \in \mathbb{R},$$

and so x is np-periodic. It remains to show that x is not constant if $\xi \neq 0$. Seeking a contradiction, suppose that $x(t) = \Phi(t, 0)\xi$ is constant for some non-zero ξ in $\ker(\Phi(p, 0) - \lambda I)$. Fixing k, $1 \leq k \leq n-1$, we have $\Phi^k(p, 0)\xi = \lambda^k \xi \neq \xi$. Since x is constant, x is kp-periodic, whence the contradiction

$$\xi = x(0) = x(kp) = \Phi(kp, 0)\xi = \Phi^k(p, 0)\xi = \lambda^k \xi \neq \xi,$$

where we have used once again (2.32). Therefore, x is non-constant, completing the proof. □

The next exercise shows that, whilst Proposition 2.22 provides sufficient conditions for the existence of a non-constant periodic solution of (2.30), this solution may have a period smaller than p.

Exercise 2.18

Let $\mathbb{F} = \mathbb{R}$, $N = 3$, $p = 2\pi$ and let $A \colon \mathbb{R} \to \mathbb{R}^{3 \times 3}$ be the continuous, p-periodic function given by

$$A(t) := \begin{pmatrix} 0 & 1/2 & 0 \\ -1/2 & 0 & 0 \\ 0 & 0 & 1 + \sin t \end{pmatrix}.$$

Show that the transition matrix function Φ generated by A is such that

$$\Phi(t, 0) = \begin{pmatrix} \cos(t/2) & \sin(t/2) & 0 \\ -\sin(t/2) & \cos(t/2) & 0 \\ 0 & 0 & \exp(1 - \cos t + t) \end{pmatrix}.$$

Verify that $\sigma(\Phi(p, 0)) = \{-1, e^p\}$, and so the hypotheses of Proposition 2.22 hold. Show that, for each non-zero $\xi \in \ker(\Phi(p, 0) + I)$, the solution of (2.30), with initial data $x(0) = \xi$, is non-constant and periodic with period $\pi < 2\pi = p$.

Proposition 2.20 (and its generalization in Exercise 2.17) and Proposition 2.22 serve to illustrate the fact that the eigenvalues of the matrix $\Phi(p, 0)$ play a crucial role in the analysis of solutions of the system (2.30): these eigenvalues are known as *Floquet multipliers* and are all non-zero, because $\Phi(p, 0)$ is non-singular.

We now consider inhomogeneous systems with piecewise continuous periodic $A \colon \mathbb{R} \to \mathbb{F}^{N \times N}$ and $b \colon \mathbb{R} \to \mathbb{F}^N$, each with period $p > 0$:

$$\dot{x}(t) = A(t)x(t) + b(t), \quad A(t+p) = A(t), \ b(t+p) = b(t) \ \forall\, t \in \mathbb{R}. \quad (2.33)$$

Exercise 2.19

Set $\eta := \int_0^p \Phi(p,s)b(s)\mathrm{d}s$. Show that (2.33) has a p-periodic solution if, and only if, $\eta \in \mathrm{im}\,(I - \Phi(p,0))$.

We proceed to investigate further the existence of p-periodic solutions of (2.33). In the following, let $\mathcal{S}_p$ denote the set of all p-periodic solutions of the homogeneous equation (2.30). It is easy to show that $\mathcal{S}_p$ is a vector space (a subspace of $\mathcal{S}_{\mathrm{hom}}$), see Exercise 2.16. The homogeneous equation

$$\dot{y}(t) = \tilde{A}(t)y(t), \quad \text{where } \tilde{A}(t) := -A^*(t) \text{ for all } t \in \mathbb{R}, \qquad (2.34)$$

is said to be the *adjoint* equation of (2.30). The transition matrix $\tilde{\Phi}$ generated by $\tilde{A}$ is given by $\tilde{\Phi}(t,s) = \Phi^*(s,t)$ for all $s,t \in \mathbb{R}$, see Exercise 2.7. The space of all p-periodic solutions of the adjoint equation (2.34) is denoted by $\tilde{\mathcal{S}}_p$. For later purposes, we state and prove the following result which shows that the dimensions of $\mathcal{S}_p$ and $\tilde{\mathcal{S}}_p$ coincide.

Lemma 2.23

$\dim \mathcal{S}_p = \dim \tilde{\mathcal{S}}_p = \dim \ker(\Phi(p,0) - I)$.

Proof

Invoking Proposition 2.7, Exercise 2.16, and Proposition 2.20 shows that

$$\dim \mathcal{S}_p = \dim \ker(\Phi(p,0) - I) \quad \text{and} \quad \dim \tilde{\mathcal{S}}_p = \dim \ker(\tilde{\Phi}(p,0) - I).$$

Therefore, it only remains to prove that

$$\dim \ker(\Phi(p,0) - I) = \dim \ker(\tilde{\Phi}(p,0) - I). \qquad (2.35)$$

Since $\tilde{\Phi}(t,s) = \Phi^*(s,t)$, it follows that

$$(\tilde{\Phi}(p,0) - I)^* = \Phi(0,p) - I = \Phi(0,p)(I - \Phi(p,0)).$$

Consequently, since $\Phi(p,0)$ is invertible,

$$\mathrm{rk}\,(\tilde{\Phi}(p,0) - I) = \mathrm{rk}\,(\tilde{\Phi}(p,0) - I)^* = \mathrm{rk}\,(\Phi(0,p)(I - \Phi(p,0)) = \mathrm{rk}\,(\Phi(p,0) - I).$$

Finally, by the dimension formula (see (A.5) in Appendix A.1),

$$\mathrm{rk}\,(\Phi(p,0)-I)+\dim\ker(\Phi(p,0)-I) = N = \mathrm{rk}\,(\tilde{\Phi}(p,0)-I)+\dim\ker(\tilde{\Phi}(p,0)-I),$$

and (2.35) follows. $\square$

The following theorem provides a necessary and sufficient condition for the existence of p-periodic solutions of the inhomogeneous equation (2.33).

Theorem 2.24

(1) There exists a p-periodic solution of the inhomogeneous equation (2.33) if, and only if,

$$\int_0^p \langle y(s), b(s) \rangle \, ds = 0 \quad \forall \, y \in \tilde{S}_p, \tag{2.36}$$

where $\langle \, , \, \rangle$ denotes the standard inner product in $\mathbb{F}^N$ (see Appendix A.1).

(2) If (2.36) does not hold, then every solution $x : \mathbb{R} \to \mathbb{F}^N$ of (2.33) is unbounded (and, *a fortiori*, non-periodic).

Proof

Set $\eta := \int_0^p \Phi(p, s) b(s) ds$. By Exercise 2.19, (2.33) has a p-periodic solution if, and only if, $\eta \in \mathrm{im}\big(I - \Phi(p, 0)\big)$. By Theorem A.1,

$$\mathrm{im}(I - \Phi(p, 0)) = \big(\ker(I - \Phi^*(p, 0))\big)^\perp.$$

Moreover, since $I - \Phi^*(p, 0) = \Phi^*(p, 0)\big(\Phi^*(0, p) - I\big)$ and $\Phi(p, 0)$ is invertible, we have $\ker\big(I - \Phi^*(p, 0)\big) = \ker\big(\Phi^*(0, p) - I\big)$. We may now infer that (2.33) has a p-periodic solution if, and only if,

$$\langle z, \eta \rangle = 0 \quad \forall \, z \in \ker\big(\Phi^*(0, p) - I\big). \tag{2.37}$$

Therefore, to establish statement (1), it suffices to prove that (2.36) and (2.37) are equivalent. With this in mind, observe that, by part (a) of Exercise 2.16 applied in the context of the adjoint equation (2.34),

$$\tilde{S}_p = \big\{\tilde{\Phi}(\cdot, 0)z : z \in \ker\big(\tilde{\Phi}(p, 0) - I\big)\big\} = \big\{\tilde{\Phi}(\cdot, 0)z : z \in \ker\big(\Phi^*(0, p) - I\big)\big\}.$$

Therefore, (2.36) is equivalent to

$$\int_0^p \langle \tilde{\Phi}(s, 0)z, b(s) \rangle ds = 0 \quad \forall \, z \in \ker\big(\Phi^*(0, p) - I\big)$$

and, noting that

$$\int_0^p \langle \tilde{\Phi}(s, 0)z, b(s) \rangle ds = \int_0^p \langle \tilde{\Phi}(s, p)\tilde{\Phi}(p, 0)z, b(s) \rangle ds$$

$$= \int_0^p \langle \Phi^*(0, p)z, \Phi(p, s)b(s) \rangle ds$$

$$= \langle \Phi^*(0, p)z, \eta \rangle = \langle z, \eta \rangle \quad \forall \, z \in \ker\big(\Phi^*(0, p) - I\big),$$

we may conclude that (2.36) holds if, and only if, (2.37) holds, completing the proof of statement (1).

To prove statement (2), let $x : \mathbb{R} \to \mathbb{F}^N$ be an arbitrary solution of (2.33). Let $k \in \mathbb{N}_0$ and define $x_k \colon \mathbb{R} \to \mathbb{F}^N$ by $x_k(t) := x(t + kp)$ for all $t \in \mathbb{R}$. It is straightforward to show that x_k is a solution of (2.33). Therefore,

$$x_k(t) = \Phi(t,0)x_k(0) + \int_0^t \Phi(t,s)b(s)\mathrm{d}s \ \ \forall\, t \in \mathbb{R}.$$

Hence, $x_k(p) = \Phi(p,0)x_k(0) + \eta$, and thus,

$$x((k+1)p) = \Phi(p,0)x(kp) + \eta \ \ \forall\, k \in \mathbb{N}_0.$$

By induction on k, we obtain

$$x(kp) = \Phi^k(p,0)x(0) + \sum_{j=0}^{k-1} \Phi^j(p,0)\eta \ \ \forall\, k \in \mathbb{N}. \tag{2.38}$$

By hypothesis, (2.36) does not hold. Since (2.36) is equivalent to (2.37), it follows that there exists $\zeta \in \ker(\Phi^*(0,p) - I)$ such that $\langle \zeta, \eta \rangle \neq 0$. Now $\Phi^*(0,p)\zeta = \zeta$, whence $\zeta = \Phi^*(p,0)\zeta$ and so $\langle \zeta, z \rangle = \langle \zeta, \Phi^n(p,0)z \rangle$ for all $z \in \mathbb{F}^N$ and all $n \in \mathbb{N}$. Invoking (2.38) leads to

$$\langle \zeta, x(kp) \rangle = \langle \zeta, x(0) \rangle + k\langle \zeta, \eta \rangle \ \ \forall\, k \in \mathbb{N}. \tag{2.39}$$

Since $\langle \zeta, \eta \rangle \neq 0$, the right-hand side of (2.39) is unbounded and, as a consequence, the sequence $(x(kp))$ is unbounded. This shows that x is unbounded. $\qquad\square$

We record consequences of Theorem 2.24 in two corollaries, the first of which is immediate and does not require a proof.

Corollary 2.25

The inhomogeneous equation (2.33) has a p-periodic solution if, and only if, it has a bounded solution $\mathbb{R} \to \mathbb{F}^N$.

Corollary 2.26

There exists a p-periodic solution of the inhomogeneous equation (2.33) for every piecewise continuous p-periodic forcing function b if, and only if, there does not exist a non-zero p-periodic solution of the homogeneous equation (2.30) (that is, 1 is not a Floquet multiplier).

Proof

To prove sufficiency, assume that the homogeneous equation (2.30) does not have a non-zero p-periodic solution. Then $\mathcal{S}_p = \{0\}$, and thus, by Lemma 2.23, $\tilde{\mathcal{S}}_p = \{0\}$. It now follows from Theorem 2.24 that the inhomogeneous equation (2.33) has a p-periodic solution for every piecewise continuous p-periodic b. Conversely, to prove necessity, assume that (2.33) has a p-periodic solution for every piecewise continuous p-periodic b. Let $y \in \tilde{\mathcal{S}}_p$. It then follows that (2.33) has a p-periodic solution for $b = y$. Consequently, by Theorem 2.24, $\int_0^p \|y(s)\|^2 \mathrm{d}s = 0$, implying that $y = 0$. Since $y \in \tilde{\mathcal{S}}_p$ was arbitrary, we conclude that $\tilde{\mathcal{S}}_p = \{0\}$, and hence, by Lemma 2.23, $\mathcal{S}_p = \{0\}$, completing the proof. $\qquad\square$

Example 2.27

Consider the harmonic oscillator with 2π-periodic forcing

$$\ddot{y}(t) + \omega^2 y(t) = \cos t, \quad \omega \in \mathbb{R}$$

which may be expressed in the form (2.33) with constant A and 2π-periodic b given by

$$A = \begin{pmatrix} 0 & 1 \\ -\omega^2 & 0 \end{pmatrix}, \quad b(t) = \begin{pmatrix} 0 \\ \cos t \end{pmatrix}.$$

By Corollary 2.26, we may conclude the existence of a 2π-periodic solution if, and only if, $\omega^2 \neq 1$. $\qquad\triangle$

We proceed with a deeper investigation into connections between Floquet multipliers and qualitative behaviour of solutions of the homogeneous equation (2.30). In order to do so, we require the concept of matrix logarithm: for matrices G and H in $\mathbb{C}^{N \times N}$, we say that G is a *logarithm* of H if $\exp(G) = H$. If G is a logarithm of H, then, by Theorem 2.19,

$$\sigma(H) = \{e^\lambda : \lambda \in \sigma(G)\}.$$

Thus, every eigenvalue of G is a logarithm of some eigenvalue of H and, conversely, every eigenvalue of H has a logarithm which is an eigenvalue of G. We say that G is a *principal logarithm* of H if G is a logarithm of H and

$$\sigma(G) = \{\operatorname{Log}\lambda : \lambda \in \sigma(H)\}, \tag{2.40}$$

where $\operatorname{Log} : \mathbb{C}\backslash\{0\} \to \mathbb{C}$ denotes the (scalar) principal logarithm, that is, for every nonzero $z \in \mathbb{C}$, $\operatorname{Log} z$ is the unique complex number with the properties that $z = e^{\operatorname{Log} z}$ and $\operatorname{Im}(\operatorname{Log} z) \in [0, 2\pi)$.

Corollary 2.28

Let $G \in \mathbb{C}^{N \times N}$ be a principal logarithm of $H \in \mathbb{C}^{N \times N}$. Then the algebraic and geometric multiplicities of each $\lambda \in \sigma(H)$ coincide with those of $\mathrm{Log}\, \lambda \in \sigma(G)$.

Proof

By hypothesis, $H = \exp(G)$ and (2.40) holds. Since, for all $z_1, z_2 \in \sigma(G)$, we have that $z_1 - z_2 \neq 2k\pi i$ for every $k \in \mathbb{Z} \backslash \{0\}$, it follows that the exponential function $\exp$ is injective on $\sigma(G)$. Furthermore, $\exp'(z) = \exp(z) \neq 0$ for all $z \in \sigma(G)$. Consequently, the claim follows from Theorem 2.19 (with $f = \exp$). $\square$

Exercise 2.20

Find a matrix H which has a logarithm G with the property that there exists $\lambda \in \sigma(G)$ such that the algebraic and geometric multiplicities of λ do not coincide with those of $e^\lambda \in \sigma(H)$.

The question of existence of principal matrix logarithms is settled by the next result.

Proposition 2.29

If $H \in \mathbb{C}^{N \times N}$ is invertible, then there exists a principal logarithm of H.

In order to avoid disrupting the investigation of qualitative features of solutions of (2.30), we relegate the proof of Proposition 2.29 to the end of the current chapter (see Section 2.4).

Returning to the context of system (2.30), we now establish the following (Floquet) representation for $\Phi(\cdot, 0)$.

Theorem 2.30

Let $G \in \mathbb{C}^{N \times N}$ be a logarithm of $\Phi(p, 0)$. There exists a piecewise continuously differentiable p-periodic function $\Theta \colon \mathbb{R} \to \mathbb{C}^{N \times N}$, with $\Theta(0) = I$ and $\Theta(t)$ non-singular for all t, such that

$$\Phi(t, 0) = \Theta(t) \exp(t p^{-1} G) \quad \forall t \in \mathbb{R}.$$

Proof

Invoking (2.32) with $n = 1$ and $\tau = 0$, we have

$$\Phi(t + p, 0) = \Phi(t, 0)\Phi(p, 0). \tag{2.41}$$

Set $F := p^{-1}G$ and define the continuous function $\Theta \colon \mathbb{R} \to \mathbb{C}^{N \times N}$ by

$$\Theta(t) := \Phi(t, 0) \exp(-tF).$$

Then $\Theta(0) = I$, $\Theta(t)$ is nonsingular for all t, and $\Phi(t, 0) = \Theta(t) \exp(tF)$ for all $t \in \mathbb{R}$. Since $\Phi(\cdot, 0)$ is piecewise continuously differentiable, it follows that Θ is also piecewise continuously differentiable. Moreover, for all $t \in \mathbb{R}$,

$$\Theta(t + p) = \Phi(t + p, 0) \exp(-(t + p)F) = \Phi(t + p, 0) \exp(-G) \exp(-tF).$$

Since $\Phi(p, 0) = \exp(G)$, we have $\Phi(0, p) = \exp(-G)$ and so, for all $t \in \mathbb{R}$,

$$\Theta(t + p) = \Phi(t + p, 0)\Phi(0, p) \exp(-tF) = \Phi(t, 0)\Phi(p, 0)\Phi(0, p) \exp(-tF),$$

where we have used (2.41) to obtain the second equation. Consequently, we have $\Theta(t + p) = \Phi(t, 0) \exp(-tF) = \Theta(t)$ for all $t \in \mathbb{R}$ and so Θ is p-periodic. $\qquad\square$

Equipped with Theorem 2.30, we are now in a position to make further connections between Floquet multipliers (eigenvalues of $\Phi(p, 0)$) and qualitative properties of solutions of (2.30). A Floquet multiplier is said to be *semisimple* if its algebraic and geometric multiplicities (as an eigenvalue of $\Phi(p, 0)$) coincide.

Theorem 2.31

(1) Every solution of (2.30) is bounded on $\mathbb{R}_+$ if, and only if, the modulus of each Floquet multiplier is not greater than 1 and any Floquet multiplier with modulus equal to 1 is semisimple.

(2) Every solution of (2.30) tends to zero at $t \to \infty$ if, and only if, the modulus of each Floquet multiplier is less than 1.

Proof

Let $(\tau, \xi) \in \mathbb{R} \times \mathbb{F}^N$ be arbitrary. The solution $x \colon \mathbb{R} \to \mathbb{F}^N$ of the initial-value problem

$$\dot{x}(t) = A(t)x(t), \quad x(\tau) = \xi,$$

is given by $x(t) = \Phi(t, \tau)\xi = \Phi(t, 0)\Phi(0, \tau)\xi = \Phi(t, 0)\zeta$, where $\zeta := \Phi(0, \tau)\xi$. Proposition 2.29 guarantees the existence of a principal logarithm G of $\Phi(p, 0)$ and, moreover, by Corollary 2.28, the algebraic and geometric multiplicities of each $\lambda \in \sigma(\Phi(p, 0))$ coincide with those of $\mathrm{Log}\,\lambda \in \sigma(G)$. Writing $F := p^{-1}G$ and invoking Theorem 2.30, we have

$$x(t) = \Theta(t) \exp(tF)\zeta \quad \forall\, t \in \mathbb{R},$$

where $\Theta\colon \mathbb{R} \to \mathbb{C}^{N\times N}$ is piecewise continuously differentiable (and hence continuous) and p-periodic, with $\Theta(0) = I$ and $\Theta(t)$ invertible for all t. Therefore, we may infer the existence of $M > 0$ such that $\|\Theta(t)\| \leq M$ and $\|\Theta^{-1}(t)\| \leq M$ for all $t \in \mathbb{R}$. Since $pF = G$, we have

$$\sigma(\Phi(p,0)) = \{e^{\mu p}\colon\ \mu \in \sigma(F)\}$$

and, moreover, the algebraic and geometric multiplicities of each $\mu \in \sigma(F)$ coincide with those of $e^{\mu p} \in \sigma(\Phi(p,0))$. We record three particular consequences.

(a) Every eigenvalue of F has non-positive real part if, and only if, every eigenvalue of $\Phi(p,0)$ has modulus not greater than 1.

(b) Every eigenvalue of F with zero real part is semisimple if, and only if, every eigenvalue of $\Phi(p,0)$ with modulus equal to 1 is semisimple.

(c) Every eigenvalue of F has negative real part if, and only if, every eigenvalue of $\Phi(p,0)$ has modulus less than 1.

Now, define $y\colon \mathbb{R} \to \mathbb{C}^N$ by $y(t) := \Theta^{-1}(t)x(t)$. Then, $\|y(t)\| \leq M\|x(t)\| \leq M^2\|y(t)\|$ and, in particular, x is bounded on $\mathbb{R}_+$ if, and only if, y is bounded on $\mathbb{R}_+$. Furthermore,

$$y(t) = \exp(tF)\zeta \quad \forall\, t \in \mathbb{R}.$$

Thus, Θ determines a one-to-one correspondence between the solutions of the nonautonomous system (2.30) and the solutions of the autonomous system

$$\dot{y} = Fy. \tag{2.42}$$

Therefore, we may conclude the following.

(d) Every solution of (2.30) is bounded on $\mathbb{R}_+$ if, and only if, every solution of (2.42) is bounded on $\mathbb{R}_+$.

(e) Every solution of (2.30) tends to zero as $t \to \infty$ if, and only if, every solution of (2.42) tends to zero as $t \to \infty$.

The conjunction of Corollary 2.13 and equivalences (a), (b) and (d) above now give statement (1). Similarly, the conjunction of Corollary 2.13 and equivalences (c) and (e) yield statement (2). □

Example 2.32

In this example, we consider Hill's equation[6]

$$\ddot{y}(t) + a(t)y(t) = 0, \quad a(t+p) = a(t) \ \forall\, t \in \mathbb{R}, \tag{2.43}$$

---

[6] George William Hill (1838-1914), US American.

where a is piecewise continuous and $p > 0$. Hill's equation describes an undamped oscillation with restoring force at time t equal to $-a(t)y(t)$. The two-dimensional first-order system associated with (2.43) is given by

$$\dot{x}(t) = A(t)x(t), \quad A(t) = \begin{pmatrix} 0 & 1 \\ -a(t) & 0 \end{pmatrix} \quad \forall t \in \mathbb{R}. \tag{2.44}$$

Let Φ be the transition matrix function generated by A. Our intention is to apply Theorem 2.31 in the context of (2.44). To this end, we calculate the Floquet multipliers. Now,

$$\det(\lambda I - \Phi(p,0)) = \lambda^2 - \lambda \operatorname{tr} \Phi(p,0) + \det \Phi(p,0),$$

and, by statement (2) of Proposition 2.7,

$$\det \Phi(p,0) = \exp\left(\int_0^p \operatorname{tr} A(s)\mathrm{d}s\right) = 1.$$

Moreover, noting that $\Phi(t,0)$ is of the form

$$\Phi(t,0) = \begin{pmatrix} \varphi_1(t) & \varphi_2(t) \\ \dot{\varphi}_1(t) & \dot{\varphi}_2(t) \end{pmatrix} \quad \forall t \in \mathbb{R},$$

where φ_1 and φ_2 are the unique solutions of (2.43) satisfying $\varphi_1(0) = 1 = \dot{\varphi}_2(0)$ and $\dot{\varphi}_1(0) = 0 = \varphi_2(0)$, respectively, it follows that

$$\operatorname{tr} \Phi(p,0) = \varphi_1(p) + \dot{\varphi}_2(p).$$

Consequently,

$$\det(\lambda I - \Phi(p,0)) = \lambda^2 - 2\gamma\lambda + 1, \quad \text{where} \quad \gamma := \tfrac{1}{2}(\varphi_1(p) + \dot{\varphi}_2(p)), \tag{2.45}$$

and the Floquet multipliers are given by

$$\lambda_{\pm} = \gamma \pm \sqrt{\gamma^2 - 1}.$$

Invoking Theorem 2.31, we draw the following conclusions.

Case 1: $|\gamma| > 1$. Then $\lambda_+ > 1$ (if $\gamma > 1$) or $\lambda_- < -1$ (if $\gamma < -1$), and hence, at least one solution of (2.44) is unbounded on $\mathbb{R}_+$.

Case 2: $|\gamma| < 1$. Then $\lambda_{\pm} = \gamma \pm i\delta$ with $\delta > 0$. Since $\lambda_+\lambda_- = 1$, it follows that $|\lambda_+| = |\lambda_-| = 1$. Moreover, λ_+ and λ_- are simple (and *a fortiori* semisimple) and hence all solutions of (2.44) are bounded on $\mathbb{R}_+$.

Case 3: $|\gamma| = 1$. Then $\gamma = \pm 1$ and $\lambda_+ = \lambda_- = \gamma$. All solutions of (2.44) are bounded on $\mathbb{R}_+$ if, and only if, γ is semisimple. Since the algebraic multiplicity of γ is two, γ is semisimple if, and only if, $\ker(\gamma I - \Phi(p,0)) = \mathbb{C}^2$. Consequently,

γ is semisimple if, and only if, $\Phi(p,0) = \gamma I$, that is, $\varphi_1(p) = \dot{\varphi}_2(p) = \gamma$ and $\dot{\varphi}_1(p) = \varphi_2(p) = 0$.

Irrespective of semisimplicity of γ, by Proposition 2.20, there exists at least one non-zero periodic solution of period p if $\gamma = 1$ and of period $2p$ if $\gamma = -1$. Furthermore, we claim that, in the case of γ being semisimple, every solution is p-periodic (if $\gamma = 1$) or $2p$-periodic (if $\gamma = -1$). To see this, assume that γ is semisimple. Then the matrix

$$G := \begin{pmatrix} \log \gamma & 0 \\ 0 & \log \gamma \end{pmatrix}.$$

is a logarithm of $\Phi(p,0) = \gamma I$. By Theorem 2.30, there exists a piecewise continuously differentiable p-periodic function $\Theta : \mathbb{R} \to \mathbb{C}^{2 \times 2}$ such that

$$\Phi(t,0) = \Theta(t) \exp(tp^{-1}G) \quad \forall\, t \in \mathbb{R}.$$

If $\gamma = 1$, then $G = 0$, and hence $\Phi(t,0) = \Theta(t)$ for all $t \in \mathbb{R}$, showing that $\Phi(t+p,0) = \Phi(t,0)$ for all $t \in \mathbb{R}$. Every solution x of (2.44) is of the form $x(t) = \Phi(t,0)x(0)$ and is therefore p-periodic. If $\gamma = -1$, then

$$G = \begin{pmatrix} i\pi & 0 \\ 0 & i\pi \end{pmatrix},$$

whence

$$\Phi(t,0) = \Theta(t) \begin{pmatrix} e^{(i\pi/p)t} & 0 \\ 0 & e^{(i\pi/p)t} \end{pmatrix} \quad \forall\, t \in \mathbb{R}.$$

Therefore, $\Phi(t+2p,0) = \Phi(t,0)$ for all $t \in \mathbb{R}$, showing that every solution x of (2.44) is $2p$-periodic.

Finally, we analyse a specific example. Assume that the function a is given by

$$a(t) = \begin{cases} \omega^2, & m \le t < m+\tau \\ 0, & m+\tau \le t < m+1, \end{cases} \quad \text{where } m \in \mathbb{Z}. \tag{2.46}$$

Here $\omega > 0$ and $\tau \in (0,1)$, Obviously, a is a piecewise continuous periodic function with period equal to 1. With this choice of a, Hill's equation (2.43) describes an undamped oscillator, the restoring force of which is switched off on the intervals $[m+\tau, m+1)$, $m \in \mathbb{Z}$. Since a is piecewise constant, $\Phi(1,0)$ can easily be determined analytically. A routine calculation yields

$$\Phi(1,0) = \begin{pmatrix} \cos(\omega\tau) - \omega(1-\tau)\sin(\omega\tau) & \omega^{-1}\sin(\omega\tau) + (1-\tau)\cos(\omega\tau) \\ -\omega\sin(\omega\tau) & \cos\omega\tau \end{pmatrix}.$$

In particular,

$$\gamma = \frac{1}{2}\big(2\cos(\omega\tau) - \omega(1-\tau)\sin(\omega\tau)\big)$$

We consider two "extreme" scenarios.

Scenario 1: τ is close to 1. In this scenario, the restoring force is switched on "most" of the time and so one might expect the behaviour of the solutions to be similar to those of the harmonic oscillator $\ddot{y} + \omega^2 y = 0$ (for which every solution is periodic, of period $2\pi/\omega$, and so *a fortiori* is bounded). However, we show that this is not the case. To this end, let $\omega = \pi/\tau$ and note that

$$\Phi(1,0) = \begin{pmatrix} -1 & \tau - 1 \\ 0 & -1 \end{pmatrix}$$

and $\gamma = -1$. Clearly, $\lambda = -1$ is an eigenvalue of $\Phi(1,0)$ and so, by Proposition 2.20, there exists a non-zero periodic solution of period 2. Since the eigenvalue $\lambda = -1$ is not semisimple, it follows from Case 3 above that there exists at least one solution which is unbounded on $\mathbb{R}_+$. A more detailed analysis (see Exercise 2.21) reveals that φ_1 is periodic of period 2 and φ_2 is unbounded on $\mathbb{R}_+$. Consequently, denoting the components of $\xi \in \mathbb{R}^2$ by ξ_1 and ξ_2 and setting $x(t) := \Phi(t,0)\xi$ for all $t \in \mathbb{R}$, the solution x is periodic of period 2 if, and only if, $\xi_2 = 0$ and, furthermore, x is unbounded on $\mathbb{R}_+$ if, and only if, $\xi_2 \neq 0$. These observations are valid for all $\tau \in (0,1)$: in particular, they hold when τ is close to 1, in which case we have $\omega \approx \pi$ and so the 2-periodic solutions do indeed mimic the behaviour of the harmonic oscillator $\ddot{y} + \omega^2 y = 0$; however, all other non-zero solutions are unbounded and so the behaviour of the system differs markedly from that of the harmonic oscillator.

Scenario 2: τ is close to 0. In this scenario, the restoring force is switched off "most" of the time and one might expect that the behaviour of the solutions is similar to those of the "double integrator" $\ddot{y} = 0$ (which has unbounded solutions, for example, $y(t) = t$). However, this is not the case. For every $\omega > 0$, we have $0 < \gamma < 1$ for all sufficiently small $\tau \in (0,1)$. Consequently, by Case 2 above, for all $\tau > 0$ sufficiently small, all solutions of (2.44) are bounded on $\mathbb{R}_+$. This behaviour differs markedly from that of the double integrator.   △

Exercise 2.21

Assume that, in Example 2.32, the periodic function a is given by (2.46) with $\tau \in (0,1)$ and $\omega = \pi/\tau$. Show that φ_1 is periodic of period 2 and φ_2 is unbounded with $\varphi_2(n) = (-1)^n n(1 - \tau)$ for all $n \in \mathbb{N}$.

Exercise 2.22

Assume that in Example 2.32 the periodic function a is even. Show that in this case $\gamma = \varphi_1(p) = \dot{\varphi}_2(p)$.

The following corollary of Theorem 2.31 provides a criterion for the existence of at least one solution of (2.30) which is unbounded on $\mathbb{R}_+$.

Corollary 2.33

If $\int_0^p \operatorname{tr} A(s)\,ds$ has positive real part, then (2.30) has a solution x with $\limsup_{t\to\infty} \|x(t)\| = \infty$.

Proof

By statement (2) of Proposition 2.7, we have

$$\det \Phi(p,0) = \exp\left(\int_0^p \operatorname{tr} A(s)\,ds\right).$$

Let λ_j, $j = 1,\ldots,d$, be the distinct eigenvalues of $\Phi(p,0)$, with algebraic multiplicities m_j, $j = 1,\ldots,d$. Then $\det\left(\Phi(p,0) - \lambda I\right) = \prod_{j=1}^d (\lambda_j - \lambda)^{m_j}$, which, upon evaluation at $\lambda = 0$, shows that $\det \Phi(p,0) = \prod_{j=1}^d \lambda_j^{m_j}$. Hence,

$$\prod_{j=1}^d \lambda_j^{m_j} = \exp\left(\int_0^p \operatorname{tr} A(s)\,ds\right).$$

Therefore, invoking the hypothesis,

$$\prod_{j=1}^d |\lambda_j|^{m_j} = \exp\left(\operatorname{Re}\int_0^p \operatorname{tr} A(s)\,ds\right) > 1.$$

Consequently, there exists $j \in \{1,\ldots,d\}$ such that $|\lambda_j| > 1$ and so, by Theorem 2.31, there must exist a solution x which is unbounded on $\mathbb{R}_+$. $\qquad\square$

Exercise 2.23

Consider (2.30) with $N = 2$, $\mathbb{F} = \mathbb{R}$ and

$$A(t) = \begin{pmatrix} 1 + \sin t & a \\ b & 1 - \cos t \end{pmatrix},$$

where $a, b \in \mathbb{R}$ are arbitrary constants. Show that there exists at least one solution which is unbounded on $\mathbb{R}_+$.

The converse of Corollary 2.33 does not hold. Specifically, if $\int_0^p \operatorname{tr} A(s)\,ds$ has negative real part, then we cannot conclude that every solution x of (2.30) is bounded on $\mathbb{R}_+$, as the following exercise shows.

Exercise 2.24

Consider (2.30) with $N = 2$, $\mathbb{F} = \mathbb{R}$ and

$$A(t) = \frac{1}{2}\begin{pmatrix} -2 + 3\cos^2 t & 2 - 3\sin t \cos t \\ -2 - 3\sin t \cos t & -2 + 3\sin^2 t \end{pmatrix}.$$

In this case, A is π-periodic and $\int_0^\pi \operatorname{tr} A(s)\,\mathrm{d}s = -\pi/2 < 0$. Show that

$$t \mapsto x(t) := e^{t/2} \begin{pmatrix} -\cos t \\ \sin t \end{pmatrix}$$

is a solution of (2.30), and is such that $\|x(t)\| \to \infty$ as $t \to \infty$.

2.4 Proof of Theorem 2.19 and Proposition 2.29

We conclude this chapter with proofs of Theorem 2.19 (the spectral mapping theorem) and Proposition 2.29.

Proof of Theorem 2.19

Let $M \in \mathbb{C}^{N \times N}$ and let $a_n \in \mathbb{C}$, with $n \in \mathbb{N}_0$, be such that the series $f(z) = \sum_{n=0}^\infty a_n z^n$ converges for all $z \in \mathbb{C}$. By Proposition A.27, $f(M) := \sum_{n=0}^\infty a_n M^n$ is a well-defined element of $\mathbb{C}^{N \times N}$. Let λ_j, $j = 1, \ldots, d$, be the distinct eigenvalues of M with associated algebraic multiplicities m_j, $j = 1, \ldots, d$. Noting that, if T is invertible, then

$$f(T^{-1}MT) = \sum_{n=0}^\infty a_n (T^{-1}MT)^n = T^{-1}\left(\sum_{n=0}^\infty M^n\right)T = T^{-1}f(M)T$$

and so, without loss of generality, we may assume that M is in Jordan[7] canonical form (see Theorem A.9) which we express as $M = \operatorname{diag}(J_1, \ldots, J_\ell)$. The generic block $J \in \{J_1, \ldots, J_\ell\}$ takes the form

$$J = \lambda I + K \quad \text{for some } \lambda \in \{\lambda_1, \ldots, \lambda_d\},$$

where, for some $r \in \mathbb{N}$, I is the $r \times r$ identity matrix and $K \in \mathbb{R}^{r \times r}$ is a matrix with every superdiagonal entry equal to 1, all other entries being 0, that is,

$$K = \begin{pmatrix} 0 & 1 & 0 & \cdots & 0 \\ 0 & 0 & 1 & \cdots & 0 \\ \vdots & \vdots & \vdots & \ddots & \vdots \\ 0 & 0 & 0 & \cdots & 1 \\ 0 & 0 & 0 & \cdots & 0 \end{pmatrix}$$

(if $r = 1$, then $K = 0$). Note that $K^n = 0$ for all $n \geq r$,

$$J^n = (\lambda I + K)^n = \sum_{k=0}^n \binom{n}{k} \lambda^{n-k} K^k \quad \forall n \in \mathbb{N}_0.$$

---

[7] Marie Ennemond Camille Jordan (1838-1922), French.

Furthermore, term-by-term differentiation of the power series yields

$$\frac{f^{(k)}(\lambda)}{k!} = \sum_{n=k}^{\infty} a_n \binom{n}{k} \lambda^{n-k} \quad \forall k \in \mathbb{N}_0,$$

where $f^{(k)}$ denotes the k-th derivative of f (with $f^{(0)} := f$). Therefore,

$$f(J) = \sum_{n=0}^{\infty} a_n J^n = \sum_{k=0}^{r-1} \sum_{n=k}^{\infty} a_n \binom{n}{k} \lambda^{n-k} K^k = \sum_{k=0}^{r-1} \frac{f^{(k)}(\lambda)}{k!} K^k.$$

In particular, $f(J)$ has the following upper triangular structure

$$f(J) = \begin{pmatrix} f(\lambda) & f'(\lambda) & * & \cdots & * & * \\ 0 & f(\lambda) & f'(\lambda) & \cdots & * & * \\ 0 & 0 & f(\lambda) & \cdots & * & * \\ \vdots & \vdots & \vdots & \ddots & \vdots & \vdots \\ 0 & 0 & 0 & \cdots & f(\lambda) & f'(\lambda) \\ 0 & 0 & 0 & \cdots & 0 & f(\lambda) \end{pmatrix} \tag{2.47}$$

and so $\sigma(f(J)) = \{f(\lambda)\}$. Of course, in the case of $r = 1$, (2.47) should be interpreted as the scalar $f(J) = f(\lambda)$. Since $f(M) = \mathrm{diag}(f(J_1), \ldots, f(J_\ell))$, it now follows that

$$\sigma(f(M)) = \{f(\lambda) : \lambda \in \sigma(M)\},$$

completing the proof of statement (1).

We proceed to prove statement (2). To this end note that the above argument also shows that, for each $\lambda \in \sigma(M)$ and every Jordan block J associated with λ, the algebraic multiplicity of λ as an eigenvalue of J coincides with the algebraic multiplicity of $f(\lambda)$ as an eigenvalue of $f(J)$. From this, we may infer that the algebraic multiplicity of $f(\lambda)$ as an eigenvalue of $f(M)$ cannot be less than the algebraic multiplicity of λ as an eigenvalue of M. Moreover, since f is injective on $\sigma(M)$, the number of distinct eigenvalues of $f(M)$ coincides with the number d of distinct eigenvalues of M. Since the algebraic multiplicities of the eigenvalues sum to N in each case, it follows that the algebraic multiplicity of each $\lambda \in \sigma(M)$ coincides with that of $f(\lambda) \in \sigma(f(M))$.

To prove statement (3), let J be any Jordan block in $M = \mathrm{diag}(J_1, \ldots, J_\ell)$ associated with $\lambda \in \sigma(M)$. If J is scalar, then trivially we have $\ker(f(J) - f(\lambda)I) = \ker(J - \lambda I) = \mathbb{C}$. If J is not scalar, then λ is not a semisimple, and so, by hypothesis, the additional property $f'(\lambda) \neq 0$ holds. By (2.47), we then have

$$\ker(f(J) - f(\lambda)I) = \mathrm{span} \begin{pmatrix} 1 \\ 0 \\ \vdots \\ 0 \end{pmatrix} = \ker(J - \lambda I).$$

Consequently, defining $\Lambda := \{k\colon J_k \text{ is associated with } \lambda\} \subset \{1,\ldots,\ell\}$, it follows that

$$\dim \ker(f(J_k) - f(\lambda)I) = \dim \ker(J_k - \lambda I) = 1 \quad \forall\, k \in \Lambda.$$

Moreover, by injectivity of f on $\sigma(M)$, $f(\lambda) \notin \sigma(f(J_k))$ for all $k \in \{1,\ldots,\ell\}\backslash\Lambda$, and thus

$$\dim \ker(f(J_k) - f(\lambda)I) = \dim \ker(J_k - \lambda I) = 0 \quad \forall\, k \in \{1,\ldots,\ell\}\backslash\Lambda.$$

Therefore, we may conclude that

$$\dim \ker(f(M) - f(\lambda)I) = \dim \ker(M - \lambda I) = \#\Lambda,$$

where $\#\Lambda$ denotes the number of elements of Λ. Finally, let $v \in \ker(M - \lambda I)$. Then $M^n v = \lambda^n v$ for all $n \in \mathbb{N}_0$ and so $f(M)v = f(\lambda)v$. Therefore, $\ker(M - \lambda I) \subset \ker(f(M) - f(\lambda)I)$ and since these subspaces have the same dimension, they must coincide. This completes the proof. $\qquad\square$

Proof of Proposition 2.29

Let λ_j, $j = 1,\ldots,d$, be the distinct eigenvalues of H with associated algebraic multiplicities m_j, $j = 1,\ldots,d$. Note that, if G is a logarithm of H and T is invertible, then $\exp(T^{-1}GT) = T^{-1}\exp(G)T = T^{-1}HT$ and so $T^{-1}GT$ is a logarithm of $T^{-1}HT$. Therefore, without loss of generality, we may assume that H is in Jordan canonical form (see Theorem A.9) which can be expressed as

$$H = \operatorname{diag}(J_1,\ldots,J_\ell), \quad \text{where } \ell \geq d.$$

The generic block $J \in \{J_1,\ldots,J_\ell\}$ takes the form

$$J = \lambda I + K \quad \text{for some } \lambda \in \{\lambda_1,\ldots,\lambda_d\},$$

where, for some $r \in \mathbb{N}$, I is the $r \times r$ identity matrix and $K \in \mathbb{R}^{r \times r}$ is a matrix with every superdiagonal entry equal to 1, all other entries being 0, that is,

$$K = \begin{pmatrix} 0 & 1 & 0 & \cdots & 0 \\ 0 & 0 & 1 & \cdots & 0 \\ \vdots & \vdots & \vdots & \ddots & \vdots \\ 0 & 0 & 0 & \cdots & 1 \\ 0 & 0 & 0 & \cdots & 0 \end{pmatrix}$$

(of course, if $r = 1$, then $K = 0$). We record that $\sigma(J) = \{\lambda\}$ and

$$\ker(J - \lambda I) = \ker(K) = \operatorname{span}\begin{pmatrix} 1 \\ 0 \\ \vdots \\ 0 \end{pmatrix} =: V. \tag{2.48}$$

Choose $\delta > 0$ sufficiently small so that $e^\delta - 1 < 3\delta/2 < 1$. Furthermore, choose $\varepsilon > 0$ sufficiently small so that $M := (\varepsilon/\lambda)K$ has norm $\|M\| \le \delta/2$. Next, we will invoke the contraction mapping theorem (Theorem A.25) to prove the existence of a logarithm of $I + M$. To this end, set $\Omega := \{X \in \mathbb{C}^{r \times r} : \|X\| \le \delta\}$ and define $F: \Omega \to \mathbb{C}^{r \times r}$ by

$$F(X) := X + I + M - \exp(X) = M - \sum_{k=2}^{\infty} \frac{1}{k!} X^k.$$

Then,

$$\|F(X)\| \le \|M\| + \sum_{k=2}^{\infty} \frac{\delta^k}{k!} \le \frac{\delta}{2} + e^\delta - 1 - \delta = e^\delta - 1 - \frac{\delta}{2} \le \delta \ \ \forall X \in \Omega$$

and so $F(\Omega) \subset \Omega$. Observing that

$$X^k - Y^k = \sum_{j=1}^{k} X^{k-j}(X - Y)Y^{j-1} \quad \forall X, Y \in \Omega, \ \forall k \in \mathbb{N} \tag{2.49}$$

(see Exercise 2.25 below for details), we obtain

$$\|F(X) - F(Y)\| = \left\| \sum_{k=2}^{\infty} \frac{1}{k!} (Y^k - X^k) \right\|$$

$$\le \sum_{k=2}^{\infty} \frac{\delta^{k-1}}{(k-1)!} \|X - Y\| = (e^\delta - 1)\|X - Y\| \ \ \forall X, Y \in \Omega.$$

Recalling that $e^\delta - 1 < 1$, it follows that F is a contraction on Ω and so, by the contraction mapping theorem (Theorem A.25), has a fixed point $Z \in \Omega$. Therefore, $I + M = \exp(Z)$, that is, Z is a logarithm of $I + M$.

Next, we will use Z to construct a principal logarithm of J. To this end, define $Q := \mathrm{diag}(\varepsilon, \varepsilon^2, \ldots, \varepsilon^r)$. Then $Q^{-1}KQ = \varepsilon K$ and so

$$Q^{-1}JQ = \lambda I + \varepsilon K = \lambda(I + M) = \lambda \exp(Z).$$

Let $\nu \in \mathbb{C}$ be a logarithm of λ, and so $e^\nu = \lambda$ (such a logarithm exists since $\lambda \ne 0$ by invertibility of H). Therefore,

$$J = \lambda Q \exp(Z) Q^{-1} = \lambda \exp(QZQ^{-1}) = \exp(\nu I + QZQ^{-1}),$$

showing that $P := \nu I + QZQ^{-1}$ is a logarithm of J. By Theorem 2.19, $e^\mu = \lambda$ for every $\mu \in \sigma(P)$. Hence, for $\mu \in \sigma(P)$ and $v \in \ker(P - \mu I)$,

$$(J - \lambda I)v = (\exp(P) - \lambda I)v = e^\mu v - \lambda v = 0,$$

and thus, by (2.48), $\ker(P - \mu I) = V$. Consequently, $\sigma(P) = \{\mu\}$ is a singleton. Setting $L := P + (2k\pi i)I$, where $k \in \mathbb{Z}$ is such that $\operatorname{Log} \lambda = \mu + 2k\pi i$, we obtain

$$\exp(L) = J \quad \text{and} \quad \sigma(L) = \{\operatorname{Log} \lambda\},$$

that is, L is a principal logarithm of J.

We have now shown that, for each $j = 1, \ldots, \ell$, there exists a principle logarithm L_j of J_j and so $G := \operatorname{diag}(L_1, \ldots, L_\ell)$ is a principal logarithm of $H = \operatorname{diag}(J_1, \ldots, J_\ell)$, completing the proof. $\qquad\square$

Exercise 2.25

Prove (2.49) by induction on k.

(*Hint.* Note that $X^{k+1} - Y^{k+1} = (X + Y)(X^k - Y^k) + XY^k - YX^k$.)

3

Introduction to linear control theory

In this chapter and with reference to Figure 3.1, we consider linear systems with *input* (*control*) u and *output* (*observation*) y. In particular, for $(0, \xi) \in J \times \mathbb{R}^N$, where $J \subset \mathbb{R}$ is an interval containing 0, we study systems of the form

$$\dot{x} = Ax + Bu, \quad x(0) = \xi \in \mathbb{R}^N, \tag{3.1a}$$

$$y = Cx, \tag{3.1b}$$

with $A \in \mathbb{R}^{N \times N}$, $B \in \mathbb{R}^{N \times M}$ and $C \in \mathbb{R}^{P \times N}$.

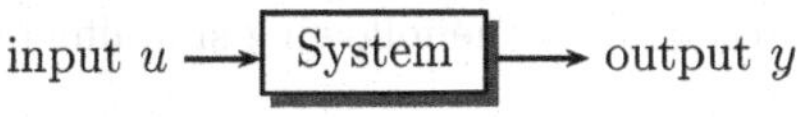

$$\text{input } u \longrightarrow \boxed{\text{System}} \longrightarrow \text{output } y$$

Figure 3.1 System with input and output

By Corollary 2.17, we know that, for each $u \in PC(J, \mathbb{R}^M)$ and $\xi \in \mathbb{R}^N$, (3.1a) has a unique solution on J, which we denote by $x(\cdot\,; \xi, u)$. Specifically, for each $(\xi, u) \in \mathbb{R}^N \times PC(J, \mathbb{R}^M)$, the unique solution of (3.1a) on J, which we denote by $x(\cdot\,; \xi, u)$, is given by the *variation of parameters* formula

$$x(t; \xi, u) := \exp(At)\xi + \int_0^t \exp(A(t - s))Bu(s) \, ds \quad \forall \, t \in J. \tag{3.2}$$

The quantity $x(t; \xi, u) \in \mathbb{R}^N$ is the *state* of the system at time t, corresponding to the initial state ξ and input u. For notational concision, in situations where the data (ξ, u) are clear from context, we may simply write $x(\cdot)$ or x in place of

H. Logemann and E. P. Ryan, *Ordinary Differential Equations*,
Springer Undergraduate Mathematics Series,
DOI: 10.1007/978-1-4471-6398-5_3, © Springer-Verlag London 2014

the more cumbersome $x(\cdot\,;\xi,u)$. The output $y(t;\xi,u) = Cx(t;\xi,u) = Cx(t)$ at time t represents the information on the current state of the system available to the outside world (for example, in the case of a mechanical system, its position may be available through measurement - but its velocity may be unavailable). Again, for concision, we may write $y(\cdot)$ in place of $y(\cdot\,;\xi,u)$ whenever the data (ξ,u) are clear from context.

Example 3.1

Consider the satellite model introduced in Section 1.1.4. The linearization of this system is described by (3.1) with $N = 4$, $M = P = 2$ and

$$A = \begin{pmatrix} 0 & 1 & 0 & 0 \\ 3\omega^2 & 0 & 0 & 2\omega \\ 0 & 0 & 0 & 1 \\ 0 & -2\omega & 0 & 0 \end{pmatrix}, \quad B = \begin{pmatrix} 0 & 0 \\ 1 & 0 \\ 0 & 0 \\ 0 & 1 \end{pmatrix}, \quad C = \begin{pmatrix} 1 & 0 & 0 & 0 \\ 0 & 0 & 1 & 0 \end{pmatrix}, \quad (3.3)$$

where we have normalized the constant σ to 1. Recall that the components u_1 and u_2 of u correspond to the radial and tangential thrust, respectively, whilst the components y_1 and y_2 of y correspond to the radial and angular coordinates respectively. $\triangle$

Example 3.2

The process of drug ingestion, distribution, and subsequent metabolism in an individual may be represented by the following simplified mathematical model:

$$\left.\begin{aligned} \dot{x}_1 &= -a_1 x_1 + u\,, \\ \dot{x}_2 &= a_1 x_1 - a_2 x_2\,, \quad a_1, a_2 > 0\,, \\ y &= x_2\,, \end{aligned}\right\} \qquad (3.4)$$

where x_1, x_2 and u, respectively, represent (as deviations from initial steady-state values) the drug mass in the gastrointestinal tract, the drug mass in the bloodstream and the drug ingestion rate. In this example, the output y is given by $y = x_2$, i.e., $y(t)$ is the the drug mass in the bloodstream at time t. The constants a_1 and a_2 reflect physiological properties of the individual in question. Setting

$$x := \begin{pmatrix} x_1 \\ x_2 \end{pmatrix}, \quad A := \begin{pmatrix} -a_1 & 0 \\ a_1 & -a_2 \end{pmatrix}, \quad B := \begin{pmatrix} 1 \\ 0 \end{pmatrix}, \quad C := (0, 1)\,,$$

system (3.4) takes the form (3.1). $\triangle$

At the core of control theory are the following two problems.

Controllability problem. For a given initial state and through suitable choice of input, can the state evolution of a system be determined so as to reach a prespecified target state in finite time?
Observability problem. Can the current state $x(t)$, $t > 0$, be determined from knowledge of the input and output signals on $[0, t]$?

These questions will be addressed in the next two sections.

3.1 Controllability

We start with definitions of the fundamental concepts of reachability and controllability.

Definition 3.3

Let $\xi, \zeta \in \mathbb{R}^N$. If there exist $T > 0$ and a control $u \in PC([0, T], \mathbb{R}^M)$ such that $x(T; \xi, u) = \zeta$, then ζ is said to be *reachable* from ξ. System (3.1a) (or system (3.1)) is said to be *controllable* if, for all $\xi, \zeta \in \mathbb{R}^N$, ζ is reachable from ξ.

We will frequently identify system (3.1a) with the pair of matrices (A, B). In this context, it is convenient to define a pair $(A, B) \in \mathbb{R}^{N \times N} \times \mathbb{R}^{N \times M}$ to be *controllable* if the associated system (3.1a) is controllable in the sense of Definition 3.3.

Exercise 3.1

Consider the problem of achieving a lunar soft landing of a spacecraft. The spacecraft moves vertically, has mass m and is equipped with thrusters capable of exerting force, in either upward or downward direction. Let $h(t)$ denote the height of the spacecraft above the surface at time t, and let $f(t)$ denote the force on the spacecraft due to the thrusters at time t. Denoting the lunar gravitational constant by g, one arrives at the initial-value problem

$$m\ddot{h}(t) = -mg + f(t), \ t \geq 0; \quad h(0) = \xi_1, \ \dot{h}(0) = \xi_2,$$

where ξ_1 and ξ_2 are the altitude and velocity of the craft at time $t = 0$.

Setting $x_1 = h$ and $x_2 = \dot{h}$, we obtain

$$\dot{x}_1 = x_2 , \quad x_1(0) = \xi_1 ,$$
$$\dot{x}_2 = -g + \frac{1}{m}f , \quad x_2(0) = \xi_2 .$$

The above system of equations can be written in the form

$$\dot{x} = Ax + bu , \quad x(0) = \xi,$$

where

$$A := \begin{pmatrix} 0 & 1 \\ 0 & 0 \end{pmatrix}, \quad b := \begin{pmatrix} 0 \\ 1 \end{pmatrix}, \quad x := \begin{pmatrix} x_1 \\ x_2 \end{pmatrix}, \quad \xi := \begin{pmatrix} \xi_1 \\ \xi_2 \end{pmatrix}.$$

and u is defined by $u(t) := f(t)/m - g$ for all $t \geq 0$.

The *soft-landing problem* consists of finding a control function $u \colon [0, T] \to \mathbb{R}$ which steers ξ to 0 in time T, that is, $x_1(T) = 0 = x_2(T)$.

(a) Show that, for every $T > 0$, there exists an affine linear control function u of the form $u(t) = \alpha + \beta t$ for all $t \in [0, T]$ which solves the soft-landing problem, where α and β are real parameters.

(b) Now consider the situation wherein the spacecraft is equipped with only one thruster which is directed towards the lunar surface (and so can exert a force only in the upward direction). Moreover, assume that the thrust is limited in magnitude by $F > 0$, that is, for all t, $f(t) \in [0, F]$. In order to achieve a soft landing, it is clear that the engine thrust must be capable of overcoming the gravitational force on the craft, whence the assumption $\alpha := (F - mg)/m > 0$.

Let $\xi_1 > 0$ and $\xi_2 = 0$. Show that there exist $T > 0$ and $S \in (0, T)$ such that the control

$$u(t) = \begin{cases} -g & \text{if } 0 \leq t < S \\ \alpha & \text{if } S \leq t \leq T. \end{cases}$$

achieves a soft landing at time T. In terms of the thruster force f, this

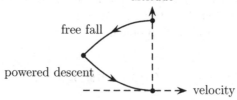

Figure 3.2   Lunar soft landing

strategy corresponds to a period of free fall under gravity ($f(t) = 0$ on $[0, S)$), followed by a period of powered descent under maximum thrust ($f(t) = F$ on $[S, T]$), as depicted in Figure 3.2.

In part (b) of the above exercise, controls were constrained to take their values in a compact set. A general study of controllability in the presence of such control constraints is beyond the scope of this book. Throughout this chapter, we consider only unconstrained controls, that is, $\mathbb{R}^M$-valued functions.

In the following, the *controllability matrix* (also called *reachability matrix*)

$$\mathcal{C}(A, B) := (B, AB, A^2B, \ldots, A^{N-1}B) \in \mathbb{R}^{N \times (MN)}$$

and the *controllability Gramian* (parameterized by $T > 0$)

$$Q_T := \int_0^T \exp(At)BB^* \exp(A^*t)dt \in \mathbb{R}^{N \times N}$$

will play important roles. Note that Q_T is symmetric and positive semi-definite.

Theorem 3.4

$\operatorname{im} \mathcal{C}(A, B) = \operatorname{im} Q_T$ for all $T > 0$.

Proof

As a preliminary, we will establish the following equivalence

$$z \in (\operatorname{im} \mathcal{C}(A, B))^\perp \iff z^* \exp(At)B = 0 \quad \forall t \in \mathbb{R}. \tag{3.5}$$

Assume $z \in (\operatorname{im} \mathcal{C}(A, B))^\perp$. Then, by Theorem A.1, $z^* \mathcal{C}(A, B) = 0$, implying that $z^* A^k B = 0$ for $k = 0, 1, \ldots, N - 1$. By the Cayley-Hamilton[1] theorem (see Theorem A.6), for every $k \in \mathbb{N}_0$, A^k is a linear combination of $I = A^0$, $A, A^2, \ldots, A^{N-1}$ and so we may conclude that $z^* A^k B = 0$ for all $k \in \mathbb{N}_0$. Consequently,

$$0 = \sum_{k=0}^\infty \frac{t^k}{k!} z^* A^k B = z^* \exp(At)B \quad \forall t \in \mathbb{R}.$$

Now assume that $z^* \exp(At)B = 0$ for all $t \in \mathbb{R}$. Evaluation at 0 and repeated differentiation at $t = 0$ gives $z^* A^k B = 0$, $k = 0, 1, \ldots, N - 1$. Therefore, $z^* \mathcal{C}(A, B) = 0$ and so, by Theorem A.1, $z \in (\operatorname{im} \mathcal{C}(A, B))^\perp$. This establishes (3.5).

---

[1] Arthur Cayley (1821-1895), British; Sir William Rowan Hamilton (1805-1865), Irish.

Next we proceed to the proof of the assertion of the theorem. The assertion is equivalent to the statement that $(\operatorname{im}\mathcal{C}(A,B))^\perp = (\operatorname{im}Q_T)^\perp$ for all $T > 0$. It is the latter statement that we will prove. Let $T > 0$ be arbitrary. We first show that $(\operatorname{im}\mathcal{C}(A,B))^\perp \subset (\operatorname{im}Q_T)^\perp$, Let $z \in (\operatorname{im}\mathcal{C}(A,B))^\perp$. Then, by (3.5), $z^*\exp(At)B = 0$ for all $t \in \mathbb{R}$ and so $z^*Q_T = \int_0^T z^*\exp(At)BB^*\exp(A^*t)\mathrm{d}t = 0$, whence, by Theorem A.1, $z \in (\operatorname{im}Q_T)^\perp$. Consequently, $(\operatorname{im}\mathcal{C}(A,B))^\perp \subset (\operatorname{im}Q_T)^\perp$. To prove the reverse inclusion, let $z \in (\operatorname{im}Q_T)^\perp$. Then, by Theorem A.1, $z^*Q_T = 0$ and, by symmetry of Q_T, $Q_Tz = 0$. Hence $\langle z, Q_Tz \rangle = 0$ and

$$0 = \int_0^T \langle z, \exp(At)BB^*\exp(A^*t)z\rangle\mathrm{d}t = \int_0^T \|B^*\exp(A^*t)z\|^2\mathrm{d}t\,.$$

Since the function $t \mapsto \|B^*\exp(A^*t)z\|^2$ is continuous and non-negative-valued, it follows that $B^*\exp(A^*t)z = 0$ for all $t \in [0,T]$. Thus, $z^*\exp(At)B = 0$ for all $t \in \mathbb{R}$ and so, by (3.5), $z \in (\operatorname{im}\mathcal{C}(A,B))^\perp$. We may now conclude that $(\operatorname{im}Q_T)^\perp \subset (\operatorname{im}\mathcal{C}(A,B))^\perp$, completing the proof of the theorem. $\qquad\square$

By Theorem 3.4, we have

$$Q_T \text{ is invertible for all } T > 0 \;\Leftrightarrow\; \operatorname{rk}\mathcal{C}(A,B) = N. \tag{3.6}$$

Also, by Theorem 3.4 in conjunction with Theorem A.4, we see that, for each $T > 0$, there exists $Q_T^\sharp \in \mathbb{R}^{N \times N}$ such that

$$Q_T Q_T^\sharp z = z \;\; \forall\, z \in \operatorname{im}\mathcal{C}(A,B), \tag{3.7}$$

If $\operatorname{rk}\mathcal{C}(A,B) = N$, then $Q_T^\sharp = Q_T^{-1}$ (the unique inverse); if $\operatorname{rk}\mathcal{C}(A,B) < N$, then $Q_T^\sharp$ is not unique.

Introduce the *input-to-state map* $\mathbf{C}_T$ (parameterized by $T > 0$)

$$\mathbf{C}_T\colon PC([0,T],\mathbb{R}^M) \to \mathbb{R}^N, \quad u \mapsto \int_0^T \exp(A(T-t))Bu(t)\,\mathrm{d}t\,. \tag{3.8}$$

It is clear that the map $\mathbf{C}_T$ is linear, that is, for $u,v \in PC([0,T],\mathbb{R}^M)$ and $\alpha,\beta \in \mathbb{R}$, we have that

$$\mathbf{C}_T(\alpha u + \beta v) = \alpha\mathbf{C}_T u + \beta\mathbf{C}_T v,$$

where, as is usual for linear maps, we have written $\mathbf{C}_T u$ for $\mathbf{C}_T(u)$.

Note that

$$x(T;\xi,u) = \exp(AT)\xi + \mathbf{C}_T u.$$

Note further that $\operatorname{im}\mathbf{C}_T$ is the set of states reachable from 0 in time $T > 0$. Let $T > 0$ be arbitrary and let $Q_T^\sharp$ be such that (3.7) holds. Introduce $\mathbf{C}_T^\sharp\colon \mathbb{R}^N \to PC([0,T],\mathbb{R}^M)$, defined by

$$\left(\mathbf{C}_T^\sharp z\right)(t) := B^*\exp(A^*(T-t))Q_T^\sharp z \;\; \forall\, t \in [0,T]. \tag{3.9}$$

Observe that $\mathbf{C}_T^\sharp$ is linear and, for all $z \in \mathbb{R}^N$,

$$\mathbf{C}_T \mathbf{C}_T^\sharp z = \int_0^T \exp(A(T-t)BB^* \exp(A^*(T-t))Q_T^\sharp z \, dt = Q_T Q_T^\sharp z,$$

and so, by (3.7),

$$\mathbf{C}_T \mathbf{C}_T^\sharp z = z \ \ \forall z \in \operatorname{im} \mathcal{C}(A,B). \tag{3.10}$$

In particular, note that, if $\operatorname{im} \mathcal{C}(A,B) = \mathbb{R}^N$, then $\mathbf{C}_T$ is right invertible (and $\mathbf{C}_T^\sharp$ is a right inverse). We record this fact in the following

$$\operatorname{rk} \mathcal{C}(A,B) = N \implies \mathbf{C}_T \text{ is right invertible for all } T > 0. \tag{3.11}$$

Let R be the set of states of system (3.1a) which are reachable from 0, that is

$$R := \cup_{T>0} \operatorname{im} \mathbf{C}_T = \{x(T;0,u) = \mathbf{C}_T u : T > 0, \ u \in PC([0,T],\mathbb{R}^M)\}. \tag{3.12}$$

Theorem 3.5

$$R = \operatorname{im} \mathcal{C}(A,B) = \operatorname{im} \mathbf{C}_T \ \forall T > 0.$$

The identity $R = \operatorname{im} \mathcal{C}(A,B)$ in the above result gives a purely algebraic description of the subspace R of all states which are reachable from 0, thereby providing the means for effective and easy computation of R. Note also that the subspaces $\operatorname{im} \mathbf{C}_T$ are identical for all $T > 0$.

Proof of Theorem 3.5

We first show that $R \subset \operatorname{im} \mathcal{C}(A,B)$. Let $z \in R$ be arbitrary. Then there exists $T > 0$ and $u \in PC([0,T],\mathbb{R}^M)$ such that $z = \mathbf{C}_T u$. Write $z = z_1 + z_2$, where $z_1 \in \operatorname{im} \mathcal{C}(A,B)$ and $z_2 \in (\operatorname{im} \mathcal{C}(A,B))^\perp$. Invoking (3.5), we have $z_2^* \exp(T-t)B = 0$ for all $t \in [0,T]$. Therefore,

$$0 = \int_0^T z_2^* \exp(A(T-t))Bu(t)dt = \langle z_2, \mathbf{C}_T u \rangle = \langle z_2, z \rangle = \|z_2\|^2,$$

and so $z = z_1 \in \operatorname{im} \mathcal{C}(A,B)$. Thus, $R \subset \operatorname{im} \mathcal{C}(A,B)$.

Now let $z \in \operatorname{im} \mathcal{C}(A,B)$. Let $T > 0$ be arbitrary and set $u = \mathbf{C}_T^\sharp z$. By (3.10),

$$z = \mathbf{C}_T \mathbf{C}_T^\sharp z = \mathbf{C}_T u = x(T;0,u)$$

and so $z \in R$ and $z \in \operatorname{im} \mathbf{C}_T$. We may now infer that $R = \operatorname{im} \mathcal{C}(A,B)$ and $R \subset \operatorname{im} \mathbf{C}_T$ for all $T > 0$. It remains only to show that, for all $T > 0$, $\operatorname{im} \mathbf{C}_T \subset R$ and this follows immediately from the fact that $R = \cup_{T>0} \operatorname{im} \mathbf{C}_T$. $\qquad \square$

Exercise 3.2

Consider the linearized (and normalized) satellite model with A and B given in (3.3).

(a) Compute R (the set of all states which are reachable from 0).

(b) Replace B by $B_1 = (0, 1, 0, 0)^*$, the first column of B (physically this means that the input u_2 is inoperative, that is, tangential thrust is lost). Compute R in this case.

Observe that Theorem 3.5 implies the following equivalence

$$\mathbf{C}_T \text{ is surjective for all } T > 0 \iff \operatorname{rk} \mathcal{C}(A, B) = N. \qquad (3.13)$$

We are now in a position to formulate various characterizations of the concept of controllability of the linear system (3.1a).

Theorem 3.6

The following statements are equivalent.

(1) (3.1a) is controllable.
(2) $\operatorname{rk} \mathcal{C}(A, B) = N$.
(3) Q_T is invertible for some $T > 0$.
(4) Q_T is invertible for all $T > 0$.
(5) $\mathbf{C}_T$ is surjective for some $T > 0$.
(6) $\mathbf{C}_T$ is surjective for all $T > 0$.
(7) $\mathbf{C}_T$ is right invertible for some $T > 0$.
(8) $\mathbf{C}_T$ is right invertible for all $T > 0$.

Of the above characterizations (2)–(8) of controllability of (3.1a), (2) is the most useful in applications: it provides a simple algebraic rank test for controllability.

Proof

The equivalences (2) $\iff$ (4) $\iff$ (6) have already been established in (3.6) and (3.13). The implication (2) $\Rightarrow$ (8) was shown in (3.11). The implications (4) $\Rightarrow$ (3), (6) $\Rightarrow$ (5) and (8) $\Rightarrow$ (7) are immediate. If statement (7) holds, then $\mathbf{C}_T \mathbf{C}_T^{\sharp} z = z$ for all $z \in \mathbb{R}^N$ and so $\operatorname{im} \mathbf{C}_T = \mathbb{R}^N$, which is statement (5). Thus, (7) $\Rightarrow$ (5). To complete the proof, it suffices to prove the implications (5) $\Rightarrow$ (1), (3) $\Rightarrow$ (1) and (1) $\Rightarrow$ (2). Assume that statement (5) holds, that is, $\mathbf{C}_T$ is surjective for some $T > 0$. Let $\xi, \zeta \in \mathbb{R}^N$ be arbitrary. Define $z := \zeta - \exp(AT)\xi$.

By surjectivity of $\mathbf{C}_T$, there exists $u \in PC([0,T], \mathbb{R}^M)$ such that $\mathbf{C}_T u = z$ or, equivalently,

$$x(T; \xi, u) = \exp(AT)\xi + \int_0^T \exp(A(T-t))Bu(t)dt = \zeta - z + \mathbf{C}_T u = \zeta,$$

and so statement (1) holds. We proceed to prove the implication (3) $\Rightarrow$ (1). Assume that statement (3) holds, that is, Q_T is invertible for some $T > 0$. Let $\xi, \zeta \in \mathbb{R}^N$. Since Q_T is invertible, it follows from Theorem 3.4 that $\mathbb{R}^N = \operatorname{im} Q_T = \operatorname{im} \mathcal{C}(A, B)$ and so $\zeta - \exp(AT)\xi \in \operatorname{im} \mathcal{C}(A, B)$. Define $u := \mathbf{C}_T^\sharp(\zeta - \exp(AT)\xi))$. Then

$$x(T; \xi, u) = \exp(AT)\xi + \mathbf{C}_T u = \exp(AT)\xi + \mathbf{C}_T \mathbf{C}_T^\sharp(\zeta - \exp(AT\xi) = \zeta,$$

wherein the final equality follows by (3.10). Hence, (3.1a) is controllable and so statement (1) holds. It remains only to establish the implication (1) $\Rightarrow$ (2). Statement (1) implies that $R = \mathbb{R}^N$, where R is the set of states reachable from 0 given by (3.12). Thus, invoking Theorem 3.5, it follows that $\operatorname{rk} \mathcal{C}(A, B) = \dim \operatorname{im} \mathcal{C}(A, B) = N$ which is statement (2). $\square$

Example 3.7

We return to the linearized (and normalized) satellite model considered in Example 3.1 with A and B given by

$$A = \begin{pmatrix} 0 & 1 & 0 & 0 \\ 3\omega^2 & 0 & 0 & 2\omega \\ 0 & 0 & 0 & 1 \\ 0 & -2\omega & 0 & 0 \end{pmatrix} \quad \text{(where } \omega \neq 0\text{)}, \qquad B = \begin{pmatrix} 0 & 0 \\ 1 & 0 \\ 0 & 0 \\ 0 & 1 \end{pmatrix}.$$

A straightforward calculation gives

$$\mathcal{C}(A, B) = (B, AB, A^2B, A^3B)$$

$$= \begin{pmatrix} 0 & 0 & 1 & 0 & 0 & 2\omega & -\omega^2 & 0 \\ 1 & 0 & 0 & 2\omega & -\omega^2 & 0 & 0 & -2\omega^3 \\ 0 & 0 & 0 & 1 & -2\omega & 0 & 0 & -4\omega^2 \\ 0 & 1 & -2\omega & 0 & 0 & -4\omega^2 & 2\omega^3 & 0 \end{pmatrix}.$$

This matrix is of rank 4, and so, by Theorem 3.6, the system is controllable.

A natural question arises: if one of the inputs becomes inoperative, is the system still controllable? Setting $u_2 = 0$ reduces B to $B_1 = (0, 1, 0, 0)^*$ and gives

$$\mathcal{C}(A, B_1) = (B_1, AB_1, A^2B_1, A^3B_1) = \begin{pmatrix} 0 & 1 & 0 & -\omega^2 \\ 1 & 0 & -\omega^2 & 0 \\ 0 & 0 & -2\omega & 0 \\ 0 & -2\omega & 0 & 2\omega^3 \end{pmatrix}.$$

This matrix is only of rank 3 (column 4 is a multiple of column 2), showing that the system, with input u_1 only, is not controllable.

Setting $u_1 = 0$ reduces B to $B_2 = (0, 0, 0, 1)^*$ and gives

$$\mathcal{C}(A, B_2) = (B_2, AB_2, A^2 B_2, A^3 B_2) = \begin{pmatrix} 0 & 0 & 2\omega & 0 \\ 0 & 2\omega & 0 & -2\omega^3 \\ 0 & 1 & 0 & -4\omega^2 \\ 1 & 0 & -4\omega^2 & 0 \end{pmatrix}.$$

This matrix is of rank 4 and we may conclude that the system, with input u_2 only, is controllable.

Since u_1 was radial thrust and u_2 was tangential thrust, we see that loss of radial thrust does not destroy controllability, whereas loss of tangential thrust does. $\triangle$

Exercise 3.3

Consider the linearized inverted pendulum given by (1.4). Show that, for all positive values of the physical parameters m, M and l, the system is controllable.

The following proposition gives a geometric characterization of $\operatorname{im} \mathcal{C}(A, B)$ which will be useful in the proof of one of the main results of linear control theory, the so-called eigenvalue-assignment theorem (see Theorem 6.3).

Proposition 3.8

The subspace $\operatorname{im} \mathcal{C}(A, B) \subset \mathbb{R}^N$ is A-invariant and contains $\operatorname{im} B$. Moreover, it is the smallest subspace of $\mathbb{R}^N$ with these properties, that is, if $\mathcal{S} \subset \mathbb{R}^N$ is an A-invariant subspace containing $\operatorname{im} B$, then $\operatorname{im} \mathcal{C}(A, B) \subset \mathcal{S}$.

Proof

Let $v \in \mathbb{R}^M$ and set

$$w := \begin{pmatrix} v \\ 0 \end{pmatrix} \in \mathbb{R}^{MN}.$$

Then $\mathcal{C}(A, B)w = Bv$, showing that $\operatorname{im} B \subset \operatorname{im} \mathcal{C}(A, B)$. Let $z \in \operatorname{im} \mathcal{C}(A, B)$. To prove A-invariance, we need to show that $Az \in \operatorname{im} \mathcal{C}(A, B)$. Since $z \in \operatorname{im} \mathcal{C}(A, B)$, there exist N vectors $v_0, v_1, \ldots, v_{N-1}$ in $\mathbb{R}^M$ such that

$$z = \sum_{k=0}^{N-1} A^k B v_k \tag{3.14}$$

Consequently,

$$Az = \sum_{k=0}^{N-1} A^{k+1} Bv_k = \sum_{k=1}^{N-1} A^k Bv_{k-1} + A^N Bv_{N-1}.$$

By the Cayley-Hamilton theorem (Theorem A.6), we can express A^N as a linear combination of $I = A^0, A, A^2, \dots, A^{N-1}$. Consequently, $Az = \sum_{k=0}^{N-1} A^k Bw_k$ for suitable $w_k \in \mathbb{R}^M$, showing that $Az \in \operatorname{im} \mathcal{C}(A, B)$. Finally, let $\mathcal{S} \subset \mathbb{R}^N$ be an A-invariant subspace containing $\operatorname{im} B$ and let $z \in \operatorname{im} \mathcal{C}(A, B)$. Then z is of the form (3.14) for suitable $v_k \in \mathbb{R}^M$. Since $\operatorname{im} B \subset \mathcal{S}$, it follows that $Bv_k \in \mathcal{S}$, and hence, by A-invariance, $A^k Bv_k \in \mathcal{S}$ for all $k = 0, \dots, N-1$. Using the subspace property of $\mathcal{S}$, we conclude that $z \in \mathcal{S}$. $\qquad\square$

Exercise 3.4

Show that $\operatorname{im} \mathcal{C}(A, B)$ is $\exp(At)$-*invariant* in the sense that $\exp(At)v \in \operatorname{im} \mathcal{C}(A, B)$ for all $v \in \operatorname{im} \mathcal{C}(A, B)$ and all $t \in \mathbb{R}$.

Exercise 3.5

Let $T > 0$. Define

$$D_T := \{\xi \in \mathbb{R}^N \colon x(T; \xi, u) = 0 \text{ for some } u \in PC([0, T], \mathbb{R}^M)\},$$

the set of states that can be driven to 0 in time T. Show that $D_T = \operatorname{im} \mathcal{C}(A, B)$.
(*Hint.* Use Exercise 3.4.)

In the proofs of Theorems 3.5 and 3.6, control functions of the form $\mathbf{C}_T^\sharp v$ for some $v \in \operatorname{im} \mathcal{C}(A, B)$ played important roles. We proceed to investigate further these particular controls.

One immediate consequence of Exercise 3.4 is the fact that, for every pair of points $\xi, \zeta \in \operatorname{im} \mathcal{C}(A, B)$ and every $T > 0$, we have $\zeta - \exp(AT)\xi \in \operatorname{im} \mathcal{C}(A, B) = \operatorname{im} \mathbf{C}_T$ and so, invoking (3.10), the control $u^\dagger := \mathbf{C}_T^\sharp(\zeta - \exp(AT)\xi)$ is such that

$$\mathbf{C}_T u^\dagger = \mathbf{C}_T \mathbf{C}_T^\sharp(\zeta - \exp(AT)\xi) = \zeta - \exp(AT)\xi \qquad (3.15)$$

or, equivalently, $\zeta = x(T; \xi, u^\dagger)$. Thus, for every pair $\xi, \zeta \in \operatorname{im} \mathcal{C}(A, B)$ and $T > 0$, there exists a control u steering ξ to ζ in time T. In this sense, $\operatorname{im} \mathcal{C}(A, B)$ is the *controllable subspace*. Denote the set of controls that achieve the transition from $\xi \in \operatorname{im} \mathcal{C}(A, B)$ to $\zeta \in \operatorname{im} \mathcal{C}(A, B)$ in time $T > 0$ by

$$\mathcal{U}_T(\xi, \zeta) := \{u \in PC([0, T], \mathbb{R}^M) \colon x(T; \xi, u) = \zeta\}.$$

Clearly, this set is non-empty since the control $u^\dagger := \mathbf{C}_T^\sharp(\zeta - \exp(AT)\xi)$ is in $\mathcal{U}_T(\xi, \zeta)$. A measure of the *energy* of a control $u \in \mathcal{U}_T(\xi, \zeta)$ is given by

$\int_0^T \|u(t\|^2 dt$. The next result asserts that the control $u^\dagger$ achieves the transition from $\xi \in \operatorname{im} \mathcal{C}(A, B)$ to $\zeta \in \operatorname{im} \mathcal{C}(A, B)$ in time T with least energy expenditure.

Proposition 3.9 (Minimum-energy control)

Let $\xi, \zeta \in \operatorname{im} \mathcal{C}(A, B)$ and $T > 0$. Define the control $u^\dagger := \mathbf{C}_T^\sharp(\zeta - \exp(AT)\xi)$. Then $u^\dagger \in \mathcal{U}_T(\xi, \zeta)$ and

$$\int_0^T \|u^\dagger(t)\|^2 dt \le \int_0^T \|u(t)\|^2 dt \ \ \forall u \in \mathcal{U}_T(\xi, \zeta).$$

Proof

Under the control $u^\dagger$ we have

$$x(T; \xi, u^\dagger) = \exp(AT)\xi + \mathbf{C}_T u^\dagger = \exp(AT)\xi + \mathbf{C}_T \mathbf{C}_T^\sharp(\zeta - \exp(AT)\xi) = \zeta,$$

and so $u^\dagger \in \mathcal{U}_T(\xi, \zeta)$. Now $x(T; \xi, u) = \zeta = x(T; \xi, u^\dagger)$ for all $u \in \mathcal{U}_T(\xi, \zeta)$ and hence

$$0 = \mathbf{C}_T(u^\dagger - u) = \int_0^T \exp(A(T-t))B(u^\dagger(t) - u(t)) \, dt.$$

Taking inner product with $Q_T^\sharp(\zeta - \exp(AT)\xi)$, we have, for all $u \in \mathcal{U}_T(\xi, \zeta)$,

$$\begin{aligned} 0 &= \int_0^T \langle B^* \exp(A^*(T-t))Q_T^\sharp(\zeta - \exp(AT)\xi), u^\dagger(t) - u(t) \rangle \, dt \\ &= \int_0^T \langle \mathbf{C}_T^\sharp(\zeta - \exp(AT)\xi), u^\dagger(t) - u(t) \rangle dt \\ &= \int_0^T \langle u^\dagger(t), u^\dagger(t) - u(t) \rangle \, dt, \end{aligned}$$

and so

$$\int_0^T \|u^\dagger(t)\|^2 dt = \int_0^T \langle u^\dagger(t), u(t) \rangle dt.$$

Therefore, for all $u \in \mathcal{U}_T(\xi, \zeta)$,

$$\begin{aligned} 0 &\le \int_0^T \|u^\dagger(t) - u(t)\|^2 dt \\ &= \int_0^T \|u^\dagger(t)\|^2 dt + \int_0^T \|u(t)\|^2 dt - 2 \int_0^T \langle u^\dagger(t), u(t) \rangle dt \\ &= \int_0^T \|(u(t)\|^2 dt - \int_0^T \|u^\dagger(t)\|^2 dt, \end{aligned}$$

whence the assertion of the proposition. □

To develop controllability theory further, some elementary facts about coordinate changes in the state space are required. To this end, consider the controlled system (3.1a) and let $S \in GL(N, \mathbb{R})$, where $GL(N, \mathbb{R})$ denotes the *general linear group*:

$$GL(N, \mathbb{R}) := \{M \in \mathbb{R}^{N \times N} : M \text{ invertible}\} = \{M \in \mathbb{R}^{N \times N} : \det M \neq 0\}.$$

Consider the coordinate change $x \mapsto S^{-1}x$. Then (3.1a) becomes

$$S^{-1}\dot{x} = S^{-1}AS(S^{-1}x) + S^{-1}Bu.$$

Setting $\tilde{x} := S^{-1}x$ and

$$\tilde{A} := S^{-1}AS, \quad \tilde{B} := S^{-1}B, \tag{3.16}$$

it follows that

$$\dot{\tilde{x}} = \tilde{A}\tilde{x} + \tilde{B}u. \tag{3.17}$$

Since

$$\mathcal{C}(\tilde{A}, \tilde{B}) = S^{-1}\mathcal{C}(A, B)$$

it follows from Theorem 3.6 that (A, B) is controllable if, and only if, $(\tilde{A}, \tilde{B})$ is controllable. This can also be easily shown from first principles (based on the definition of controllability, without using Theorem 3.6), see Exercise 3.6.

Exercise 3.6

Show from first principles (using only the definition of controllability) that (A, B) is controllable if, and only if, $(\tilde{A}, \tilde{B})$ is controllable.

Next, we state and prove a useful result, referred to as the Kalman[2] controllability decomposition lemma.

Lemma 3.10 (Kalman controllability decomposition)

Assume that $(A, B) \in \mathbb{R}^{N \times N} \times \mathbb{R}^{N \times M}$ is not controllable and $B \neq 0$, in which case,

$$0 < K := \operatorname{rk}\mathcal{C}(A, B) = \dim \operatorname{im} \mathcal{C}(A, B) < N.$$

Then there exists $S \in GL(N, \mathbb{R})$ such that $\tilde{A} := S^{-1}AS$ and $\tilde{B} := S^{-1}B$ have the block structure

$$\tilde{A} = \begin{pmatrix} A_1 & A_2 \\ 0 & A_3 \end{pmatrix}, \quad \tilde{B} = \begin{pmatrix} B_1 \\ 0 \end{pmatrix}, \tag{3.18}$$

where $A_1 \in \mathbb{R}^{K \times K}$, $B_1 \in \mathbb{R}^{K \times M}$ and (A_1, B_1) is controllable.

---

[2] Rudolf Emil Kalman (born 1930), US American.

Proof

Pick any subspace $V \subset \mathbb{R}^N$ such that

$$\operatorname{im}\mathcal{C}(A,B) \oplus V = \mathbb{R}^N . \tag{3.19}$$

Let $\{v_1, \ldots, v_K\}$ and $\{v_{K+1}, \ldots, v_N\}$ be bases of $\operatorname{im}\mathcal{C}(A,B)$ and V, respectively. Define

$$S := (v_1, v_2, \ldots, v_N) \in GL(N, \mathbb{R}).$$

Then, for $z = (z_1, \ldots, z_N)^* \in \mathbb{R}^N$, we have $Sz = \sum_{j=1}^N z_j v_j$, and it follows from (3.19) that

$$Sz \in \operatorname{im}\mathcal{C}(A,B) \quad \Longleftrightarrow \quad z_{K+1} = \ldots = z_N = 0 . \tag{3.20}$$

Writing

$$\tilde{B} := S^{-1}B = \begin{pmatrix} B_1 \\ B_2 \end{pmatrix}, \quad B_1 \in \mathbb{R}^{K \times M},$$

it follows that

$$S \begin{pmatrix} B_1 v \\ B_2 v \end{pmatrix} = Bv \in \operatorname{im}\mathcal{C}(A,B) \quad \forall v \in \mathbb{R}^M .$$

Hence, by (3.20), $B_2 v = 0$ for all $v \in \mathbb{R}^M$ and thus $B_2 = 0$. Now write

$$\tilde{A} := S^{-1}AS = \begin{pmatrix} A_1 & A_2 \\ A_4 & A_3 \end{pmatrix}, \quad A_1 \in \mathbb{R}^{K \times K}.$$

We proceed to show that $A_4 = 0$. Let $w \in \operatorname{im}\mathcal{C}(A,B) \subset \mathbb{R}^N$. Since $\operatorname{im}\mathcal{C}(A,B)$ is A-invariant (by Proposition 3.8), we have $z := Aw \in \operatorname{im}\mathcal{C}(A,B)$. Moreover,

$$\tilde{A}S^{-1}w = S^{-1}z .$$

Since $w, z \in \operatorname{im}\mathcal{C}(A,B)$, it follows from (3.20) that

$$S^{-1}w = \begin{pmatrix} w_1 \\ 0 \end{pmatrix}, \quad S^{-1}z = \begin{pmatrix} z_1 \\ 0 \end{pmatrix}$$

for some $w_1, z_1 \in \mathbb{R}^K$. Consequently,

$$\begin{pmatrix} A_1 & A_2 \\ A_4 & A_3 \end{pmatrix} \begin{pmatrix} w_1 \\ 0 \end{pmatrix} = \begin{pmatrix} z_1 \\ 0 \end{pmatrix},$$

which implies, in particular, that $A_4 w_1 = 0$. The above argument holds for every $w \in \operatorname{im}\mathcal{C}(A,B)$ and so

$$A_4 w_1 = 0 \quad \forall w_1 \in \mathbb{R}^K ,$$

whence $A_4 = 0$. Finally, it is not difficult to show that (A_1, B_1) is controllable, see Exercise 3.7. $\qquad \square$

Exercise 3.7

Show that the pair (A_1, B_1) is controllable.

The next result – the Hautus[3] criterion – provides another algebraic test for controllability.

Theorem 3.11 (Hautus criterion for controllability)

System (3.1a) is controllable if, and only if, $\operatorname{rk}(sI - A, B) = N$ for all $s \in \mathbb{C}$.

Note that, if $s \in \mathbb{C} \backslash \sigma(A)$ (that is, if s is *not* an eigenvalue of A), then $N = \operatorname{rk}(sI - A) = \operatorname{rk}(sI - A, B)$. Therefore, by Theorem 3.11, system (3.1a) is controllable if, and only if, $\operatorname{rk}(sI - A, B) = N$ for all $s \in \sigma(A)$.

Proof of Theorem 3.11

We prove the theorem using contraposition. To this end, assume first that $\operatorname{rk}(\lambda I - A, B) < N$ for some $\lambda \in \mathbb{C}$ (an eigenvalue of A). Then there exists $z \in \mathbb{C}^N$, $z \neq 0$ such that $z^*(\lambda I - A, B) = 0$. Thus, $z^* A = \lambda z^*$ and $z^* B = 0$. As a consequence,

$$z^* A^k B = \lambda^k z^* B = 0 \quad \forall k \in \mathbb{N}_0,$$

implying that $z^* \mathcal{C}(A, B) = 0$. Since $z \neq 0$, this shows that $\operatorname{rk} \mathcal{C}(A, B) < N$. Hence, by Theorem 3.6, system (3.1a) is not controllable.

Conversely, assume that system (3.1a) is not controllable. If $B = 0$, then $\operatorname{rk}(sI - A, B) = \operatorname{rk}(sI - A) < N$ for all $s \in \sigma(A)$. If $B \neq 0$, then it follows from Proposition 3.10 that there exists $S \in GL(N, \mathbb{R})$ such that

$$\tilde{A} := S^{-1} A S = \begin{pmatrix} A_1 & A_2 \\ 0 & A_3 \end{pmatrix}, \quad \tilde{B} := S^{-1} B = \begin{pmatrix} B_1 \\ 0 \end{pmatrix},$$

where $A_1 \in \mathbb{R}^{K \times K}$, $B_1 \in \mathbb{R}^{K \times M}$ and $K < N$. Let $\lambda \in \mathbb{C}$ and $v \in \mathbb{C}^{N-K}$ be an eigenvalue/eigenvector pair of A_3^*. Then

$$v \neq 0, \quad v^*(\bar{\lambda} I - A_3) = 0.$$

Setting

$$w := \begin{pmatrix} 0 \\ v \end{pmatrix} \in \mathbb{C}^N,$$

it follows that

$$w^*(\bar{\lambda} I - \tilde{A}) = (0, v^*) \begin{pmatrix} \bar{\lambda} I - A_1 & -A_2 \\ 0 & \bar{\lambda} I - A_3 \end{pmatrix} = 0, \quad w^* \tilde{B} = (0, v^*) \begin{pmatrix} B_1 \\ 0 \end{pmatrix} = 0.$$

---

[3] Malo Hautus (1940 –), Dutch.

Hence, $z = (S^*)^{-1}w \neq 0$ satisfies

$$z^*(\bar{\lambda}I - A)S = w^*(\bar{\lambda}I - \tilde{A}) = 0, \quad z^*B = w^*\tilde{B} = 0,$$

implying that

$$z^*(\bar{\lambda}I - A) = 0, \quad z^*B = 0.$$

Consequently, $z^*(\bar{\lambda}I - A, B) = 0$ and hence, $\mathrm{rk}\,(\bar{\lambda}I - A, B) < N$. $\qquad\square$

Example 3.12

Once more we return to the linearized (and normalized) satellite model considered in Examples 3.1 and 3.7, with A and B given by

$$A = \begin{pmatrix} 0 & 1 & 0 & 0 \\ 3\omega^2 & 0 & 0 & 2\omega \\ 0 & 0 & 0 & 1 \\ 0 & -2\omega & 0 & 0 \end{pmatrix} \quad \text{(where } \omega \neq 0\text{)}, \quad B = \begin{pmatrix} 0 & 0 \\ 1 & 0 \\ 0 & 0 \\ 0 & 1 \end{pmatrix}.$$

In Example 3.7, it was shown that (A, B) is controllable by invoking Theorem 3.6. Here we wish to re-derive controllability of (A, B) by using Theorem 3.11.

The matrix $(sI - A, B)$ is given by

$$(sI - A, B) = \begin{pmatrix} s & -1 & 0 & 0 & 0 & 0 \\ -3\omega^2 & s & 0 & -2\omega & 1 & 0 \\ 0 & 0 & s & -1 & 0 & 0 \\ 0 & 2\omega & 0 & s & 0 & 1 \end{pmatrix}.$$

A routine calculation reveals that, if $s = \sqrt{3}\,\omega$, then columns 1, 3, 4 and 6 of this matrix are linearly independent, whilst, for all $s \neq \sqrt{3}\,\omega$, columns 1, 2, 4 and 6 are linearly independent. Therefore, the matrix has full rank 4 for all $s \in \mathbb{C}$. By the Hautus criterion, it follows that (A, B) is controllable. Note that, in the above calculation, column 5 plays no role. Therefore, if control component u_1 becomes inoperative, the system remains controllable (which was also shown in Example 3.7).

On the other hand, if control component u_2 becomes inoperative, then, replacing B by $B_1 = (0, 1, 0, 0)^*$, we obtain

$$(sI - A, B_1) = \begin{pmatrix} s & -1 & 0 & 0 & 0 \\ -3\omega^2 & s & 0 & -2\omega & 1 \\ 0 & 0 & s & -1 & 0 \\ 0 & 2\omega & 0 & s & 0 \end{pmatrix}.$$

Setting $s = 0$ gives

$$(-A, B_1) = \begin{pmatrix} 0 & -1 & 0 & 0 & 0 \\ -3\omega^2 & 0 & 0 & -2\omega & 1 \\ 0 & 0 & 0 & -1 & 0 \\ 0 & 2\omega & 0 & 0 & 0 \end{pmatrix}$$

with $\text{rk}\,(-A, B_1) = 3 < 4$. Hence, Theorem 3.11 shows that (A, B_1) is not controllable (which was also shown in Example 3.7). $\triangle$

Exercise 3.8

Consider the pair of matrices (A, B) given by

$$A = \begin{pmatrix} 0 & 1 & 0 & 0 & 0 \\ 0 & 0 & 1 & 0 & 0 \\ 0 & 0 & 0 & 0 & 1 \\ \alpha & 0 & 0 & 0 & 1 \\ 0 & 0 & 0 & 0 & 0 \end{pmatrix}, \qquad B = \begin{pmatrix} 0 \\ -1 \\ 0 \\ \beta \\ 1 \end{pmatrix},$$

where α and β are real parameters. Find all pairs (α, β) such that (A, B) is controllable.

3.2 Observability

In this section, we address the observability problem which was introduced, in an informal way, earlier in this chapter.

For $\xi \in \mathbb{R}^N$, $T > 0$ and $u \in PC([0, T], \mathbb{R}^M)$, we denote the output of system (3.1) on $[0, T]$ by

$$y(t; \xi, u) := Cx(t; \xi, u) = C \exp(At)\xi + C \int_0^t \exp(A(t - \tau)Bu(\tau)\mathrm{d}\tau \quad \forall t \in [0, T],$$

where we have explicitly indicated the dependence of the output on the initial state ξ and input u. We assume given the following data: (A, B, C), $T > 0$, $u \in PC([0, T], \mathbb{R}^M)$ and $y(\cdot; \xi, u) \in C([0, T], \mathbb{R}^P)$. The initial state $\xi \in \mathbb{R}^N$ is unknown and so $x(t) = x(t; \xi, u)$, $t \in [0, T]$ may not, in general, be computable from the given data. Observability investigates circumstances wherein such a computation is feasible. In particular, observability seeks to identify conditions under which the initial state ξ (and hence the state $x(t) = x(t; \xi, u)$, $t \in [0, T]$) is uniquely determined by the given data.

Let $\mathbf{O}_T$, parameterized by $T > 0$, denote the *state-to-output map* $\mathbb{R}^N \mapsto C([0,T],\mathbb{R}^P)$ given by

$$(\mathbf{O}_T\xi)(t) := C\exp(At)\xi = y(t;\xi,0) \quad \forall\, t \in [0,T].$$

It is clear that the map $\mathbf{O}_T$ is linear, that is, for $\xi, \zeta \in \mathbb{R}^N$ and $\alpha, \beta \in \mathbb{R}$, we have $\mathbf{O}_T(\alpha\xi + \beta\zeta) = \alpha\mathbf{O}_T\xi = \beta\mathbf{O}_T\zeta$, where, as is usual for linear maps, we have written $\mathbf{O}_T\xi$ for $\mathbf{O}_T(\xi)$. Furthermore, we note that, for each $\xi \in \mathbb{R}^N$, the function $\mathbf{O}_T\xi$ is computable from the given data:

$$(\mathbf{O}_T\xi)(t) = y(t;\xi,u) - y(t;0,u)$$

$$= y(t;\xi,u) - C\int_0^t \exp(A(t-\tau))Bu(\tau)\mathrm{d}\tau, \quad \forall\, t \in [0,T].$$

A map $\mathbf{O}_T^\sharp \colon C([0,T],\mathbb{R}^P) \to \mathbb{R}^N$ is a left inverse of $\mathbf{O}_T$ if the following holds: $\mathbf{O}_T^\sharp(\mathbf{O}_T\xi) = \xi$ for all $\xi \in \mathbb{R}^N$. Note that a left inverse of $\mathbf{O}_T$ exists if, and only if, $\mathbf{O}_T$ is injective. To see this, first assume that $\mathbf{O}_T$ is injective and choose any map $\mathbf{O}_T^\sharp \colon C([0,T],\mathbb{R}^P) \to \mathbb{R}^N$ with the property that, for each $y \in \mathrm{im}\,\mathbf{O}_T$, $\mathbf{O}_T^\sharp y$ is defined to be the unique vector $\xi \in \mathbb{R}^N$ such that $\mathbf{O}_T\xi = y$: this choice ensures that the requisite property $\mathbf{O}_T^\sharp(\mathbf{O}_T\xi) = \xi$ holds for all $\xi \in \mathbb{R}^N$ and so $\mathbf{O}_T$ is left invertible. Conversely, we note that the existence of a left inverse of $\mathbf{O}_T$ immediately implies the injectivity of $\mathbf{O}_T$.

Thus, the question of observability can be identified as that of left invertibility or, equivalently, injectivity of the map $\mathbf{O}_T$. If a left inverse $\mathbf{O}_T^\sharp$ exists, then the initial state ξ of (3.1) is given by

$$\xi = \mathbf{O}_T^\sharp(y(\,\cdot\,;\xi,0))$$

and, once ξ is determined, the state $x(t;\xi,u)$ is computable for all $t \in [0,T]$ via the variation of parameters formula (3.2). Note that $y(\,\cdot\,;\xi,0)$ is computable from the given data, since $y(\,\cdot\,;\xi,0) = y(\,\cdot\,;\xi,u) - y(\,\cdot\,;0,u)$.

Next, we note that, for each $T > 0$ and $\xi \in \mathbb{R}^N$, $\mathbf{O}_T\xi = 0$ if, and only if, $C\exp(At)\xi = 0$ for all $t \in \mathbb{R}_+$: that the latter identity implies the former is clear; to see that the former implies the latter, assume $\mathbf{O}_T\xi = 0$, then $C\exp(At)\xi = 0$ for all $t \in [0,T]$ and evaluation at 0 and repeated differentiation at 0 gives $CA^k\xi = 0$ for all $k \in \mathbb{N}_0$, whence $C\exp(At)\xi = \sum_{k=0}^\infty (t^k/k!)CA^k\xi = 0$ for all $t \in \mathbb{R}_+$. Furthermore, we note that, by linearity, $\mathbf{O}_T$ is injective if, and only if, the following holds

$$\mathbf{O}_T\xi = 0 \Rightarrow \xi = 0$$

and this, in turn, is equivalent to

$$y(t;\xi,0) = C\exp(At)\xi = 0 \ \ \forall\, t \in \mathbb{R}_+ \ \Rightarrow \ \xi = 0.$$

The above discussion underpins the following definition and establishes the ensuing proposition.

Definition 3.13

System (3.1) is said to be *observable* if, for all $\xi \in \mathbb{R}^N$, the following holds

$$\left(y(t;\xi,0) = C\exp(At)\xi = 0 \ \forall t \in \mathbb{R}_+ \right) \quad \Rightarrow \quad \xi = 0. \tag{3.21}$$

Proposition 3.14

The following statements are equivalent.
(1) System (3.1) is observable.
(2) $\mathbf{O}_T$ is injective for some $T > 0$.
(3) $\mathbf{O}_T$ is injective for all $T > 0$.
(4) $\mathbf{O}_T$ is left invertible for some $T > 0$.
(5) $\mathbf{O}_T$ is left invertible for all $T > 0$.

In Definition 3.13, the matrix B does not play a role: observability of (3.1) depends only on the matrices A and C. It is therefore convenient to define a pair of matrices $(C, A) \in \mathbb{R}^{P \times N} \times \mathbb{R}^{N \times N}$ to be *observable* if, for all $\xi \in \mathbb{R}^N$, the implication (3.21) holds. The next assertion is an immediate consequence of Definition 3.13.

Lemma 3.15

Define

$$U := \{\xi \in \mathbb{R}^N : C\exp(At)\xi = 0 \ \forall t \in \mathbb{R}_+\}. \tag{3.22}$$

The pair (C, A) is observable if, and only if, $U = \{0\}$.

With a view to convenient characterizations of the set U, we introduce the *observability matrix* defined by

$$\mathcal{O}(C,A) := \begin{pmatrix} C \\ CA \\ \vdots \\ CA^{N-1} \end{pmatrix} \in \mathbb{R}^{(PN) \times N}$$

and the *observability Gramian* (parameterized by $T > 0$) defined by

$$S_T := \int_0^T \exp(A^*t) C^* C \exp(At) dt \in \mathbb{R}^{N \times N}.$$

Note that S_T is symmetric and positive semi-definite.

Theorem 3.16

$\ker \mathcal{O}(C, A) = \ker S_T$ for all $T > 0$.

Proof

First note that $(\mathcal{O}(C, A))^* = \mathcal{C}(A^*, C^*)$ and that S_T is the controllability Gramian associated with the pair (A^*, C^*). Hence, by Theorem 3.4, $(\operatorname{im} S_T)^{\perp} = (\operatorname{im} \mathcal{C}(A^*, C^*))^{\perp}$. Invoking Theorem A.1, it follows that, for $z \in \mathbb{R}^N$ and $T > 0$,

$$\mathcal{O}(C, A)z = 0 \;\Leftrightarrow\; z^* \mathcal{C}(A^*, C^*) = 0 \;\Leftrightarrow\; z \in (\operatorname{im} \mathcal{C}(A^*, C^*))^{\perp} \;\Leftrightarrow\; z \in (\operatorname{im} S_T)^{\perp}.$$

Therefore, using symmetry of S_T and invoking Theorem A.1,

$$\mathcal{O}(C, A)z = 0 \;\Leftrightarrow\; z^* S_T = 0 \;\Leftrightarrow\; S_T z = 0.$$

Thus, $\ker \mathcal{O}(A, B) = \ker S_T$, completing the proof of the theorem. $\qquad\square$

By Theorem 3.16, in conjunction with Theorem A.4, we see that, for each $T > 0$, there exists $S_T^{\sharp} \in \mathbb{R}^{N \times N}$ such that

$$S_T^{\sharp} S_T z = z \quad \forall\, z \in \left(\ker S_T \right)^{\perp} = \left(\ker \mathcal{O}(C, A) \right)^{\perp}. \tag{3.23}$$

Introduce the map

$$\mathbf{O}_T^{\sharp} : C([0, T], \mathbb{R}^P) \to \mathbb{R}^N, \quad w \mapsto S_T^{\sharp} \int_0^T \exp(A^* t) C^* w(t)\, dt. \tag{3.24}$$

Note that

$$\mathbf{O}_T^{\sharp}(\mathbf{O}_T \xi) = S_T^{\sharp} \int_0^T \exp(A^* t) C^* C \exp(At) \xi\, dt = S_T^{\sharp} S_T \xi \quad \forall\, \xi \in \mathbb{R}^N$$

and so, by (3.23),

$$\mathbf{O}_T^{\sharp}(\mathbf{O}_T \xi) = \xi \quad \forall\, \xi \in \left(\ker S_T \right)^{\perp} = \left(\ker \mathcal{O}(C, A) \right)^{\perp}. \tag{3.25}$$

By the dimension formula (see (A.5) in Appendix A.1), $\operatorname{rk} \mathcal{O}(C, A) = N - \dim \ker \mathcal{O}(C, A)$, and so a particular consequence of (3.25) is the fact that, if $\operatorname{rk} \mathcal{O}(C, A) = N$, then $\mathbf{O}_T$ is left invertible (and $\mathbf{O}_T^{\sharp}$ is a left inverse). We record this in the following

$$\operatorname{rk} \mathcal{O}(C, A) = N \implies \mathbf{O}_T \text{ is left invertible for all } T > 0. \tag{3.26}$$

In view of (3.25) and in the context of the observability question, we may infer that, if the initial state ξ is in $\left(\ker \mathcal{O}(C, A) \right)^{\perp}$, then ξ can be computed, via the known function $\mathbf{O}_T \xi =: y$, by evaluating $\mathbf{O}_T^{\sharp} y$. In this sense, $\left(\ker \mathcal{O}(C, A) \right)^{\perp}$ is the *observable subspace*. The next result establishes that the set U, defined by (3.22), coincides with $\ker \mathcal{O}(C, A)$, the orthogonal complement of the observable subspace. In the following, U is referred to as the *unobservable subspace*.

Theorem 3.17

$\ker \mathcal{O}(C, A) = U$, where U is given by (3.22).

The significance of this result is that the set U can be computed by purely algebraic means.

Proof

Note that, for $\xi \in \mathbb{R}^N$, $\mathcal{O}(C, A)\xi = 0$ is equivalent to

$$C\xi = CA\xi = \cdots = CA^{N-1}\xi = 0.$$

By the Cayley-Hamilton theorem (see Theorem A.6), this is equivalent to

$$CA^k\xi = 0, \quad \forall k \in \mathbb{N}_0,$$

which, in turn, is equivalent to

$$Ce^{At}\xi = \sum_{k=0}^{\infty} \frac{t^k}{k!} CA^k\xi = 0, \quad \forall t \in \mathbb{R}_+.$$

Hence, $\ker \mathcal{O}(A, B) = U$. $\square$

We now formulate a result which, in conjunction with Proposition 3.14, provides an "observability" counterpart to the "controllability" Theorem 3.6 and which, via the equivalence of (1) and (2), provides an easily checkable algebraic rank condition for observability.

Theorem 3.18

The following statements are equivalent.
(1) (3.1) is observable.
(2) $\operatorname{rk} \mathcal{O}(C, A) = N$.
(3) S_T is invertible for some $T > 0$.
(4) S_T is invertible for all $T > 0$.

Proof

Note that $\operatorname{rk} \mathcal{O}(C, A) = \operatorname{rk}(\mathcal{O}(C, A))^* = \operatorname{rk} \mathcal{C}(A^*, C^*)$ and, for each $T > 0$, the observability Gramian S_T for (C, A) S_T coincides with the controllability Gramian associated with the pair (A^*, C^*). Therefore, the equivalences (2) $\Leftrightarrow$ (3) $\Leftrightarrow$ (4) are immediate consequences of Theorem 3.6. To complete the proof, it suffices to show that (1) $\Rightarrow$ (2) and (3) $\Rightarrow$ (1). To prove (1) $\Rightarrow$ (2), note that, by Lemma 3.15, statement (1) is equivalent to $U = \{0\}$ which in turn, by Theorem 3.17, is equivalent to $\ker \mathcal{O}(C, A) = \{0\}$. The latter is equivalent to the linear independence of the N columns of $\mathcal{O}(C, A)$. Hence, $\operatorname{rk} \mathcal{O}(C, A) = N$, showing that $(1) \Rightarrow (2)$.

Assume that statement (3) holds and so S_T is invertible for some $T > 0$. Note that

$$\mathbf{O}_T^{\sharp}\big(y(\cdot\,;\xi,0)\big) = S_T^{-1} \int_0^T \exp(A^*t)C^*C \exp(At)\xi dt = S_T^{-1} S_T \xi = \xi \ \ \forall \xi \in \mathbb{R}^N.$$

Thus, if $C \exp(At)\xi = y(\cdot\,;\xi,0) = 0$ for all $t \in \mathbb{R}_+$, then $\xi = \mathbf{O}_T^{\sharp}(0) = 0$. Therefore, (3.1) is observable. This establishes the implication $(3) \Rightarrow (1)$. $\qquad\square$

We now have the following scenario. Assume that (3.1) is observable and let $T > 0$. Furthermore, assume that the initial-state ξ is unknown, but that the data A, B and C are known and that, for all $t \in [0, T]$, the input and output signals $u(t)$ and $y(t; \xi, u)$ are known (and hence $\mathbf{O}_T\xi = y(\cdot\,;\xi, u) - y(\cdot\,; 0, u)$ is a known function). By observability, it follows from Theorem 3.18 that $\ker S_T = \{0\}$ and so, by (3.25), the map $\mathbf{O}_T^{\sharp}$, given by (3.24) with $S_T^{\sharp} = S_T^{-1}$, is a left inverse of the map $\mathbf{O}_T$. Therefore, $\mathbf{O}_T^{\sharp}(\mathbf{O}_T\xi) = \xi$, and the initial state is determined by evaluating $\mathbf{O}_T^{\sharp}(\mathbf{O}_T\xi)$. Once ξ has been determined, the state $x(t; \xi, u)$, for $t \in [0, T]$, can be obtained from the variation of parameters formula (3.2).

Example 3.19

In this example (and in Exercise 3.9 below), we consider the observability properties of the linearized (and normalized) satellite model described in Example 3.1, with A and C given by

$$A = \begin{pmatrix} 0 & 1 & 0 & 0 \\ 3\omega^2 & 0 & 0 & 2\omega \\ 0 & 0 & 0 & 1 \\ 0 & -2\omega & 0 & 0 \end{pmatrix} \quad \text{(where } \omega \neq 0\text{)}, \qquad C = \begin{pmatrix} 1 & 0 & 0 & 0 \\ 0 & 0 & 1 & 0 \end{pmatrix}.$$

A routine calculation shows that

$$\mathcal{O}(C, A) = \begin{pmatrix} 1 & 0 & 0 & 0 \\ 0 & 0 & 1 & 0 \\ 0 & 1 & 0 & 0 \\ 0 & 0 & 0 & 1 \\ 3\omega^2 & 0 & 0 & 2\omega \\ 0 & -2\omega & 0 & 0 \\ 0 & -\omega^2 & 0 & 0 \\ -6\omega^3 & 0 & 0 & -4\omega^2 \end{pmatrix}.$$

This matrix has rank 4, and so, by Theorem 3.18, the system is observable. $\triangle$

Exercise 3.9

Here, we continue the investigation of the observability properties of the linearized (and normalized) satellite model started in Example 3.19. To minimize the measurements required, we might consider not measuring y_2 (angle measurement) or y_1 (radial measurement). This means replacing the matrix C in Example 3.19 by $C_1 = (1, 0, 0, 0)$ or $C_2 = (0, 0, 1, 0)$, respectively. Determine whether or not (C_1, A) and (C_2, A) are observable.

The next result is an immediate consequence of Theorems 3.6 and 3.18.

Corollary 3.20

The pair (A, B) is controllable if, and only if, the pair (B^*, A^*) is observable. Similarly, the pair (C, A) is observable if, and only if, the pair (A^*, C^*) is controllable.

The above corollary (which is sometimes encapsulated as "controllability and observability are *dual* concepts") can be used to derive results on observability from corresponding controllability results and vice versa. The proof of the following theorem is a typical example.

Theorem 3.21 (Hautus criterion for observability)

System (3.1) is observable if, and only if,

$$\mathrm{rk} \begin{pmatrix} sI - A \\ C \end{pmatrix} = N \quad \forall\, s \in \mathbb{C}. \tag{3.27}$$

Proof

By Corollary 3.20, (C, A) is observable if, and only if, (A^*, C^*) is controllable. By the Hautus criterion for controllability (Theorem 3.11), the latter is the case if, and only if, $\mathrm{rk}\,(\bar{s}I - A^*, C^*) = N$ for all $s \in \mathbb{C}$, which, in turn, is equivalent to (3.27). $\qquad\square$

Exercise 3.10

Revisit Example 3.19 and Exercise 3.9: investigate the observability properties of the satellite example by invoking Theorem 3.21.

Lemma 3.22 (Kalman observability decomposition)

Assume that $(C, A) \in \mathbb{R}^{P \times N} \times \mathbb{R}^{N \times N}$ is not observable and $C \neq 0$, in which case,

$$0 < K := \operatorname{rk} \mathcal{O}(C, A) = \dim \operatorname{im} \mathcal{O}(C, A) < N.$$

Then there exists $S \in GL(N, \mathbb{R})$ such that $\tilde{A} := S^{-1}AS$ and $\tilde{C} := CS$ have the block structure

$$\tilde{A} = \begin{pmatrix} A_1 & 0 \\ A_2 & A_3 \end{pmatrix}, \quad \tilde{C} = (C_1, 0) \tag{3.28}$$

where $A_1 \in \mathbb{R}^{K \times K}$, $C_1 \in \mathbb{R}^{P \times K}$ and (C_1, A_1) is observable.

Exercise 3.11

Prove Lemma 3.22. (*Hint.* Use Lemma 3.10 and Corollary 3.20.)

Exercise 3.12

Let $(A, B, C) \in \mathbb{R}^{N \times N} \times \mathbb{R}^{N \times M} \times \mathbb{R}^{P \times N}$ with $C \neq 0$ and $B \neq 0$.
(a) Assume that (C, A) is observable and (A, B) is not controllable. Show that there exists $S \in GL(N, \mathbb{R})$ such that the matrices $\tilde{A} = S^{-1}AS$, $\tilde{B} = S^{-1}B$ and $\tilde{C} = CS$ have the block structure

$$\tilde{A} = \begin{pmatrix} A_1 & A_2 \\ 0 & A_3 \end{pmatrix}, \quad \tilde{B} = \begin{pmatrix} B_1 \\ 0 \end{pmatrix}, \quad \tilde{C} = (C_1, C_2)$$

where the pair $(C_1, A_1) \in \mathbb{R}^{P \times K} \times \mathbb{R}^{K \times K}$ is observable, the pair $(A_1, B_1) \in \mathbb{R}^{K \times K} \times \mathbb{R}^{K \times M}$ is controllable, and $K := \operatorname{rk} \mathcal{C}(A, B)$.
(b) Assume that (C, A) is not observable and (A, B) is controllable. Show that there exists $S \in GL(N, \mathbb{R})$ such that the matrices $\tilde{A} = S^{-1}AS$, $\tilde{B} = S^{-1}B$ and $\tilde{C} = CS$ have the block structure

$$\tilde{A} = \begin{pmatrix} A_1 & 0 \\ A_2 & A_3 \end{pmatrix}, \quad \tilde{B} = \begin{pmatrix} B_1 \\ B_2 \end{pmatrix}, \quad \tilde{C} = (C_1, 0)$$

where the pair $(C_1, A_1) \in \mathbb{R}^{P \times K} \times \mathbb{R}^{K \times K}$ is observable, the pair $(A_1, B_1) \in \mathbb{R}^{K \times K} \times \mathbb{R}^{K \times M}$ is controllable, and $K := \operatorname{rk} \mathcal{O}(C, A)$.

Exercise 3.13

Give an alternative proof of Theorem 3.21 by invoking arguments akin to those in the proof of Theorem 3.11 and by making use of Lemma 3.22.

The following result is the observability counterpart of Proposition 3.8. The proof is left to the reader (see Exercise 3.14).

Proposition 3.23

The subspace $\ker \mathcal{O}(C, A) \subset \mathbb{R}^N$ is A-invariant and is contained in $\ker C$. Moreover, it is the largest subspace in $\mathbb{R}^N$ with these properties, that is, if $\mathcal{S} \subset \mathbb{R}^N$ is an A-invariant subspace contained in $\ker C$, then $\mathcal{S} \subset \ker \mathcal{O}(C, A)$.

Exercise 3.14

Prove Proposition 3.23.

3.3 Impulse response and transfer function

This section is devoted to two fundamental results relating to impulse responses and transfer functions, concepts which will be introduced below. Setting

$$G(t) := C \exp(At)B \quad \forall t \in \mathbb{R}_+, \tag{3.29}$$

the output y of the system (3.1) can be written in the form

$$y(t) = C \exp(At)\xi + (G \star u)(t) \quad \forall t \in \mathbb{R}_+,$$

where $G \star u$ denotes the convolution of G and u, that is,

$$(G \star u)(t) = \int_0^t G(t - \tau)u(\tau)\mathrm{d}\tau \quad \forall t \in \mathbb{R}_+.$$

The function G defined in (3.29) is $\mathbb{R}^{P \times M}$-valued. We denote its entries by g_{kl}, $k = 1, \ldots, P$, $l = 1, \ldots, M$, and write $G(t) = (g_{kl}(t))$. A sequence of functions (d_n) in $PC(\mathbb{R}_+, \mathbb{R})$ is said to be a *Dirac sequence* if $d_n(t) \geq 0$ for all $n \in \mathbb{N}$ and all $t \in \mathbb{R}_+$ and there exists a sequence of positive numbers τ_n with $\tau_n \to 0$ as $n \to \infty$ and such that, for all $n \in \mathbb{N}$, $d_n(t) = 0$ for all $t > \tau_n$ and $\int_0^\infty d_n(\tau)\mathrm{d}\tau = 1$. The sequence (d_n) should be thought of as a sequence of approximants of an "ideal" impulse at $t = 0$.

Example 3.24

For $n \in \mathbb{N}$, define d_n by

$$d_n(t) = n \ \text{ if } t \in [0, 1/n] \quad \text{and} \quad d_n(t) = 0 \ \text{ if } t > 1/n.$$

Then (d_n) is a Dirac sequence (with $\tau_n = 1/n$), see Figure 3.3.   △

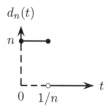

Figure 3.3 Typical element of a Dirac sequence

Proposition 3.25

Let (d_n) be a Dirac sequence and define $\delta_{l,n} \in PC(\mathbb{R}_+, \mathbb{R}^M)$ by $\delta_{l,n}(t) = d_n(t)e_l$, where e_l denotes the l-th canonical basis vector of $\mathbb{R}^M$. Then, for all $t > 0$,

$$\lim_{n \to \infty} y_k(t; 0, \delta_{l,n}) = g_{kl}(t),$$

where $y_k(t; 0, \delta_{l,n})$ denotes the k-th component of $y(t; 0, \delta_{l,n})$.

The above result says that, under zero initial conditions, the k-th component of the response of the system to the input $u = \delta_{l,n}$ can be approximated by g_{kl} arbitrarily closely by taking n is sufficiently large. As remarked above, for large n, $\delta_{l,n}$ models an "ideal" impulse and therefore G is called the *impulse response* of system (3.1).

Proof of Proposition 3.25

Let $t > 0$ be fixed but arbitrary and choose an integer m such that $\tau_n \leq t$ for all $n \geq m$. Then, for all $n \geq m$,

$$y_k(t; 0, \delta_{l,n}) = \int_0^t g_{kl}(t - \tau)d_n(\tau)d\tau = \int_0^{\tau_n} g_{kl}(t - \tau)d_n(\tau)d\tau.$$

By the mean-value theorem for integrals (see Theorem A.29), there exists $t_n \in [t - \tau_n, t]$ such that, for all $n \geq m$,

$$y_k(t; 0, \delta_{l,n}) = g_{kl}(t_n)\int_0^{\tau_n} d_n(\tau)d\tau = g_{kl}(t_n)\int_0^\infty d_n(\tau)d\tau = g_{kl}(t_n).$$

The claim now follows, because g_{kl} is continuous and $t_n \to t$ as $n \to \infty$. $\square$

By Proposition A.38, we know that the matrix exponential function $t \mapsto \exp(At)$ has Laplace[4] transform given by $(sI - A)^{-1}$. Therefore, the impulse

---

[4] Pierre-Simon Laplace (1749-1827), French.

response G of system (3.1) is Laplace transformable, with Laplace transform $\hat{G}$ given by

$$\hat{G}(s) = C(sI - A)^{-1}B. \tag{3.30}$$

The (matrix-valued) function $\hat{G}$ is called the *transfer function matrix* or simply *transfer function* of system (3.1).

If the input u is exponentially bounded, then the corresponding output y of (3.1) is exponentially bounded and the Laplace transform $\hat{y}$ of y is given by

$$\hat{y}(s) = C(sI - A)^{-1}\xi + \hat{G}(s)\hat{u}(s),$$

see Appendix A.4. Furthermore, if $\xi = 0$, then the formula for $\hat{y}$ simplifies to

$$\hat{y}(s) = \hat{G}(s)\hat{u}(s).$$

By Cramer's[5] rule (see Theorem A.5),

$$\hat{G}(s) = \frac{1}{\det(sI - A)} C\mathrm{adj}(sI - A)B,$$

where $\mathrm{adj}(sI - A)$ denotes the adjugate of the matrix $sI - A$. Consequently, the entries $\hat{g}_{kl}(s)$ of $\hat{G}(s)$ are rational functions of s, that is ratios of two polynomials.

Example 3.26

In this example we compute the transfer function and the impulse response of the linearized (and normalized) satellite model described in Example 3.1, with A, B and C given by

$$A = \begin{pmatrix} 0 & 1 & 0 & 0 \\ 3\omega^2 & 0 & 0 & 2\omega \\ 0 & 0 & 0 & 1 \\ 0 & -2\omega & 0 & 0 \end{pmatrix}, \quad B = \begin{pmatrix} 0 & 0 \\ 1 & 0 \\ 0 & 0 \\ 0 & 1 \end{pmatrix}, \quad C = \begin{pmatrix} 1 & 0 & 0 & 0 \\ 0 & 0 & 1 & 0 \end{pmatrix},$$

where $\omega > 0$. The transfer function matrix $\hat{G}$ of the system is given by $\hat{G}(s) = C(sI - A)^{-1}B$. In view of the structure of the matrices B and C, it is clear that only four entries of $(sI - A)^{-1}$ are relevant, namely the entries in the first and third rows which are in the second and fourth columns. A routine calculation invoking Cramer's rule shows that

$$(sI - A)^{-1} = \frac{1}{s^2(s^2 + \omega^2)} \begin{pmatrix} * & s^2 & * & 2\omega s \\ * & * & * & * \\ * & -2\omega s & * & s^2 - 3\omega^2 \\ * & * & * & * \end{pmatrix}.$$

---

[5] Gabriel Cramer (1704-1752), Swiss.

Consequently,

$$\hat{G}(s) = \frac{1}{s^2(s^2 + \omega^2)} \begin{pmatrix} s^2 & 2\omega s \\ -2\omega s & s^2 - 3\omega^2 \end{pmatrix} = \begin{pmatrix} \dfrac{1}{s^2 + \omega^2} & \dfrac{2\omega}{s(s^2 + \omega^2)} \\ \dfrac{-2\omega}{s(s^2 + \omega^2)} & \dfrac{4}{s^2 + \omega^2} - \dfrac{3}{s^2} \end{pmatrix}.$$

Using elementary properties of the Laplace transform (see Appendix A.4), we obtain the following expression for the impulse response G:

$$G(t) = \frac{1}{\omega} \begin{pmatrix} \sin(\omega t) & 2(1 - \cos(\omega t)) \\ 2(\cos(\omega t) - 1) & 4\sin(\omega t) - 3\omega t \end{pmatrix}.$$

$\triangle$

Exercise 3.15

With reference to Figure 3.4, applying output feedback of the form $u = v - Ky$ to (3.1), where $K \in \mathbb{R}^{m \times p}$ and $v \in PC(\mathbb{R}_+, \mathbb{R}^m)$ represents an external signal, leads to the following feedback (or closed-loop) system

$$\dot{x} = (A - BKC)x + Bv, \quad x(0) = x^0; \quad y = Cx.$$

Show that the transfer function $\hat{G}_K$ of the feedback system, with input

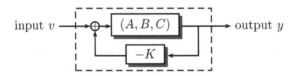

input v ——⊕—→ (A, B, C) ——→ output y

$-K$

Figure 3.4 Feedback system

v and output y, satisfies $\hat{G}_K = \hat{G}(I + K\hat{G})^{-1}$ and $\hat{G}_K = (I + \hat{G}K)^{-1}\hat{G}$.

The transfer function of a system (3.1) contains information relating to the asymptotic behaviour of the response to sinusoidal inputs. This is made precise in the following result.

Proposition 3.27

Assume that

$$\text{Re}\,\lambda < 0 \quad \forall\,\lambda \in \sigma(A). \tag{3.31}$$

For $\omega \in \mathbb{R}$, set $u_{l,\omega}(t) := \sin(\omega t)e_l$, where e_l denotes the l-th canonical basis vector of $\mathbb{R}^M$. Then, for every $\xi \in \mathbb{R}^N$,

$$\lim_{t \to \infty} \left(y_k(t; \xi, u_{l,\omega}) - |\hat{g}_{kl}(i\omega)| \sin(\omega t + \varphi_{kl}) \right) = 0,$$

where φ_{kl} is the argument of $\hat{g}_{kl}(i\omega)$ in $[0, 2\pi)$ and $y_k(t; \xi, u_{l,\omega})$ denotes the k-th component of $y(t; \xi, u_{l,\omega})$.

Recall that, by Theorem 2.13, (3.31) is equivalent to $\exp(At) \to 0$ as $t \to \infty$ for all $\xi \in \mathbb{R}^N$. Furthermore, anticipating the stability theory to be developed in Chapter 5, we mention that (3.31) is equivalent to the global asymptotic stability of the uncontrolled system (the system with control $u = 0$).

Proof of Proposition 3.27

Theorem 2.11 in conjunction with (3.31) shows that $\hat{g}_{kl}(i\omega)$ is well-defined. Moreover, since $\exp(At) \to 0$ as $t \to \infty$ (by Theorem 2.13), it is sufficient to prove the claim for $\xi = 0$. Setting

$$v(t) := \exp(i\omega t)e_l,$$

we have that

$$y_k(t; 0, v) = \int_0^t g_{kl}(\tau)\exp(i\omega(t - \tau))\mathrm{d}\tau = \exp(i\omega t)\int_0^t g_{kl}(\tau)\exp(-i\omega\tau)\mathrm{d}\tau,$$

and thus,

$$y_k(t; 0, v) = \hat{g}_{kl}(i\omega)\exp(i\omega t) - \int_t^\infty g_{kl}(\tau)\exp(i\omega(t - \tau))\mathrm{d}\tau.$$

Since

$$\left| \int_t^\infty g_{kl}(\tau)\exp(i\omega(t - \tau))\mathrm{d}\tau \right| \leq \int_t^\infty |g_{kl}(\tau)|\mathrm{d}\tau \to 0 \quad \text{as } t \to \infty,$$

we obtain that

$$\lim_{t \to 0} \left(y_k(t; 0, v) - |\hat{g}_{kl}(i\omega)|\exp(i(\omega t + \varphi_{kl})) \right) = 0.$$

Since $\operatorname{Im} y_k(t; 0, v) = y_k(t; 0, \operatorname{Im} v) = y_k(t; 0, u_{l,\omega})$, taking imaginary parts in the above identity yields the claim. $\qquad\square$

Proposition 3.27 says that if (3.31) holds, then, for large t, the k-th component of the response corresponding to an arbitrary initial state ξ and an oscillating input of the form $u_{l,\omega}$ is approximately equal to an oscillation of the same frequency ω, with amplitude $|\hat{g}_{kl}(i\omega)|$ and phase shift equal to the argument of $\hat{g}_{kl}(i\omega)$ (in $[0, 2\pi)$).

Exercise 3.16

Consider system (3.1) with $M = P = 1$. Assume that $\operatorname{Re} \lambda < 0$ for all $\lambda \in \sigma(A)$ and that $\hat{G}$ is of the form $\hat{G}(s) = \alpha/(s + \beta)$, where $\alpha, \beta > 0$. Moreover, let u be given by $u(t) = \sin t$ for all $t \geq 0$ and assume that

$$\lim_{t \to \infty} \left(y(t; 0, u) - \sqrt{2} \sin(t - \pi/4) \right) = 0.$$

Determine α and β.

3.4 Realization theory

We conclude this introduction to linear control theory with some fundamental results on *realization theory*. These results will play an important role in Section 6.3. Recall that $\mathbb{R}(s)$ denotes the set of all rational functions with real coefficients, that is, all functions which can be written as the quotient of two polynomials with real coefficients. The set of matrices of format $P \times M$ with entries in $\mathbb{R}(s)$ is denoted by $\mathbb{R}(s)^{P \times M}$. A matrix $R \in \mathbb{R}(s)^{P \times M}$ is said to be *proper* if the limit

$$\lim_{|s| \to \infty} R(s) =: D \in \mathbb{R}^{P \times M}$$

exists. If $D = 0$, then R is said to be *strictly proper*. We have seen that the transfer function matrix $\hat{G}$ (see (3.30)) of the system (3.1) is in $\mathbb{R}(s)^{P \times M}$. Moreover,

$$\lim_{|s| \to \infty} \hat{G}(s) = \lim_{|s| \to \infty} s^{-1} C(I - s^{-1} A)^{-1} B = 0,$$

showing that $\hat{G}$ is strictly proper.

Conversely, given a strictly proper matrix $R \in \mathbb{R}(s)^{P \times M}$, we now ask the following question: do there exist $N \in \mathbb{N}$ and matrices $A \in \mathbb{R}^{N \times N}$, $B \in \mathbb{R}^{N \times M}$ and $C \in \mathbb{R}^{P \times N}$ such that R is the transfer matrix of a system of the form (3.1), that is, $R(s) = C(sI - A)^{-1} B$? Any such triple of matrices (A, B, C) (and the associated system (3.1)) is called a *state-space realization* or simply a *realization* of R and the integer N is called the *dimension* of the realization. A realization is said to be *minimal* if its dimension is less than or equal to the dimension of any other realization.

The question of *existence* of realizations is settled by the following result.

Proposition 3.28

Every strictly proper $R \in \mathbb{R}(s)^{P \times M}$ has a realization, that is, there exist $N \in \mathbb{N}$ and matrices $A \in \mathbb{R}^{N \times N}$, $B \in \mathbb{R}^{N \times M}$ and $C \in \mathbb{R}^{P \times N}$ such that

$$R(s) = C(sI - A)^{-1}B.$$

Proof

Let $R \in \mathbb{R}(s)^{P \times M}$ be strictly proper and let d be the least common multiple of all the denominator polynomials of the entries of R. Without loss of generality, we may assume that the coefficient of the highest power of d is equal to 1, that is, d is of the form

$$d(s) = s^l + d_{l-1}s^{l-1} + \cdots + d_1 s + d_0.$$

Since R is strictly proper, it follows that

$$d(s)R(s) = N_{l-1}s^{l-1} + \cdots + N_1 s + N_0,$$

where the N_j are real matrices of format $P \times M$. In the following, let 0_M and I_M denote the zero and identity matrices, respectively, of format $M \times M$. Setting $N := lM$ and defining

$$A := \begin{pmatrix} 0_M & I_M & 0_M & \cdots & 0_M \\ 0_M & 0_M & I_M & \cdots & 0_M \\ \vdots & & & & \vdots \\ 0_M & 0_M & 0_M & \cdots & I_M \\ -d_0 I_M & -d_1 I_M & -d_2 I_m & \cdots & -d_{l-1} I_M \end{pmatrix} \in \mathbb{R}^{N \times N},$$

$$B := \begin{pmatrix} 0_M \\ 0_M \\ \vdots \\ 0_M \\ I_M \end{pmatrix} \in \mathbb{R}^{N \times M} \quad \text{and} \quad C := (N_0, N_1, \ldots, N_{l-1}) \in \mathbb{R}^{P \times N},$$

we will show that (A, B, C) is a realization of G. To this end, set

$$H(s) := (sI - A)^{-1}B \tag{3.32}$$

and partition the $N \times M$ matrix H into l blocks H_j, each of format $M \times M$. Multiplying (3.32) by $sI - A$ and expressing the result in terms of the submatrices H_j gives

$$H_{j+1}(s) = sH_j(s), \quad j = 1, 2, \ldots, l - 1 \tag{3.33}$$

and

$$sH_l(s) + d_0H_1(s) + d_1H_2(s) + \cdots + d_{l-1}H_l(s) = I_M. \qquad (3.34)$$

By (3.33), $H_j(s) = s^{j-1}H_1(s)$ for $j = 1,\ldots,l$. Inserting this into (3.34) then gives $H_1 = (1/d)I_M$. Therefore,

$$H(s) = \frac{1}{d(s)} \begin{pmatrix} I_M \\ sI_M \\ \vdots \\ s^{l-1}I_M \end{pmatrix}.$$

Multiplying this identity by C from the left yields

$$C(sI - A)^{-1}B = CH(s) = \frac{1}{d(s)}\left(N_0 + N_1 s + \cdots + N_{l-1}s^{s-1}\right) = R(s).$$

$\square$

Realizations which are controllable and observable are of particular interest. Proposition 3.28 together with the following exercise shows that there always exist such realizations.

Exercise 3.17

Let (A, B, C) be a realization of the strictly proper matrix $R \in \mathbb{R}(s)^{P \times M}$ and let N denote the dimension of the realization. We assume that $R(s)$ is not identically equal to the zero matrix.

(a) Show that there exists $T \in GL(N, \mathbb{R})$ such that $T^{-1}AT$, $T^{-1}B$ and CT have the block structure

$$T^{-1}AT = \begin{pmatrix} A_{11} & A_{12} & 0 \\ 0 & A_{22} & 0 \\ A_{31} & A_{32} & A_{33} \end{pmatrix}, \quad T^{-1}B = \begin{pmatrix} B_{11} \\ 0 \\ B_{31} \end{pmatrix}, \quad CT = (C_{11}, C_{12}, 0),$$

where the pair (A_{11}, B_{11}) is controllable and the pair (C_{11}, A_{11}) is observable.

(b) Prove that $CA^kB = C_{11}A_{11}^k B_{11}$ for all $k \in \mathbb{N}_0$ and conclude that (A_{11}, B_{11}, C_{11}) is a realization of R.

Whilst the realization constructed in the proof of Proposition 3.28 is in general not minimal, the following proposition shows that, in the single-input single-output case (that is, $M = P = 1$), this realization is indeed minimal. Recall that two polynomials are said to be *coprime* if they have no common zeros.

Proposition 3.29

Let $R = n/d \in \mathbb{R}(s)$ be strictly proper, where n and d are coprime real polynomials given by

$$d(s) = s^l + d_{l-1}s^{l-1} + \cdots + d_1 s + d_0, \quad n(s) = n_{l-1}s^{l-1} + \cdots + n_1 s + n_0.$$

Then the dimension of a minimal realization of R is equal to l. Moreover, $(A, B, C) \in \mathbb{R}^{l \times l} \times \mathbb{R}^{l \times 1} \times \mathbb{R}^{1 \times l}$, given by

$$A := \begin{pmatrix} 0 & 1 & 0 & \cdots & 0 \\ 0 & 0 & 1 & \cdots & 0 \\ \vdots & & & & \vdots \\ 0 & 0 & 0 & \cdots & 1 \\ -d_0 & -d_1 & -d_2 & \cdots & -d_{l-1} \end{pmatrix}, \quad B := \begin{pmatrix} 0 \\ 0 \\ \vdots \\ 0 \\ 1 \end{pmatrix}, \quad C := (n_0, n_1, \ldots, n_{l-1}),$$

is a minimal realization of R.

Proof

It has already been shown in the proof of Proposition 3.28 that (A, B, C) is a realization. It remains to show that there does not exist a realization with dimension smaller than l. To this end, let $(\tilde{A}, \tilde{B}, \tilde{C})$ be another realization of R of dimension $\tilde{l}$. Then, using Cramer's rule (see Theorem A.5),

$$\frac{n(s)}{d(s)} = R(s) = \tilde{C}(sI - \tilde{A})^{-1}\tilde{B} = \frac{1}{\det(sI - \tilde{A})}\tilde{C}\mathrm{adj}(sI - \tilde{A})\tilde{B},$$

and hence,

$$\frac{n(s)\det(sI - \tilde{A})}{d(s)} = \tilde{C}\mathrm{adj}(sI - \tilde{A})\tilde{B}. \tag{3.35}$$

Since the right-hand side of (3.35) is a polynomial and since n and d are coprime, it follows that d must divide $\det(sI - \tilde{A})$, the characteristic polynomial of $\tilde{A}$ which has degree $\tilde{l}$. Since the degree of d is l, we obtain that $l \leq \tilde{l}$. □

The next result shows that minimality is equivalent to joint controllability and observability,

Theorem 3.30

Let $R \in \mathbb{R}(s)^{P \times M}$ be strictly proper. A realization $(A, B, C) \in \mathbb{R}^{N \times N} \times \mathbb{R}^{N \times M} \times \mathbb{R}^{P \times N}$ of R is minimal if, and only if, (A, B) is controllable and (C, A) is observable.

Proof

We prove the two implications by contraposition.

First assume that the realization (A, B, C) is not minimal. Then there exists a realization $(\tilde{A}, \tilde{B}, \tilde{C})$ of dimension $\tilde{N} < N$. We have to show that (A, B, C) is not jointly controllable and observable. Since $C(sI - A)^{-1}B = R(s) = \tilde{C}(sI - \tilde{A})^{-1}\tilde{B}$, it follows that $Ce^{At}B = \tilde{C}e^{\tilde{A}t}\tilde{B}$ for all $t \in \mathbb{R}$. Evaluation and repeated differentiation at $t = 0$ shows that

$$CA^k B = \tilde{C}\tilde{A}^k\tilde{B} \quad \forall\, k \in \mathbb{N}_0.$$

Setting

$$\tilde{\mathcal{O}} := \begin{pmatrix} \tilde{C} \\ \tilde{C}\tilde{A} \\ \vdots \\ \tilde{C}\tilde{A}^{N-1} \end{pmatrix} \in \mathbb{R}^{(PN)\times \tilde{N}}, \quad \tilde{\mathcal{C}} := (\tilde{B}, \tilde{A}\tilde{B}, \dots, \tilde{A}^{N-1}\tilde{B}) \in \mathbb{R}^{\tilde{N}\times (MN)},$$

it follows that

$$\mathcal{O}(C, A)\mathcal{C}(A, B) = \tilde{\mathcal{O}}\tilde{\mathcal{C}}. \tag{3.36}$$

Note that, since $N > \tilde{N}$, $\tilde{\mathcal{O}}$ is not the observability matrix of $(\tilde{C}, \tilde{A})$ and $\tilde{\mathcal{C}}$ is not the controllability matrix of $(\tilde{A}, \tilde{B})$.

If (C, A) is not observable, there is nothing to prove. Thus, let us assume that (C, A) is observable. Then $\operatorname{rk}\mathcal{O}(C, A) = N$ and so $\mathcal{O}(C, A)$ has a left inverse $L \in \mathbb{R}^{N \times NP}$. Multiplying (3.36) from the left by L gives

$$\mathcal{C}(A, B) = L\tilde{\mathcal{O}}\tilde{\mathcal{C}}.$$

Now $\operatorname{rk}\tilde{\mathcal{O}} \leq \tilde{N}$ and therefore, by Proposition A.3, we have $\operatorname{rk}(L\tilde{\mathcal{O}}\tilde{\mathcal{C}}) \leq \tilde{N}$. Consequently, $\operatorname{rk}\mathcal{C}(A, B) \leq \tilde{N} < N$, showing that (A, B) is not controllable.

Conversely, assume that (A, B, C) is not jointly controllable and observable. If (C, A) is not observable, then there exists $z \in \mathbb{R}^N$ such that $z \neq 0$ and $\mathcal{O}(C, A)z = 0$. Let $S \in GL(N, \mathbb{R})$ be such that z is the N-th column of S. Then

$$CS = (C_1, 0), \quad \text{where } C_1 \in \mathbb{R}^{P\times (N-1)}.$$

Partition the matrices $S^{-1}AS$ and $S^{-1}B$ accordingly, that is,

$$S^{-1}AS = \begin{pmatrix} A_1 & A_2 \\ A_3 & A_4 \end{pmatrix}, \quad S^{-1}B = \begin{pmatrix} B_1 \\ B_2 \end{pmatrix},$$

where $A_1 \in \mathbb{R}^{(N-1)\times(N-1)}$ and $B_1 \in \mathbb{R}^{(N-1)\times M}$. Since $CA^k z = 0$ for all $k \in \mathbb{N}_0$, it follows that the last column of $CA^k S$ is equal to zero for all $k \in \mathbb{N}_0$. Combining this with a routine calculation then shows that

$$(CS)(S^{-1}AS)^k = CA^k S = (C_1 A_1^k, 0) \quad \forall\, k \in \mathbb{N}_0,$$

so that $CA^kB = (CS)(S^{-1}AS)^k(S^{-1}B) = C_1A_1^kB_1$ for all $k \in \mathbb{N}_0$. This in turn leads to

$$Ce^{At}B = \sum_{k=0}^{\infty} \frac{t^k}{k!}CA^kB = \sum_{k=0}^{\infty} \frac{t^k}{k!}C_1A_1^kB_1 \quad \forall t \in \mathbb{R}.$$

Applying Laplace transform yields,

$$R(s) = C(sI - A)^{-1}B = C_1(sI - A_1)^{-1}B_1.$$

Thus (A_1, B_1, C_1) is a realization of R. The dimension of this realization is $N - 1$, showing that the realization (A, B, C) is not minimal. Finally, if (A, B) is not controllable, then the above argument applies *mutatis mutandis* to show that (A, B, C) is not a minimal realization (see Exercise 3.18). $\square$

Exercise 3.18

Complete the proof of Theorem 3.30 by showing that if (A, B) is not controllable, then (A, B, C) is not a minimal realization.

Exercise 3.19

Consider two single-input single-output systems

$$\dot{x}_j = A_jx_j + Bu_j, \quad y_j = C_jx_j, \quad j = 1, 2,$$

where $A_j \in \mathbb{R}^{N_j \times N_j}$, $j = 1, 2$. The transfer functions are given by

$$\hat{G}_j(s) = C_j(sI - A_j)^{-1}B_j, \quad j = 1, 2.$$

Define matrices

$$A := \begin{pmatrix} A_1 & B_1C_2 \\ 0 & A_2 \end{pmatrix}, \quad B := \begin{pmatrix} 0 \\ B_2 \end{pmatrix}, \quad C := (C_1, 0).$$

In the following, we consider the series interconnection of these two systems obtained by setting $u_1 = y_2$, see Figure 3.5.

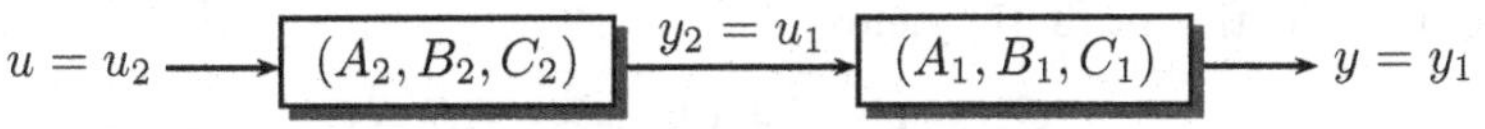

$$u = u_2 \longrightarrow \boxed{(A_2, B_2, C_2)} \xrightarrow{y_2 = u_1} \boxed{(A_1, B_1, C_1)} \longrightarrow y = y_1$$

Figure 3.5 Cascade of two linear systems

(a) Show that the series interconnection is described by

$$\dot{x} = Ax + Bu, \quad y = Cx, \quad \text{where } x := \begin{pmatrix} x_1 \\ x_2 \end{pmatrix}, \quad u := u_2, \quad y := y_1.$$

(b) Show that the transfer function $\hat{G}$ of the series interconnection is given by $\hat{G} = \hat{G}_1\hat{G}_2$.

(c) Assume that, for $j = 1, 2$, (A_j, B_j) is controllable and (C_j, A_j) is observable. Show that (A, B, C) is a minimal realization of $\hat{G}$ if, and only if, no zero (pole) of $\hat{G}_1$ is a pole (zero) of $\hat{G}_2$ (no pole/zero cancellation in the product $\hat{G}_1\hat{G}_2$).

We conclude this introduction to realization theory with a "uniqueness" result: minimal realizations are "essentially" unique in the sense that any two minimal realizations are related by a coordinate transformation (in the state space).

Theorem 3.31

Let $R \in \mathbb{R}(s)^{P \times M}$ be strictly proper. If (A, B, C) and $(\tilde{A}, \tilde{B}, \tilde{C})$ are minimal realizations (of dimension N) of R, then there exists $S \in GL(N, \mathbb{R})$ such that

$$\tilde{A} = S^{-1}AS, \quad \tilde{B} = S^{-1}B, \quad \tilde{C} = CS.$$

Proof

As in the proof of Theorem 3.30, it can be shown that $CA^kB = \tilde{C}\tilde{A}^k\tilde{B}$ for all $k \in \mathbb{N}_0$. Therefore, setting

$$\mathcal{C} := \mathcal{C}(A, B), \quad \mathcal{O} := \mathcal{O}(C, A), \quad \tilde{\mathcal{C}} := \mathcal{C}(\tilde{A}, \tilde{B}), \quad \tilde{\mathcal{O}} := \mathcal{O}(\tilde{C}, \tilde{A}),$$

it follows that

$$\mathcal{O}\mathcal{C} = \tilde{\mathcal{O}}\tilde{\mathcal{C}} \quad \text{and} \quad \mathcal{O}A\mathcal{C} = \tilde{\mathcal{O}}\tilde{A}\tilde{\mathcal{C}}. \tag{3.37}$$

Invoking minimality, Theorem 3.30 guarantees that the realizations (A, B, C) and $(\tilde{A}, \tilde{B}, \tilde{C})$ are jointly controllable and observable. Consequently, invoking Theorem 3.6, Theorem 3.18 and Theorem A.4, we conclude that the matrices $\mathcal{O}$ and $\tilde{\mathcal{O}}$ have left inverses which will be denoted by $\mathcal{O}^\sharp$ and $\tilde{\mathcal{O}}^\sharp$, respectively, and $\mathcal{C}$ and $\tilde{\mathcal{C}}$ have right inverses which will be denoted by $\mathcal{C}^\sharp$ and $\tilde{\mathcal{C}}^\sharp$, respectively. By (3.37), $\tilde{\mathcal{O}}^\sharp\mathcal{O}\mathcal{C}\tilde{\mathcal{C}}^\sharp = I$, showing that the $N \times N$ matrices $\tilde{\mathcal{O}}^\sharp\mathcal{O}$ and $\mathcal{C}\tilde{\mathcal{C}}^\sharp$ are invertible and

$$S := \mathcal{C}\tilde{\mathcal{C}}^\sharp = (\tilde{\mathcal{O}}^\sharp\mathcal{O})^{-1} \in GL(N, \mathbb{R}).$$

The first identity in (3.37) yields

$$\mathcal{O}S = \tilde{\mathcal{O}}\tilde{\mathcal{C}}\tilde{\mathcal{C}}^\sharp = \tilde{\mathcal{O}} \quad \Rightarrow \quad CS = \tilde{C}$$

and, furthermore,

$$S^{-1}\mathcal{C} = \tilde{\mathcal{O}}^\sharp\mathcal{O}\mathcal{C} = \tilde{\mathcal{C}} \quad \Rightarrow \quad S^{-1}B = \tilde{B}.$$

Finally, invoking the second identity in (3.37), we obtain

$$S^{-1}AS = \tilde{\mathcal{O}}^\sharp\mathcal{O}A\mathcal{C}\tilde{\mathcal{C}}^\sharp = \tilde{A}.$$

$\square$

4
Nonlinear differential equations

We now turn our attention to the initial-value problem for a nonlinear differential equation of the form

$$\dot{x}(t) = f(t, x(t)), \quad x(\tau) = \xi, \quad (\tau, \xi) \in J \times G,$$

where $J \subset \mathbb{R}$ is an interval, G is a non-empty open subset of $\mathbb{R}^N$ and $f \colon J \times G \to \mathbb{R}^N$. As discussed in Chapter 1, in order to make progress in the development of an existence theory for the initial-value problem, it is necessary to impose some regularity on the function f. We will treat two cases separately, under the respective assumptions:

- the function f is jointly continuous;

- the function f is locally Lipschitz with respect to its second argument (in a sense to be made precise) and, for each continuous $y \colon J \to G$, the function $t \mapsto f(t, y(t))$ is piecewise continuous.

Before doing so, we briefly digress to make a useful observation pertaining to the fact that no *a priori* assumption is imposed on the nature of the interval J, which may be bounded or unbounded, open or closed, neither open nor closed. The following result is a straightforward consequence of material presented in Example A.12 of Appendix A.2.

Proposition 4.1

Let J be an interval. For every $t, \tau \in J$, there exists an interval $J_0 \subset J$ that contains both t and τ, and has the following properties:

H. Logemann and E. P. Ryan, *Ordinary Differential Equations*,
Springer Undergraduate Mathematics Series,
DOI: 10.1007/978-1-4471-6398-5_4, © Springer-Verlag London 2014

(1) J_0 is relatively open in J;

(2) the closure $\bar{J}_0$ of J_0 is compact and contained in J.

4.1 Peano existence theory

We first treat the case wherein $f\colon J \times G \to \mathbb{R}^N$ is continuous and consider the initial value problem

$$\dot{x}(t) = f(t, x(t)), \quad x(\tau) = \xi, \quad (\tau, \xi) \in J \times G, \quad f \text{ continuous} \qquad (4.1)$$

We will develop an existence theory in the spirit of Peano[1]. By a *solution* of (4.1) we mean a continuously differentiable function $x\colon I \to G$ on some interval $I \subset J$ containing τ such that $x(\tau) = \xi$ and the differential equation in (4.1) holds for all $t \in I$. Clearly, $x\colon I \to G$ is a solution of (4.1) if, and only if,

$$x(t) = \xi + \int_\tau^t f(s, x(s))\mathrm{d}s \quad \forall\, t \in I.$$

Theorem 4.2 (Peano existence theorem)

For each $(\tau, \xi) \in J \times G$, there exists a solution of (4.1).

Proof

Let $(\tau, \xi) \in J \times G$ be arbitrary. The proof consists of two steps: first, we show (by construction) that, for each $\varepsilon > 0$, there exists an "ε-approximate" solution of (4.1); then we show that there exists a sequence of "approximate solutions" that converges to a solution.

Step 1. Existence of an ε-approximate solution. We start by making precise what we mean by an approximate solution: for $\varepsilon > 0$, an *ε-approximate solution* of (4.1) on an interval $I \subset J$ is a piecewise continuously differentiable (see Appendix A.3) function $y\colon I \to G$ such that $\tau \in I$, $y(\tau) = \xi$ and

$$\|\dot{y}(t) - f(t, y(t))\| \le \varepsilon \quad \forall\, t \in I \setminus E,$$

where $E \subset I$ denotes a (finite) set of points t at which y may fail to be differentiable.

Choose $\gamma > 0$ sufficiently small so that the closed ball $B := \bar{\mathbb{B}}(\xi, \gamma)$ is contained in G (such a γ exists since G is open and contains ξ). Let $J_0 \subset J$

---

[1] Giuseppe Peano (1858-1932), Italian.

be a compact interval with $\tau \in J_0$. Let $\mu > 1$ be such that $1 + \|f(t,z)\| \le \mu$ for all $(t,z) \in \bar{J}_0 \times B$ (such a μ exists since f is continuous on $J \times G$ and $\bar{J}_0 \times B$ is a compact subset of $J \times G$). Now let $[\alpha, \beta] =: I \subset \bar{J}_0$ be any compact interval containing τ and such that its length does not exceed γ/μ, that is, $0 < \beta - \alpha \le \gamma/\mu$.

Let $\varepsilon > 0$ be arbitrary. We will construct an ε-approximate solution y on I. Since f is continuous, it is uniformly continuous on the compact set $I \times B$ and so there exists $\delta > 0$ such that, for all $(t,z),(t',z') \in I \times B$,

$$\|(t,z) - (t',z')\| \le \delta \implies \|f(t,z) - f(t',z')\| \le \varepsilon. \tag{4.2}$$

Let $E := \{t_{\kappa^-}, \ldots, t_0, \ldots, t_{\kappa^+}\}$, where $\kappa^-, \kappa^+ \in \mathbb{Z}$, $\kappa^- \le 0$ and $\kappa^+ \ge 0$ (not both zero), be a finite partition of $I = [\alpha, \beta]$ with $t_0 = \tau$ and $\max_k |t_k - t_{k-1}| < \delta/\mu$. In particular,

$$t_{\kappa^-} = \alpha, \; t_0 = \tau, \; t_{\kappa^+} = \beta, \; 0 < t_k - t_{k-1} < \delta/\mu \quad \text{where } \kappa^- < k \le \kappa^+.$$

Let $x_0 := \xi$ and, for all $k \ne 0$ with $\kappa^- \le k \le \kappa^+$, define x_k via the following recursions, the first (respectively, second) of which is vacuous if $\tau = \kappa^+$ (respectively, $\tau = \kappa^-$):

$$x_k = x_{k-1} + f(t_{k-1}, x_{k-1})(t_k - t_{k-1}), \quad k = 1, \ldots, \kappa^+$$
$$x_k = x_{k+1} + f(t_{k+1}, x_{k+1})(t_k - t_{k+1}), \quad k = -1, \ldots, \kappa^-.$$

These recursions, which are inspired by Euler's method from numerical analysis, give

$$x_k = x_0 + \sum_{j=0}^{k-1} f(t_j, x_j)(t_{j+1} - t_j), \quad k = 1, \ldots, \kappa^+$$
$$x_k = x_0 + \sum_{j=k+1}^{0} f(t_j, x_j)(t_{j-1} - t_j), \quad k = -1, \ldots, \kappa^-.$$

We claim that $x_k \in B$ for all k, $\kappa^- \le k \le \kappa^+$. Suppose that the claim is false. Then one (or both) of the following must hold: (a) there exists $k \in \{1, \ldots, \kappa^+\}$ such that $x_k \notin B$ and $x_j \in B$ for all $j = 0, \ldots, k-1$, or (b) there exists $k \in \{\kappa^-, \ldots, -1\}$ such that $x_k \notin B$ and $x_j \in B$ for all $j = k+1, \ldots, 0$. In either case, we arrive at a contradiction. In particular, case (a) yields the contradiction

$$\gamma < \|x_k - x_0\| \le \sum_{j=0}^{k-1} \|f(t_j, x_j)\|(t_{j+1} - t_j) \le \mu(t_k - t_0) \le \mu(\beta - \alpha) \le \gamma$$

and case (b) yields the contradiction

$$\gamma < \|x_k - x_0\| \le \sum_{j=k+1}^{0} \|f(t_j, x_j)\|(t_j - t_{j-1}) \le \mu(t_0 - t_k) \le \mu(\beta - \alpha) \le \gamma.$$

Therefore, $(t_k, x_k) \in I \times B$ for all $\kappa^- \le k \le \kappa^+$. Let $y \colon I \to B$ to be the linear interpolant of these points, that is, the continuous piecewise linear function given by

$$y(t) = x_{k-1} + f(t_{k-1}, x_{k-1})(t - t_{k-1}), \quad t_{k-1} \le t \le t_k, \ \kappa^- < k \le \kappa^+$$

with graph as shown Figure 4.1. Next, we show that y is an ε-approximate

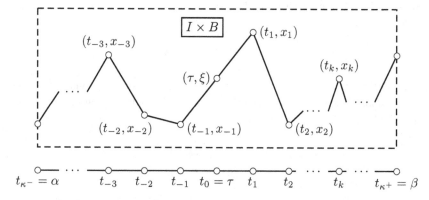

Figure 4.1 Piecewise linear interpolant y of points (t_k, x_k)

solution on I. It is clear that $y(\tau) = \xi$. It is also clear that y is piecewise continuously differentiable. Furthermore, for each k with $\kappa^- < k \le \kappa^+$, we have $\|x_{k-1} - y(t)\| \le \|f(t_{k-1}, x_{k-1})\|(t - t_{k-1})$ for all $t \in (t_{k-1}, t_k)$ and so, for all $\kappa^- < k \le \kappa^+$ and all $t \in (t_{k-1}, t_k)$,

$$\|(t_{k-1}, x_{k-1}) - (t, y(t))\| \le (1 + \|f(t_{k-1}, x_{k-1})\|)(t - t_{k-1}) \le \mu(\delta/\mu) = \delta.$$

Since $(t_{k-1}, x_{k-1}) \in I \times B$ and $(t, y(t))) \in I \times B$ for all $t \in (t_{k-1}, t_k)$, we may invoke (4.2) to conclude that, for each k with $\kappa^- < k \le \kappa^+$,

$$\|\dot{y}(t) - f(t, y(t))\| = \|f(t_{k-1}, x_{k-1}) - f(t, y(t))\| \le \varepsilon \ \forall t \in (t_{k-1}, t_k).$$

Therefore, recalling that $E := \{t_{\kappa^-}, \ldots, t_{\kappa^+}\}$, we have $\|\dot{y}(t) - f(t, y(t))\| \le \varepsilon$ for all $t \in I \setminus E$, and so y is an ε-approximate solution on I.

Step 2. Existence of an "exact" solution. The aim is to construct a sequence of approximate solutions converging to an "exact" solution. Let (ε_n) be a sequence in $(0,1]$ such that $\varepsilon_n \to 0$ as $n \to \infty$. By Step 1, for each $n \in \mathbb{N}$, there exists an ε_n-approximate solution $y_n \colon I \to B$ with $y_n(\tau) = \xi$. For each $n \in \mathbb{N}$, let $\rho_n \colon I \to \mathbb{R}^N$ be any piecewise continuous function such that $\rho_n(t) = \dot{y}_n(t) - f((t, y_n(t))$ for all $t \in I \backslash E_n$, where E_n denotes the (finite) set of points at which y_n fails to be differentiable. We now have

$$\dot{y}_n(t) = f(t, y_n(t)) + \rho_n(t) \quad \text{and} \quad \|\rho_n(t)\| \le \varepsilon_n \le 1 \ \forall t \in I \backslash E_n \ \forall n \in \mathbb{N}$$

and an application of the (generalized) fundamental theorem of calculus (see Theorem A.31) yields

$$y_n(s_2) = y_n(s_1) + \int_{s_1}^{s_2} f(s, y_n(t)) \mathrm{d}t + \int_{s_1}^{s_2} \rho_n(t) \mathrm{d}t \ \forall s_1, s_2 \in I \ \forall n \in \mathbb{N}. \quad (4.3)$$

Therefore,

$$\|y_n(s_2) - y_n(s_1)\| \le \left| \int_{s_1}^{s_2} \|f(t, y_n(t))\| \, \mathrm{d}t \right| + |s_2 - s_1| \le \mu |s_2 - s_1|$$

$$\forall s_1, s_2 \in I \ \forall n \in \mathbb{N}.$$

Therefore, the set $\{y_n \colon n \in \mathbb{N}\} \subset C(I, \mathbb{R}^N)$ is equicontinuous (see Appendix A.2). Moreover, since, for every $n \in \mathbb{N}$, y_n maps into B, the set $\{y_n \colon n \in \mathbb{N}\}$ is evidently bounded. By the Arzelà-Ascoli[2] theorem (Theorem A.24), it follows that the sequence (y_n) has a subsequence – which we do not relabel – converging uniformly to a continuous function x which, by compactness of B, takes its values in B, that is, $x \colon I \to B$. Invoking (4.3) with $s_1 = \tau$ and $s_2 = t \in I$, we obtain

$$\left\| y_n(t) - \xi - \int_\tau^t f(s, y_n(s)) \, \mathrm{d}s \right\| = \left\| \int_\tau^t \rho_n(s) \, \mathrm{d}s \right\| \le \varepsilon_n (\beta - \alpha) \ \forall t \in I \ \forall n \in \mathbb{N}.$$

By uniform continuity of f on $I \times B$ and since $y_n \to x$ uniformly, it follows that $f(t, y_n(t)) \to f(t, x(t))$ uniformly on I as $n \to \infty$. Therefore, as $n \to \infty$, we may take the limit under the integral and obtain

$$x(t) - \xi - \int_\tau^t f(s, x(s)) \, \mathrm{d}s = 0 \ \forall t \in I.$$

By the fundamental theorem of calculus, $\dot{x}(t) = f(t, x(t))$ for all $t \in I$. Furthermore, $x(\tau) = \xi$. It follows that $x \colon I \to B \subset G$ solves the initial-value problem. □

---

[2] Cesare Arzelà (1847-1912), Italian; Giulio Ascoli (1843-1896), Italian.

Corollary 4.3

For each $(\tau, \xi) \in J \times G$, there exists $a > 0$ such that:

(1) if $\inf J = \tau < \sup J$, then (4.1) has a solution on $[\tau, \tau + a)$;

(2) if $\inf J < \tau = \sup J$, then (4.1) has a solution on $(\tau - a, \tau]$;

(3) if $\inf J < \tau < \sup J$, then (4.1) has a solution on $(\tau - a, \tau + a)$.

Proof

By Proposition 4.1, there exists an interval J_1 such that $\tau \in J_1$, J_1 is relatively open in J and $\bar{J}_1$ is compact and contained in J. Note that τ is an interior point of J_1 if, and only if, τ is an interior point of J. Inspection of the proof of Theorem 4.2 with $J_0 = \bar{J}_1$ shows that $I \subset J_0$ can be chosen so that statements (1) – (3) hold.   □

4.2 Maximal interval of existence

Theorem 4.2 establishes that continuity of f ensures the existence of a solution of (4.1) on some interval containing τ. A question naturally arises: is it possible to extend a solution to a larger interval and, if so, is there a "largest" interval that supports a solution? We proceed to formulate this issue more precisely. Throughout this section, it is again assumed that $f: J \times G \to \mathbb{R}^N$ is continuous, where J is an interval and G is an open subset of $\mathbb{R}^N$. Let $I \subset J$ be an interval and assume that $x: I \to G$ is a solution of (4.1). An *extension* of x is a function $\tilde{x}: \tilde{I} \to G$ such that $I \subset \tilde{I} \subset J$ and $\tilde{x}(t) = x(t)$ for all $t \in I$; an extension $\tilde{x}: \tilde{I} \to G$ of x is *proper* if $\tilde{I} \neq I$. The interval I is a *maximal interval of existence*, and x is said to be a *maximally defined solution* or simply a *maximal solution*, if x does not have a proper extension which is also a solution of (4.1) (that is, if there does not exist a solution $\tilde{x}: \tilde{I} \to G$ of (4.1) such that $I \subset \tilde{I}$, $I \neq \tilde{I}$ and $\tilde{x}(t) = x(t)$ for all $t \in I$).

Proposition 4.4

Let $x: I \to G$ be a maximal solution of (4.1). Then the interval I is relatively open in J, that is, $I = O \cap J$ for some open set $O \subset \mathbb{R}$.

Proof

Define $\alpha := \inf I$ and $\omega := \sup I$. Seeking a contradiction, suppose that I is *not* relatively open in J. Then one (or both) of the following must hold: (i) $\inf J < \alpha \in I$, or (ii) $\sup J > \omega \in I$. If (i) holds, then, by statement (2) of Corollary 4.3, the initial-value problem $\dot{y}(t) = f(t, y(t))$, $y(\alpha) = x(\alpha)$, has a solution y on $I_\alpha = (\alpha - a, \alpha]$ for some $a > 0$. The function $z : I_\alpha \cup I \to G$ defined by

$$z(t) = \begin{cases} y(t) & t \in (\alpha - a, \alpha) \\ x(t) & t \in I, \end{cases}$$

is such that

$$\xi + \int_\tau^t f(s, z(s))\mathrm{d}s = \xi + \int_\tau^t f(s, x(t))\mathrm{d}s = x(t) = z(t) \ \forall\, t \in I$$

and

$$\xi + \int_\tau^t f(s, z(s))\mathrm{d}s = \xi + \int_\tau^\alpha f(s, x(s))\mathrm{d}s + \int_\alpha^t f(s, y(s))\mathrm{d}s$$

$$= x(\alpha) + \int_\alpha^t f(s, y(s))\mathrm{d}s = y(t) = z(t) \ \forall\, t \in (\alpha - a, \alpha).$$

Therefore, z a solution of (4.1) on the interval $I_\alpha \cup I$ which extends x to the left, thereby contradicting the maximality of x. If (ii) holds, then, by statement (1) of Corollary 4.3, the initial-value problem $\dot{y}(t) = f(t, y(t))$, $y(\omega) = x(\omega)$, has a solution y on $I_\omega = [\omega, \omega + a)$ for some $a > 0$. The above argument applies *mutatis mutandis* to conclude that the function

$$I \cup I_\omega \to G, \quad t \mapsto \begin{cases} x(t) & t \in I \\ y(t) & t \in (\omega, \omega + a) \end{cases}$$

is a solution of (4.1) which extends x to the right, again contradicting the maximality of x. Therefore, I is relatively open in J.   □

Example 4.5

Consider (4.1) with $J = (-\infty, 2)$, $G = \mathbb{R}$, $(\tau, \xi) = (0, 2)$ and f given by $f(t, z) := (z/(2 - t))^2$ for all $(t, z) \in J \times G$. The function

$$x: (-\infty, 1) \to G, \quad t \mapsto (2 - t)/(1 - t)$$

solves this initial-value problem. The solution "blows up" in forwards time in the sense that $x(t) \to \infty$ as $t \uparrow 1$ and so has no proper extension. Therefore, x is a maximal solution and $I = (-\infty, 1)$ is the corresponding maximal interval of existence. Note that I is open (and so *a fortiori* is relatively open in J). △

Example 4.6

Consider (4.1) with $J = [-1,1]$, $G = \mathbb{R}$, $(\tau, \xi) = (0,1)$ and f given by $f(t,z) :=$ $\left(3z^2\sqrt{1-|t|}\right)/2$ for all $(t,z) \in J \times G$. The function

$$x\colon [-1,1) \to G, \ t \mapsto \begin{cases} 1/(1-t)^{3/2}, & t \in [0,1) \\ 1/\left(2-(1+t)^{3/2}\right), & t \in [-1,0) \end{cases}$$

solves the initial-value problem. As in the previous example, the solution "blows up" in forwards time and has no proper extension. Therefore, x is a maximal solution and $I = [-1,1)$ is the corresponding maximal interval of existence. Observe that I is relatively open in J. $\triangle$

In the above two examples, the set G coincided with $\mathbb{R}$. The next example, treats a case wherein G is a proper subset of $\mathbb{R}$.

Example 4.7

Consider (4.1) with $J = (-\infty, 1) = G$, $(\tau, \xi) = (0,0)$ and f given by $f(t,z) :=$ $1/\sqrt{(1-t)(1-z)}$ for all $(t,z) \in J \times G$. The function

$$x\colon (-\infty, 5/9) \to G, \ t \mapsto 1 - \left(3\sqrt{1-t} - 2\right)^{2/3}$$

solves the initial-value problem. Notice that $x(t) \to 1$ as $t \uparrow 5/9$ and so the solution approaches the boundary of G and has no proper extension. Therefore, x is a maximal solution with maximal interval of existence $I = (-\infty, 5/9)$. Note that I is open (and so *a fortiori* is relatively open in J). $\triangle$

Next, we provide an affirmative answer to an earlier question by showing that every solution can be maximally extended.

Theorem 4.8

Every solution of (4.1) can be extended to a maximal solution.

Proof

Let $x\colon I \to G$ be a solution of (4.1) on an interval $I \subset J$. Define the set of extensions of x:

$$\mathcal{E} := \{y\colon I_y \to G \,|\, I_y \text{ is an interval}, \ I \subset I_y \subset J,$$
$$y \text{ solves } (4.1), \ y(t) = x(t) \ \forall t \in I\} \,.$$

Note that $x \in \mathcal{E}$, and so $\mathcal{E} \neq \emptyset$. We define a partial ordering $\preceq$ on $\mathcal{E}$ by

$$w \preceq y \quad \Longleftrightarrow \quad I_w \subset I_y \text{ and } w(t) = y(t) \ \forall t \in I_w.$$

To prove the theorem, it suffices to show that $\mathcal{E}$ has a maximal element (that is, an element $z \in \mathcal{E}$ such that, if $y \in \mathcal{E}$ and $z \preceq y$, than $y = z$). This we do by an application of Zorn's[3] lemma (Lemma A.39). To this end, let $\mathcal{T}$ be any totally ordered subset of $\mathcal{E}$. Let $I_z := \cup_{y \in \mathcal{T}} I_y$ and define the function $z \colon I_z \to G$ by the property

$$z|_{I_y} = y \ \forall y \in \mathcal{T}.$$

Since $\mathcal{T}$ is totally ordered it is easy to see that I_z is an interval and z is well-defined (see Exercise 4.1). Moreover, z is in $\mathcal{E}$ and is an upper bound for $\mathcal{T}$. By Zorn's lemma, it follows that $\mathcal{E}$ has a maximal element. $\qquad\square$

Exercise 4.1

Consider the function $z \colon I_z \to G$ defined in the proof of Theorem 4.8. Show that I_z is an interval contained in J and z is well-defined.

Exercise 4.2

Let $J = \mathbb{R} = G$. For $f \colon \mathbb{R} \to \mathbb{R}, z \mapsto z^2$, consider the initial-value problem

$$\dot{x}(t) = f(x(t)) = x^2(t), \quad x(\tau) = \xi,$$

where $(\tau, \xi) \in \mathbb{R} \times \mathbb{R}$. For each of the following choices of (τ, ξ) find a maximal solution:

(a) $(\tau, \xi) = (0, 1)$,   (b) $(\tau, \xi) = (1, 0)$,   (c) $(\tau, \xi) = (1, 1)$.

Exercise 4.3

Let $J = \mathbb{R} = G$. For $f \colon \mathbb{R} \times \mathbb{R} \to \mathbb{R}, z \mapsto t^3 z^2$, consider the initial-value problem

$$\dot{x}(t) = f(t, x(t)) = t^3(x(t))^2, \quad x(\tau) = \xi,$$

where $(\tau, \xi) \in \mathbb{R} \times \mathbb{R}$.

(a) Find $\tau \in \mathbb{R}$ and $\xi \in \mathbb{R}$ such that the initial-value problem has a maximal solution with a bounded maximal interval of existence.

(b) Find $\tau \in \mathbb{R}$ and $\xi \neq 0$ such that the initial-value problem has a maximal solution with maximal interval of existence $I = \mathbb{R}$.

---

[3] Max August Zorn (1906-1993), German.

Exercise 4.4

Let $J = \mathbb{R}$, let $f\colon G \to \mathbb{R}^N$ be continuous, and let $x\colon (\alpha, \omega) \to \mathbb{R}^N$ be a maximal solution of $\dot{x} = f(x)$, where $G \subset \mathbb{R}^N$ is open. Assume that $\omega = \infty$ and $\lim_{t\to\infty} x(t) =: x^\infty$ exists and $x^\infty \in G$. Show that $f(x^\infty) = 0$ (that is, x^∞ is an equilibrium point of $\dot{x} = f(x)$).

Exercise 4.5

Let $(\tau, \xi) \in J \times G$ and let $f\colon J \times G \to \mathbb{R}^N$ be continuous. Define $g\colon J \times G \to \mathbb{R}^{N+1}$ by

$$g(z_1, z_2) = (1, f(z_1, z_2)) \quad \forall\, (z_1, z_2) \in J \times G.$$

Investigate the relationship between (maximal) solutions of the initial-value problems

$$\dot{x}(t) = f(t, x(t)), \quad x(\tau) = \xi$$

and

$$\dot{y}(t) = g(y(t)), \quad y(0) = (\tau, \xi).$$

The next result identifies conditions on a solution x of (4.1) that are sufficient to ensure that x has a proper extension which is also a solution (loosely speaking, the hypotheses ensure that there is enough "room" in $J \times G$ to contain the graph of an extension of a solution x).

Lemma 4.9

Let $x\colon I \to G$ be a solution of (4.1) and write $a := \inf I$, $b := \sup I$.

(1) If $b \in J\backslash I$ and the closure of the set $\{x(t)\colon t \in [\tau, b)\}$ is a compact subset of G, then there exists a solution $y\colon I \cup \{b\} \to G$ of (4.1) such that $y(t) = x(t)$ for all $t \in I$.

(2) If $a \in J\backslash I$ and the closure of the set $\{x(t)\colon t \in (a, \tau]\}$ is a compact subset of G, then there exists a solution $y\colon \{a\} \cup I \to G$ of (4.1) such that $y(t) = x(t)$ for all $t \in I$.

Proof

We only prove (1) (the proof of (2) is similar). Let C denote the closure of the set $\{x(t)\colon t \in [\tau, b)\}$. Assume that $b \in J\backslash I$ and C is compact with $C \subset G$. By boundedness of f on the compact set $[0, b] \times C$, the following is well defined

$$\eta := \xi + \lim_{t\to b} \int_\tau^t f(s, (x(s))\mathrm{d}s.$$

Extend x to a continuous function $y\colon I \cup \{b\} \to G$ by setting $y(t) := x(t)$ for all $t \in I$ and $y(b) := \eta$. Then

$$y(t) = \xi + \int_\tau^t f(s, y(s))\mathrm{d}s \quad \forall\, t \in [0, b]$$

and so y is a solution of (4.1) with $y(t) = x(t)$ for all $t \in I$. $\qquad\square$

The following corollary is a consequence of Lemma 4.9.

Corollary 4.10

Let $x\colon I \to G$ be a maximal solution of (4.1).

(1) If the closure of the set $\{x(t)\colon t \in I,\, t \geq \tau\}$ is compact and contained in G, then $[\tau, \infty) \cap I = [\tau, \infty) \cap J$.

(2) If the closure of the set $\{x(t)\colon t \in I,\, t \leq \tau\}$ is compact and contained in G, then $(-\infty, \tau] \cap I = (-\infty, \tau] \cap J$.

(3) If the closure of the set $x(I)$ is compact and contained in G, then $I = J$.

Proof

(1) Write $\alpha := \inf I$, $\omega := \sup I$ and $\tilde\omega := \sup J$. Seeking a contradiction, suppose that $[\tau, \infty) \cap I \neq [\tau, \infty) \cap J$. Then either (i) $\omega < \tilde\omega$ or (ii) $\omega = \tilde\omega \in J \backslash I$. If the former case (i) holds, then, since I is relatively open in J (recall Proposition 4.4), we see that $\omega \notin I$. Thus, in each case, we have $\omega \in J \backslash I$. By statement (1) of Lemma 4.9, it follows that x has a proper extension to a solution on $I \cup \{\omega\}$, contradicting maximality of I.

(2) An analogous argument to the above, invoking statement (2) of Lemma 4.9 yields the requisite result.

(3) This follows immediately from (1) and (2). $\qquad\square$

We remark that, in many circumstances (for example, in our study of periodic solutions in Section 4.5, and in our treatment of autonomous differential equations in Section 4.6), the time domain J underlying (4.1) is the real line $\mathbb{R}$. In this case, Proposition 4.4 simply says that every maximal interval of existence is open, and so takes the form $I = (\alpha, \omega)$, with $-\infty \leq \alpha < \omega \leq \infty$. Continuing to consider the case wherein $J = \mathbb{R}$, let $x\colon (\alpha, \omega) \to G$ be a maximal solution of (4.1) and let $\tau \in (\alpha, \omega)$. In this setting, the assertions of Corollary 4.10 are:
(1) if the closure of the set $x([\tau, \omega))$ is a compact subset of G, then $\omega = \infty$;
(2) if the closure of the set $x(\alpha, \tau])$ is a compact subset of G, then $\alpha = -\infty$;

(3) if the closure of the set $x(\alpha, \omega))$ is a compact subset of G, then $\alpha = -\infty$
and $\omega = \infty$ (that is, the solution x is globally defined).

Returning to the general case wherein J is an arbitrary interval, we proceed
to use Corollary 4.10 to show that, if $x \colon I \to G$ is maximal solution with $I \neq J$,
then its trajectory $\{x(t) \colon t \in I\}$ is not contained in any compact subset of G,
that is, roughly speaking, $x(t)$ must approach the boundary ∂G of G or its
norm must "blow up" to infinity in at least one of the cases $t \uparrow \sup I$ or
$t \downarrow \inf I$. Recalling that I is relatively open in J, it is straightforward to show
(see Exercise 4.6) that $I \neq J$ if, and only if, at least one of the following holds:
$\inf I \in J \backslash I$ or $\sup I \in J \backslash I$.

Exercise 4.6

Let I and J be intervals and assume that I is relatively open in J. Prove
that

$$I \neq J \iff \{\inf I, \sup I\} \cap (J \backslash I) \neq \emptyset.$$

Recall that the *distance* $\operatorname{dist}(u, V)$ of a point $u \in \mathbb{R}^N$ to a non-empty set
$V \subset \mathbb{R}^N$ is defined by

$$\operatorname{dist}(u, V) := \inf\{\|u - v\| \colon v \in V\}. \tag{4.4}$$

Exercise 4.7

Let $V \subset \mathbb{R}^N$ be non-empty. Prove that

$$|\operatorname{dist}(x, V) - \operatorname{dist}(y, V)| \leq \|x - y\| \quad \forall\, x, y \in \mathbb{R}^N.$$

Theorem 4.11

Let $x \colon I \to G$ be a maximal solution of (4.1) with maximal interval of existence
$I \subset J$ and assume that $I \neq J$. Write $\alpha := \inf I$ and $\omega := \sup I$. Then, either
$\omega \in J \backslash I$ or $\alpha \in J \backslash I$ and the following hold.

(1) If $\omega \in J \backslash I$, then, for each compact set $C \subset G$, there exists $\sigma \in I$, with
$\sigma < \omega$, such that $x(t) \notin C$ for all $t \in (\sigma, \omega)$; in particular,

$$\left.\begin{array}{l} \lim_{t \to \omega} \min\{\operatorname{dist}(x(t), \partial G), 1/\|x(t)\|\} = 0 \ \text{ if } G \neq \mathbb{R}^N, \\ \|x(t)\| \to \infty \text{ as } t \to \omega \ \text{ if } G = \mathbb{R}^N. \end{array}\right\} \tag{4.5}$$

(2) If $\alpha \in J \backslash I$, then, for each compact set $C \subset G$, there exists $\sigma \in I$, with
$\sigma > \alpha$, such that $x(t) \notin C$ for all $t \in (\alpha, \sigma)$; in particular,

$$\begin{array}{l} \lim_{t \to \alpha} \min\{\operatorname{dist}(x(t), \partial G), 1/\|x(t)\|\} = 0 \ \text{ if } G \neq \mathbb{R}^N, \\ \|x(t)\| \to \infty \text{ as } t \to \alpha \ \text{ if } G = \mathbb{R}^N. \end{array}$$

Before proving this theorem, some remarks on particular cases are warranted. In many situations, the underlying time interval J does not contain its supremum (for example, if $\sup J = \infty$), in which case we may infer the following: if $\omega \in J \backslash I$, then $\omega < \sup J$; conversely, if $\omega < \sup J$, then $\omega \in J$ and, since I is relatively open in J, we have $\omega \notin I$ and so $\omega \in J \backslash I$. In summary,

whenever $\sup J \notin J$, the following holds: $\omega \in J \backslash I$ if, and only if, $\omega < \sup J$.

Examples 4.5 and 4.7 conform to this observation and provide illustrations of (4.5) in the case $G = \mathbb{R}$ (Example 4.5) and in the case $G \subset \mathbb{R}$ with $G \neq \mathbb{R}$ (Example 4.7). An analogous observation holds in the situation where $\inf J \notin J$, namely,

whenever $\inf J \notin J$, the following holds: $\alpha \in J \backslash I$ if, and only if, $\alpha > \inf J$.

Example 4.6 considers a case wherein $\inf J \in J$ and $\sup J \in J$ with $J \backslash I = \{\omega\}$, and provides a further illustration of (4.5) with $G = \mathbb{R}$.

Proof of Theorem 4.11

That $\{\alpha, \omega\} \cap (J \backslash I) \neq \emptyset$ is a direct consequence of Exercise 4.6. We proceed to prove (1) only (the proof of (2) is similar). Assume that $\omega \in J \backslash I$ and so $[\tau, \infty) \cap I = [\tau, \omega)$. Let $C \subset G$ be compact and define $T := \{t \in [\tau, \omega) \colon x(t) \in C\}$. If $T = \emptyset$, then $x(t) \notin C$ for all $t \in [\tau, \omega)$ and the claim holds with $\sigma = \tau$. Assume $T \neq \emptyset$ and define $\sigma := \sup T \leq \omega$. Seeking a contradiction, suppose $\sigma = \omega$. Then $x(t) \in C$ for all $t \in [\tau, \omega)$ and so, by compactness of $C \subset G$, it follows that $x([\tau, \omega))$ has compact closure in G. By Corollary 4.10, it follows that $[\tau, \omega) = [\tau, \infty) \cap J$, which is impossible since $\omega \in [\tau, \infty) \cap J$. Therefore, $\sigma < \omega$ and $x(t) \notin C$ for all $t \in (\sigma, \omega)$.

To prove (4.5), we again invoke a contradiction argument. Suppose that (4.5) does not hold. Then there exist $\varepsilon > 0$ and a sequence (t_n) in (α, ω) such that $t_n \to \omega$ as $n \to \infty$ and either (i) $1/\|x(t_n)\| \geq \varepsilon$ for all $n \in \mathbb{N}$ (if $G = \mathbb{R}^N$), or (ii) $\min\{\operatorname{dist}(x(t_n), \partial G), 1/\|x(t_n)\|\} \geq \varepsilon$ for all $n \in \mathbb{N}$ (if $G \neq \mathbb{R}^N$). In each case, the set $C := \operatorname{cl}\{x(t_n) \colon n \in \mathbb{N}\}$ is compact and contained in G and has the property that $x(t_n) \in C$ for all $n \in \mathbb{N}$, yielding a contradiction, because, by what has already been proved, we know that $x(t_n) \notin C$ for all sufficiently large n. $\qquad \square$

Next, in the case $G = \mathbb{R}^N$, we identify an extra condition on $f \colon J \times \mathbb{R}^N \to \mathbb{R}^N$ which is sufficient to ensure that every maximal solution of (4.1) has interval of existence J. In particular, we will show that, if, for every compact interval $K \subset J$, there exists $L > 0$ such that

$$\|f(t, z)\| \leq L(1 + \|z\|) \quad \forall (t, z) \in K \times \mathbb{R}^N, \tag{4.6}$$

then every maximal solution of (4.1) has interval of existence J.

Proposition 4.12

Assume that $G = \mathbb{R}^N$ and, for every compact interval $K \subset J$, there exists $L > 0$ such that (4.6) holds. If $x \colon I \to \mathbb{R}^N$ is a maximal solution of (4.1), then $I = J$.

Proof

Write $\alpha := \inf I$, $\omega := \sup I$. Seeking a contradiction, suppose that the claim is not true. Then, since I is relatively open in J, one (or both) of the following two cases holds: $\omega \in J \backslash I$ or $\alpha \in J \backslash I$ (recall Exercise 4.6).
Case 1. Assume $\omega \in J$ and $\omega \notin I$, and so $\tau < \omega$. Let $L > 0$ be such that (4.6) holds with $K := [\tau, \omega]$. Then

$$\|x(t)\| \le \|x(\tau)\| + \int_\tau^t \|f(s, x(s))\| ds \le \|x(\tau)\| + L \int_\tau^t (1 + \|x(s)\|) ds \ \forall t \in [\tau, \omega),$$

whence

$$\|x(t)\| \le \|x(\tau)\| + L(\omega - \tau) + L \int_\tau^t \|x(s)\| ds \quad \forall t \in [\tau, \omega).$$

It follows from Gronwall's lemma (Lemma 2.4) that x is bounded on $[\tau, \omega)$. Consequently, the closure of the set $\{x(t) \colon t \in [\tau, \omega)\}$ is compact (and contained in $G = \mathbb{R}^N$). Part (1) of Lemma 4.9 shows that there exists a solution of (4.1) extending x to $I \cup \{\omega\}$, contradicting the maximality of x.
Case 2. Assume $\alpha \in J$ and $\alpha \notin I$, and so $\alpha < \tau$. The argument used in Case 1 above applies *mutatis mutandis* to arrive (via part (2) of Lemma 4.9) at a contradiction to the maximality of x.   □

Exercise 4.8

Set $J = \mathbb{R}$ and $G = \mathbb{R}^N$. Let $A \in \mathbb{R}^{N \times N}$ and let $b \colon \mathbb{R} \times \mathbb{R}^N \to \mathbb{R}^N$ be continuous. Assume that there exists a continuous function $\gamma \colon \mathbb{R} \to [0, \infty)$ such that

$$\|b(t, z)\| \le \gamma(t) \|z\| \ \forall t \in \mathbb{R}, \ \forall z \in \mathbb{R}^N.$$

Define $f \colon \mathbb{R} \times \mathbb{R}^N \to \mathbb{R}^N$ by $f(t, z) := Az + b(t, z)$ and consider the differential equation

$$\dot{x}(t) = f(t, x(t)) = Ax(t) + b(t, x(t)).$$

(a) Let $x\colon I \to \mathbb{R}^N$ be a maximal solution. Show that $I = \mathbb{R}$.

(b) Let $\mu \in \mathbb{R}$ be such that $\mathrm{Re}\lambda < \mu$ for every eigenvalue of A. Let x be a maximal solution. Show that

$$\|x(t)\| \leq M\|x(0)\| \exp\left(\mu t + M \int_0^t \gamma(s)\,\mathrm{d}s\right) \quad \forall\, t \geq 0,$$

where $M \geq 1$ is such that $\|e^{At}z\| \leq Me^{\mu t}\|z\|$ for all $t \geq 0$ and all $z \in \mathbb{R}^N$.

(*Hint.* Use the variation of parameters formula to derive an integral inequality for the function g defined by $g(t) = \|x(t)\|e^{-\mu t}$ for all $t \geq 0$. Then apply Gronwall's lemma.)

(c) Find conditions on μ and γ which guarantee that $x(t) \to 0$ as $t \to \infty$ for every maximal solution x.

4.3 The Lipschitz condition and uniqueness of solutions

We now have a theory of existence of solutions of (4.1): if $f\colon J \times G \to \mathbb{R}^N$ is continuous, then, for each $(\tau, \xi) \in J \times G$, there exists at least one solution; moreover, every solution can be maximally extended. However, we have no reason to expect only one (maximal) solution, as Example 1.1 in Chapter 1 serves to illustrate. Our next goal is to identify a condition on f under which *uniqueness* of (maximal) solutions is assured for the initial-value problem (4.1): that is, for each $(\tau, \xi) \in J \times G$, there exists one, and only one, maximal solution.

We introduce the so-called Lipschitz condition. Let $D \subset \mathbb{R}^Q$ be a non-empty set. A function $g\colon D \to \mathbb{R}^M$ is said to be *locally Lipschitz* if, for every $z \in D$, there exist a set $U \subset D$ containing z and relatively open in D and a number $L \geq 0$ (which may depend on U) such that

$$\|g(u) - g(v)\| \leq L\|u - v\| \quad \forall\, u, v \in U. \tag{4.7}$$

This condition is equivalent to the "difference quotient" $\|g(u) - g(v)\|/\|u - v\|$ being bounded by L for all $u, v \in U$ with $u \neq v$. In the single-variable case (that is, $Q = M = 1$) this means that, for all $u, v \in U$ with $u \neq v$, the absolute value of the slope of the "secant line" through the points $(u, g(u))$ and $(v, g(v))$ is bounded by L (see also Figure 4.2). Note that, if a function is locally Lipschitz, then it is continuous. If (4.7) holds with $U = D = \mathbb{R}^Q$, then $g\colon \mathbb{R}^Q \to \mathbb{R}^M$ is said to be *globally Lipschitz* with Lipschitz constant L.

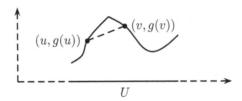

Figure 4.2 Local Lipschitz property: for all $u, v \in U$, the absolute value of slope of secant line through $(u, g(u))$ and $(v, g(v))$ bounded by L.

Example 4.13

In view of Exercise 4.7, we see that, for non-empty $V \subset \mathbb{R}^N$, the function $\mathbb{R}^N \to \mathbb{R}$, $u \mapsto \text{dist}(u, V)$ is globally Lipschitz with Lipschitz constant 1.   △

Exercise 4.9

Are the following functions $\mathbb{R} \to \mathbb{R}$ locally Lipschitz?

(a) $g : z \mapsto \begin{cases} \sqrt{z} & z \geq 0 \\ 0 & z < 0, \end{cases}$ (b) $g : z \mapsto \begin{cases} z \ln z & z > 0 \\ 0 & z \leq 0. \end{cases}$

The next result shows that continuously differentiable functions are locally Lipschitz.

Proposition 4.14

Let $V \subset \mathbb{R}^Q$ be a non-empty open set. If $g : V \to \mathbb{R}^M$ is continuously differentiable in the sense that the first-order partial derivatives of the components of g exist and are continuous, then g is locally Lipschitz.

The proof is relegated to the following exercise.

Exercise 4.10

Prove Proposition 4.14. (*Hint*. Let $z \in V$ and let $u, v \in B$, where $B \subset V$ is a ball centred at z. Apply the mean-value theorem of differentiation to the functions $[0, 1] \to \mathbb{R}$, $t \mapsto g_j((1 - t)u + tv)$, where g_j denotes the j-th component of g).

Exercise 4.11

Is the following function $\mathbb{R} \to \mathbb{R}$ locally Lipschitz?

$$g : z \mapsto \begin{cases} (1/z) \sin z & z > 0 \\ 1 & z \leq 0. \end{cases}$$

Now let $D \subset \mathbb{R}^P \times \mathbb{R}^Q$ be non-empty. A function $g \colon D \to \mathbb{R}^M$ is said to be *locally Lipschitz with respect to its second argument* if, for every $(r, z) \in D$, there exist sets S, U and a number $L \geq 0$ (in general, depending on S and U) such that $(r, z) \in S \times U$, $S \times U$ is relatively open in D, and

$$\|g(s, u) - g(s, v)\| \leq L\|u - v\| \quad \forall s \in S, \ \forall u, v \in U.$$

Proposition 4.15

Let $D \subset \mathbb{R}^P \times \mathbb{R}^Q$ be non-empty. Let $g \colon D \to \mathbb{R}^M$ be locally Lipschitz with respect to its second argument and let $C \subset D$ be compact. If g is bounded on C, then there exists a constant $L \geq 0$ such that

$$\|g(s, u) - g(s, v)\| \leq L\|u - v\| \quad \forall (s, u), (s, v) \in C.$$

Proof

Seeking a contradiction, suppose that the claim is false. Then, for each $n \in \mathbb{N}$, there exist $(s_n, u_n), (s_n, v_n) \in C$ such that

$$\|g(s_n, u_n) - g(s_n, v_n)\| > n\|u_n - v_n\|.$$

Then $u_n \neq v_n$ for all $n \in \mathbb{N}$ and moreover, by boundedness of g on C, we have $\|u_n - v_n\| \to 0$ as $n \to \infty$. By compactness of C, and passing to a subsequence if necessary, we may assume that the sequences $\big((s_n, u_n)\big)$ and $\big((s_n, v_n)\big)$ in C are convergent. We may now infer the existence of $(t, z) \in C$ such that $\lim_{n \to \infty}(s_n, u_n) = (t, z) = \lim_{n \to \infty}(s_n, v_n)$. By the Lipschitz property of g, there exist sets S, U and a number $L \geq 0$ such that $(t, z) \in S \times U$, $S \times U$ is relatively open in D, and

$$\|g(s, u) - g(s, v)\| \leq L\|u - v\| \quad \forall (s, u), (s, v) \in S \times U.$$

Choose $N \in \mathbb{N}$ sufficiently large to that $(s_n, u_n), (s_n, v_n) \in S \times U$ for all $n \geq N$. We now arrive at a contradiction:

$$n < \frac{\|g(s_n, u_n) - g(s_n, v_n)\|}{\|u_n - v_n\|} \leq L \quad \forall n \geq N.$$

This completes the proof. $\qquad\square$

Since a continuous function is bounded on compact sets, the following is an immediate consequence.

Corollary 4.16

Let $D \subset \mathbb{R}^P \times \mathbb{R}^Q$ be non-empty. Let $g \colon D \to \mathbb{R}^N$ be continuous and locally Lipschitz with respect to its second argument. For every compact set $C \subset D$, there exists constant $L \geq 0$ such that

$$\|g(s, u) - g(s, v)\| \leq L\|u - v\| \quad \forall (s, u), (s, v) \in C.$$

We now arrive at a uniqueness result for (4.1): if the continuous function $f \colon J \times G \to \mathbb{R}^N$ is locally Lipschitz with respect to its second argument, then, for each (τ, ξ), the initial-value problem (4.1) has one and only one maximal solution.

Theorem 4.17

Let $(\tau, \xi) \in J \times G$ and let $f \colon J \times G \to \mathbb{R}^N$ be continuous. Furthermore, let $x \colon I_x \to G$ and $y \colon I_y \to G$ be maximal solutions of (4.1) with $x(\tau) = \xi = y(\tau)$. If f is locally Lipschitz with respect to its second argument, then $x = y$.

Proof

Clearly, $\tau \in I_x \cap I_y$. Furthermore, by Proposition 4.4, the intervals I_x and I_y are relatively open in J and so their intersection $I_x \cap I_y$ is an interval. To prove the theorem, it suffices to show that $x(t) = y(t)$ for all $t \in I_x \cap I_y$ (in which case $I_x = I_y$ by maximality of the solutions): this we do by showing that, for every compact subinterval $I \subset I_x \cap I_y$ containing τ, $x(t) = y(t)$ for all $t \in I$. Let I be any such compact subinterval (containing τ) of the interval $I_x \cap I_y$. By continuity of x and y and compactness of I, the set $K := x(I) \cup y(I) \subset G$ is compact. Consequently, the set $I \times K \subset J \times G$ is compact, and thus, by the local Lipschitz assumption together with Corollary 4.16, there exists $L \geq 0$ such that

$$\|f(s, z_1) - f(s, z_2)\| \leq L\|z_1 - z_2\|, \quad \forall (s, z_1), (s, z_2) \in I \times K.$$

We now conclude that, for all $t \in I$,

$$\|x(t) - y(t)\| \leq \left| \int_\tau^t \|f(s, x(s)) - f(s, y(s))\| ds \right| \leq \left| \int_\tau^t L\|x(s) - y(s)\| ds \right|,$$

and so, by Gronwall's lemma (Lemma 2.4), $\|x(t) - y(t)\| = 0$ for all $t \in I$. $\square$

The next result follows immediately from Theorems 4.2, 4.8 and 4.17.

Theorem 4.18

Let $f\colon J \times G \to \mathbb{R}^N$ be continuous and locally Lipschitz with respect to its second argument. Then, for each $(\tau, \xi) \in J \times G$, there exists a unique maximal solution x of (4.1).

4.4 Contraction-mapping approach to existence and uniqueness

The existence (and uniqueness) theory developed in the previous section is predicated on the assumption that the function f is continuous. However, as discussed previously in Section 1.2.2, in many situations wherein the dependence on t may arise through inputs impinging on the system, the assumption of continuity with respect to its first argument t is difficult to justify. These inputs, in one scenario, may be extraneous disturbances/perturbations or, in another scenario, may be appropriately chosen controls. There are many circumstances in which it is unnatural to impose continuity on such input functions. For this reason, we will work with the notion of piecewise continuity and, in particular, will consider the initial-value problem under the following assumption.

Assumption A

1. $f\colon J \times G \to \mathbb{R}^N$ is locally Lipschitz with respect to its second argument;

2. for every continuous function $y\colon J \to G$, the function $t \mapsto f(t, y(t))$ is piecewise continuous.

Recall that J is an interval and $G \subset \mathbb{R}^N$ is a non-empty open set such that $J \times G$ contains the initial data (τ, ξ).

In summary, throughout this section, we assume that $f\colon J \times G \to \mathbb{R}^N$ satisfies Assumption **A** and consider the initial-value problem

$$\dot{x}(t) = f(t, x(t)), \ x(\tau) = \xi, \ (\tau, \xi) \in J \times G, \ f \text{ satisfies Assumption } \mathbf{A} \quad (4.8)$$

By a *solution* of (4.8) we mean a continuous function $x\colon I \to G$ such that

$$x(t) = \xi + \int_{\tau}^{t} f(s, x(s)) \, ds \ \ \forall t \in I,$$

where $I \subset J$ is an interval with $\tau \in I$. As before, the interval I is a *maximal interval of existence*, and x is said to be a *maximal solution*, if x does not have a proper extension which is also a solution of (4.8). We record that, in view of Theorems A.30 and A.31, $x\colon I \to G$ is a solution of (4.8), if, and only if, x is piecewise continuously differentiable and, for arbitrary points $a, b \in I$ with $a < b$,

$$\dot{x}(t) = f(t, x(t)) \quad \forall t \in [a, b]\backslash E,$$

where E is the finite set of points $t \in [a, b]$ at which the piecewise continuous function $t \mapsto f(t, x(t))$ fails to be continuous (see Appendix A.3 for the concepts of piecewise continuity and piecewise continuous differentiability). Moreover,

$$\frac{\mathrm{d}^+ x}{\mathrm{d}t}(t) = f(t^+, x(t^+)) \quad \forall t \in I,\ t < \sup I,$$

$$\frac{\mathrm{d}^- x}{\mathrm{d}t}(t) = f(t^-, x(t^-)) \quad \forall t \in I,\ t > \inf I.$$

where $\mathrm{d}^+/\mathrm{d}t$ and $\mathrm{d}^-/\mathrm{d}t$ denote the right and left derivative, respectively, whilst $f(t^+, x(t^+))$ and $f(t^-, x(t^-))$ denote the right and left limits of $s \mapsto f(s, x(s))$ at $s = t$ (see also Appendix A.3).

We emphasize that Assumption **A** is not very restrictive. The following example gives an important class of functions satisfying Assumption **A**.

Example 4.19

With $J = \mathbb{R}$ and $G = \mathbb{R}^N$, define $f\colon J \times G \to \mathbb{R}^N$ by $f(t, z) = g(z) + k(t)h(z)$ for all $(t, z) \in \mathbb{R} \times \mathbb{R}^N$, where $g, h\colon \mathbb{R}^N \to \mathbb{R}^N$ are locally Lipschitz and $k\colon \mathbb{R} \to \mathbb{R}$ is piecewise continuous. We show that f satisfies Assumption **A**. To this end, let $(t, z) \in \mathbb{R} \times \mathbb{R}^N$ and let T be a bounded open interval containing t. Piecewise continuity of k implies that $K := \sup_{s \in T} |k(s)| < \infty$. Furthermore, since g and h are locally Lipschitz, there exist an open neighbourhood $U \subset \mathbb{R}^N$ of z and a positive constant L such that

$$\|g(u) - g(v)\| \le L\|u - v\| \quad \text{and} \quad \|h(u) - h(v)\| \le L\|u - v\| \ \ \forall u, v \in U.$$

It follows that, for all $s \in T$ and all $u, v \in U$,

$$\|f(s, u) - f(s, v)\| \le \|g(u) - g(v)\| + |k(s)|\|h(u) - h(v)\| \le (1 + K)L\|u - v\|,$$

implying that f is locally Lipschitz with respect to its second argument. Finally, it is clear that for every continuous function $y : \mathbb{R} \to \mathbb{R}^N$, the function $t \mapsto f(t, y(t))$ is piecewise continuous, showing that the function f satisfies Assumption **A**. $\triangle$

The above example is a special case of the following result which identifies a large class of functions f satisfying Assumption **A**.

Proposition 4.20

Assume that $f: J \times G \to \mathbb{R}^N$ is given by $f(t,z) = f_3(f_1(t), f_2(z))$ where $f_1: J \to \mathbb{R}^P$ is piecewise continuous, $f_2: G \to \mathbb{R}^Q$ is locally Lipschitz, and $f_3: D \subset \mathbb{R}^P \times \mathbb{R}^Q \to \mathbb{R}^N$ is such that the following hold: f_3 is continuous and locally Lipschitz with respect to its second argument, $D \subset \mathbb{R}^P \times \mathbb{R}^Q$ is open and, for every $(t_0, z_0) \in J \times G$, there exist neighbourhoods J_0 and $G_0 \subset G$ of t_0 and z_0, respectively, such that D contains the closure of the set $\{(f_1(t), f_2(z)) : (t, z) \in (J_0 \cap J) \times G_0\}$. Then f satisfies Assumption **A**.

The proof of the proposition is routine and is therefore left as an exercise.

Exercise 4.12

(a) Prove Proposition 4.20.

(b) Let f be the function defined in Example 4.19, where it was proved, from first principles, that f satisfies Assumption **A**. Provide an alternative proof of this result by invoking Proposition 4.20.

We now proceed to develop an existence and uniqueness theory for (4.8). First, we record a technicality.

Lemma 4.21

Let $f: J \times G \to \mathbb{R}^N$ satisfy Assumption **A**. Then f is bounded on every compact set $C \subset J \times G$.

Proof

Let $C \subset J \times G$ be compact and, for contradiction, suppose that f is not bounded on C. Then there exists a sequence $((t_n, z_n))$ in C such that

$$\|f(t_n, z_n)\| \to \infty \quad \text{as } t \to \infty. \tag{4.9}$$

By compactness of C, and passing to a subsequence if necessary, we may assume that $((t_n, z_n))$ is convergent and we denote its limit by $(t, z) \in C$. By Assumption **A**, the function f is locally Lipschitz with respect to its second argument, and thus, there exists an open neighbourhood T of t, an open neighbourhood $U \subset G$ of z and a constant $L \geq 0$ such that

$$\|f(s, u) - f(s, v)\| \leq L\|u - v\| \quad \forall (s, u), (s, v) \in (T \cap J) \times U.$$

Let $I \subset T \cap J$ be a compact interval containing t and such that, if t is an interior point of J, then t is an interior point of I. By Assumption **A**, the function

$s \mapsto f(s, z)$ is piecewise continuous and so is bounded on I: let $K > 0$ be such that $\|f(s, z)\| \leq K$ for all $s \in I$. Choosing $k \in \mathbb{N}$ such that $(t_n, z_n) \in I \times U$ and $\|z_n - z\| \leq 1$ for all $n \geq k$, we conclude that

$$\|f(t_n, z_n)\| \leq \|f(t_n, z)\| + \|f(t_n, z_n) - f(t_n, z)\| \leq K + L\|z_n - z\| \leq K + L$$

for all $n \geq k$, contradicting (4.9). $\square$

We now arrive at the existence and uniqueness result for (4.8).

Theorem 4.22

For each $(\tau, \xi) \in J \times G$, there exists a unique maximal solution of the initial-value problem (4.8). Moreover, the maximal interval of existence is relatively open in J.

Proof

Let $(\tau, \xi) \in J \times G$ be arbitrary. Let $I \subset J$ be a compact interval containing τ and such that, if τ is an interior point of J, then τ is an interior point of I (Proposition 4.1 ensures the existence of such a compact interval I). Let $\gamma > 0$ be sufficiently small so that the closed ball $B := \overline{\mathbb{B}}(\xi, \gamma)$ is contained in G. By Lemma 4.21, we have

$$M := \sup \{\|f(s, z)\|: \ (s, z) \in I \times B\} < \infty.$$

Using Proposition 4.15 and Lemma 4.21 in conjunction with the Lipschitz property of f (guaranteed by Assumption **A**) and the compactness of $I \times B$, we conclude that there exists $L > 0$ such that

$$\|f(s, z_1) - f(s, z_2)\| \leq L\|z_1 - z_2\| \quad \forall\, s \in I, \ \forall\, z_1, z_2 \in B.$$

Choose ε, with

$$0 < \varepsilon < \min\{\gamma/M, 1/L\},$$

sufficiently small so that

$$I \supset I_\varepsilon := \begin{cases} [\tau, \tau + \varepsilon] & \text{if } \tau = \min J, \\ [\tau - \varepsilon, \tau] & \text{if } \tau = \max J, \\ [\tau - \varepsilon/2, \tau + \varepsilon/2] & \text{if } \inf J < \tau < \sup J. \end{cases} \tag{4.10}$$

Let $\mathcal{M} := C(I_\varepsilon, B)$ denote the space of continuous functions $I_\varepsilon \to B$ which, equipped with the metric d given by

$$d(y_1, y_2) = \|y_1 - y_2\|_\infty = \sup_{s \in I_\varepsilon} \|y_1(s) - y_2(s)\|,$$

is a metric space. We proceed in five steps.

Step 1. Any solution of (4.8) *defined on* I_ε *is in* $\mathcal{M}$. Let $x\colon I_\varepsilon \to G$ be a solution of (4.8). We claim that $x(t) \in B$ for all $t \in I_\varepsilon$. Seeking a contradiction, suppose that this is not true. Then the set $T := \{t \in I_\varepsilon \colon x(t) \notin B\}$ is non-empty. Define $\sigma := \inf T$, and so $\|x(\sigma) - \xi\| = \gamma$ and $x(s) \in B$ for all $s \in [\sigma, \tau]$ (if $\sigma < \tau$) or for all $s \in [\tau, \sigma]$ (if $\sigma > \tau$). We arrive at the contradiction

$$\gamma = \|x(\sigma) - \xi\| \le \left| \int_\tau^\sigma \|f(s, x(s))\| \mathrm{d}s \right| \le M|\sigma - \tau| \le M\varepsilon < \gamma\,.$$

Step 2. Existence and uniqueness on I_ε. Consider the map Γ defined on $\mathcal{M}$ by

$$(\Gamma(y))(t) = \xi + \int_\tau^t f(s, y(s))\,\mathrm{d}s \quad \forall\, t \in I_\varepsilon,\ \forall\, y \in \mathcal{M}\,.$$

Then $x \in \mathcal{M}$ is a solution of (4.8) if, and only if, $\Gamma(x) = x$. Therefore, by the contraction mapping principle (Theorem A.25), it suffices to show that Γ maps into $\mathcal{M}$ and has the contraction property, that is, for some $c < 1$, $d(\Gamma(y_1), \Gamma(y_2)) \le c\,d(y_1, y_2)$ for all $y_1, y_2 \in \mathcal{M}$. To see that Γ maps into $\mathcal{M}$, simply note that, for each $y \in \mathcal{M}$,

$$\|(\Gamma(y))(t) - \xi\| \le \left| \int_\tau^t \|f(s, y(s))\| \mathrm{d}s \right| \le M|t - \tau| \le M\varepsilon < \gamma \quad \forall\, t \in I_\varepsilon\,,$$

and so $(\Gamma(y))(t) \in B$ for all $t \in I_\varepsilon$. To show that the contraction property also holds, note that, for $y_1, y_2 \in \mathcal{M}$ and $t \in I_\varepsilon$,

$$\begin{aligned}
\|(\Gamma(y_1))(t) - (\Gamma(y_2))(t)\| &\le \left| \int_\tau^t \|f(s, y_1(s)) - f(s, y_2(s))\| \mathrm{d}s \right| \\
&\le L \left| \int_\tau^t \|y_1(s) - y_2(s)\| \mathrm{d}s \right| \\
&\le L|t - \tau| \sup_{s \in I_\varepsilon} \|y_1(s) - y_2(s)\| \le (L\varepsilon)d(y_1, y_2)\,.
\end{aligned}$$

Consequently, writing $c := L\varepsilon < 1$, we have

$$d(\Gamma(y_1), \Gamma(y_2)) = \sup_{t \in I_\varepsilon} \|(\Gamma(y_1))(t) - (\Gamma(y_2))(t)\| \le c\,d(y_1, y_2) \quad \forall\, y_1, y_2 \in \mathcal{M}\,.$$

and so Γ is a contraction. Therefore, by Theorem A.25, Γ has a unique fixed point in $\mathcal{M}$ and thus, within the class $\mathcal{M}$ of continuous functions $I_\varepsilon \to B$, (4.8) has a unique solution. Step 1 ensures that there is no other solution of (4.8) on the interval I_ε. In summary, we have now shown that, for each $(\tau, \xi) \in J \times G$ and all $\varepsilon > 0$ sufficiently small, (4.8) has a unique solution on I_ε, where I_ε is a compact interval of length ε and, if τ is an interior point of J, then τ is an interior point of I_ε.

Step 3. Extended uniqueness. Next, we show that any two solutions of (4.8) must coincide on the intersection of their domains. In particular, if $x \colon J_x \to G$ and $y \colon J_y \to G$ are two solutions of (4.8), then we claim that

$$x(t) = y(t) \ \forall\, t \in J_{xy} := J_x \cap J_y\,.$$

Seeking a contradiction, suppose that the claim is false. Then there exists $t_0 \in J_{xy}$ such that $x(t_0) \neq y(t_0)$. Evidently, either $t_0 > \tau$ or $t_0 < \tau$. Define

$$\sigma := \begin{cases} \inf\{t \in J_{xy} \cap [\tau, \infty) \colon x(t) \neq y(t)\} & \text{if } t_0 > \tau\,, \\ \sup\{t \in J_{xy} \cap (-\infty, \tau] \colon x(t) \neq y(t)\} & \text{if } t_0 < \tau\,. \end{cases}$$

First consider the case wherein $t_0 > \tau$. Then, $\sigma > \tau$ (by Step 2) and $\sigma < t_0$ (by continuity of x and y), and so σ is in the interior of J_{xy}. Moreover, $x(t) = y(t)$ for all $t \in [\tau, \sigma]$. Setting $\eta := x(\sigma) = y(\sigma)$, it is clear that both $x|_{J_{xy}}$ and $y|_{J_{xy}}$ solve the initial value problem

$$\dot{z}(t) = f(t, z(t))\,, \quad z(\sigma) = \eta$$

But by Step 2, we know that, for all sufficiently small $\varepsilon > 0$, this initial-value problem has a unique solution on a compact interval I_ε of length ε containing σ in its interior. Since σ is an interior point of J_{xy}, we may assume that $\varepsilon > 0$ is sufficiently small so that $I_\varepsilon \subset J_{xy}$. Now choose $\delta > 0$ sufficiently small so that $\sigma + \delta \in I_\varepsilon$. Since $x(t) = y(t)$ for all $t \in I_\varepsilon$, it follows that $x(t) = y(t)$ for all $t \in [\tau, \sigma + \delta]$, contradicting the definition of σ.

Now consider the case wherein $t_0 < \tau$. In this case we have $\sigma \in (t_0, \tau)$ and so σ is again an interior point of J. Moreover, $x(t) = y(t)$ for all $t \in [\sigma, \tau]$. The argument adopted in the previous case again applies – *mutatis mutandis* – to ensure the existence of $\delta > 0$ such that $x(t) = y(t)$ for all $t \in [\sigma - \delta, \tau]$, thereby contradicting the definition of δ.

Step 4. Existence of a unique maximal solution. Define

$$\mathcal{S} := \{y \colon J_y \to G \,|\, J_y \text{ is an interval}, \ \tau \in J_y \subset J, \ y \text{ solves } (4.8)\}\,.$$

Set $I(\tau, \xi) := \cup_{y \in \mathcal{S}} J_y$ and let $x \colon I(\tau, \xi) \to G$ be defined by the property that, for every $y \in \mathcal{S}$, $x(t) = y(t)$ for all $t \in J_y$. The function x is well-defined, since, by Step 3, for any two functions $y, z \in \mathcal{S}$, $y(t) = z(t)$ for all $t \in J_y \cap J_z$. It is clear that x is a solution of (4.8). Furthermore, x does not have a proper extension which is also a solution of (4.8). Uniqueness of x follows by Step 3.

Step 5. The maximal interval of existence is relatively open in J. Write $\alpha := \inf I(\tau, \xi)$ and $\omega := \sup I(\tau, \xi)$. Seeking a contradiction, suppose that $I(\tau, \xi)$ is not relatively open in J. Then one of the following must hold (see also Example A.12): either (i) $\inf J < \alpha \in I(\tau, \xi)$, or $\sup J > \omega \in I(\tau, \xi)$. If (i) holds, then, by

Step 2, the initial-value problem $\dot{z}(t) = f(t, z(t))$, $z(\alpha) = x(\alpha)$, has a solution on $[\alpha - \varepsilon, \alpha]$ for some $\varepsilon > 0$: the concatenation of this solution with x is a left extension of x. If (ii) holds, then, by Step 2, the initial-value problem $\dot{z}(t) = f(t, z(t))$, $z(\omega) = x(\omega)$, has a solution on $[\omega, \omega + \varepsilon]$ for some $\varepsilon > 0$: the concatenation of this solution with x is a right extension of x. Thus, in each case, maximality of x is contradicted, and so $I(\tau, \xi)$ is relatively open in J. $\square$

Perusal of the proof of Lemma 4.9 reveals that the same argument is valid in the present context of (4.8). Therefore, the assertions of Lemma 4.9 hold for the initial-value problem (4.8). Equipped with this fact, inspection of the proofs of Corollary 4.10 and Proposition 4.12 confirms that the arguments used therein apply *mutatis mutandis* in the context of (4.8). We may therefore infer the following two results (counterparts of Corollary 4.10 and Proposition 4.12).

Corollary 4.23

Let $x \colon I \to G$ be the maximal solution of (4.8).

(1) If the closure of the set $\{x(t) \colon t \in I, t \geq \tau\}$ is compact and contained in G, then $[\tau, \infty) \cap I = [\tau, \infty) \cap J$.

(2) If the closure of the set $\{x(t) \colon t \in I, t \leq \tau\}$ is compact and contained in G, then $(-\infty, \tau] \cap I = (-\infty, \tau] \cap J$.

(3) If the closure of the set $x(I)$ is compact and contained in G, then $I = J$.

Proposition 4.24

Let $G = \mathbb{R}^N$. Assume that $f \colon J \times \mathbb{R}^N \to \mathbb{R}^N$ satisfies Assumption **A** and, for every compact interval $K \subset J$, there exists $L > 0$ such that

$$\|f(t, z)\| \leq L(1 + \|z\|) \quad \forall (t, z) \in K \times \mathbb{R}^N,$$

If $x \colon I \to \mathbb{R}^N$ is the maximal solution of (4.8), then $I = J$.

A study of the proof of Theorem 4.11 reveals that the same argument (but now invoking Corollary 4.23 in lieu of Corollary 4.10) may be adopted to establish the following counterpart to Theorem 4.11.

Theorem 4.25

Let $x \colon I \to G$ be a maximal solution of (4.8) with maximal interval of existence $I \subset J$ and assume that $I \neq J$. Write $\alpha := \inf I$ and $\omega := \sup I$. Then, either $\omega \in J \backslash I$ or $\alpha \in J \backslash I$ and the following hold.

(1) If $\omega \in J\backslash I$, then, for each compact set $C \subset G$, there exists $\sigma \in I(\tau, \xi)$, with $\sigma < \omega$), such that $x(t) \notin C$ for all $t \in (\sigma, \omega)$; in particular,

$$\lim_{t \to \omega} \min \left\{ \text{dist}(x(t), \partial G), 1/\|x(t)\| \right\} = 0 \quad \text{if } G \neq \mathbb{R}^N,$$
$$\|x(t)\| \to \infty \text{ as } t \to \omega \quad \text{if } G = \mathbb{R}^N.$$

(2) If $\alpha \in J\backslash I$, then, for each compact set $C \subset G$, there exists $\sigma \in I$, with $\sigma > \alpha$, such that $x(t) \notin C$ for all $t \in (\alpha, \sigma)$; in particular,

$$\lim_{t \to \alpha} \min \left\{ \text{dist}(x(t), \partial G), 1/\|x(t)\| \right\} = 0 \quad \text{if } G \neq \mathbb{R}^N,$$
$$\|x(t)\| \to \infty \text{ as } t \to \alpha \quad \text{if } G = \mathbb{R}^N.$$

The remarks following the statement of Theorem 4.11 remain valid in the present context of the initial-value problem (4.8).

Next, we proceed to develop the notion of the *transition map* associated with (4.8). In what follows, for an interval I and $\tau \in \mathbb{R}$, we adopt the notation

$$I + \tau := \{t + \tau : t \in I\}.$$

If Assumption **A** holds, then, for each $(\tau, \xi) \in J \times G$, Theorem 4.22 guarantees that there exists a unique maximal solution of the initial-value problem (4.8): we denote the associated maximal interval of existence by $I(\tau, \xi)$ and define the map $\psi \colon \text{dom}(\psi) \subset J \times J \times G \to G$ by the property that, for each $(\tau, \xi) \in J \times G$, $t \mapsto \psi(t, \tau, \xi)$ is the unique maximal solution of the initial-value problem (4.8). The domain of ψ is

$$\text{dom}(\psi) := \{(t, \tau, \xi) \in J \times J \times G \colon t \in I(\tau, \xi)\}.$$

We say that the system (4.8) is *autonomous* if $J = \mathbb{R}$ and f does not depend on t, that is, there exists a function $\tilde{f} \colon G \to \mathbb{R}^N$ such that $f(t, z) = \tilde{f}(z)$ for all $(t, z) \in \mathbb{R} \times G$.

Theorem 4.26

Let $f : J \times G \to \mathbb{R}^N$ satisfy Assumption **A**.

(1) Let $(\tau, \xi) \in J \times G$ and $\sigma \in I(\tau, \xi)$. Then $I(\sigma, \psi(\sigma, \tau, \xi)) = I(\tau, \xi)$ and

$$\psi(t, \sigma, \psi(\sigma, \tau, \xi)) = \psi(t, \tau, \xi) \quad \forall t \in I(\tau, \xi). \qquad (4.11)$$

(2) Assume that the system is autonomous. Then, for arbitrary $\tau, \sigma \in \mathbb{R}$ and $\xi \in G$, $I(\tau, \xi) = I(\sigma, \xi) - \sigma + \tau$ and

$$\psi(t, \tau, \xi) = \psi(t + \sigma - \tau, \sigma, \xi) \quad \forall t \in I(\tau, \xi). \qquad (4.12)$$

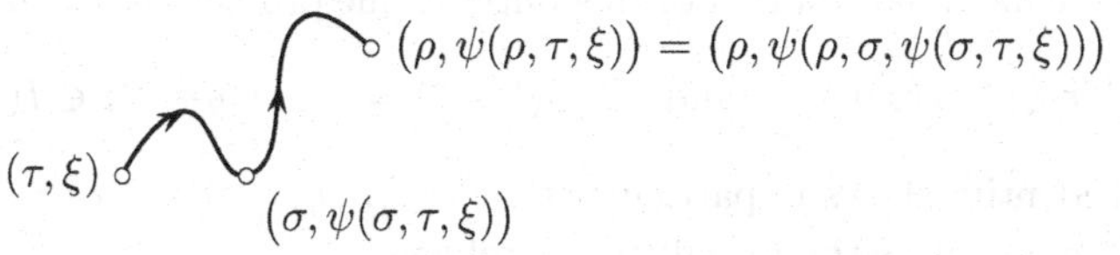

$$(\rho, \psi(\rho, \tau, \xi)) = (\rho, \psi(\rho, \sigma, \psi(\sigma, \tau, \xi)))$$

(τ, ξ)

$(\sigma, \psi(\sigma, \tau, \xi))$

Figure 4.3 The transition property

We will refer to the identity (4.11) as the *transition property* (also known in the literature as the *co-cycle property*) of the solution $\psi(\cdot, \tau, \xi)$, whilst identity (4.12) expresses the so-called *translation invariance* of autonomous systems. We will refer to ψ as the *transition map* generated by f. With reference to Figure 4.3, the transition property (4.11) formalizes an intuitive dynamical interpretation. To describe this interpretation, let $\tau, \sigma, \rho \in J$ with $\tau < \sigma < \rho$ and assume that $\rho \in I(\tau, \xi)$. Consider the evolution of $\psi(t, \tau, \xi)$ as t increases from τ to ρ. Obviously, "stopping" at time $t = \sigma$ and "restarting" the evolution at time $t = \sigma$ with initial value $\psi(\sigma, \tau, \xi) \in G$, we "arrive" at time $t = \rho$ at $\psi(\rho, \sigma, \psi(\sigma, \tau, \xi))$, which, by the transition property (4.11), is the same as $\psi(\rho, \tau, \xi)$, the point at time $t = \rho$ of the evolution "starting" at time $t = \tau$ with initial value ξ. The translation invariance identity (4.12) is a formal and precise version of the statement that, for an autonomous system, a translation (with respect to time) of a solution is again a solution.

Example 4.27

Set $J = \mathbb{R}$, $G = \mathbb{R}^N$ and let Φ be the transition matrix function generated by the piecewise continuous function $A \colon \mathbb{R} \to \mathbb{R}^{N \times N}$. If $f \colon \mathbb{R} \times \mathbb{R}^N \to \mathbb{R}^N$, $(t, z) \mapsto A(t)z$ (that is, if the system is linear), then the transition map is

$$\psi \colon \mathbb{R} \times \mathbb{R} \times \mathbb{R}^N \to \mathbb{R}^N, \quad (t, \tau, \xi) \mapsto \Phi(t, \tau)\xi$$

and the transition property is a consequence of properties of Φ. In particular,

$$\psi(t, \sigma, \psi(\sigma, \tau, \xi)) = \Phi(t, \sigma)\Phi(\sigma, \tau)\xi = \Phi(t, \tau)\xi = \psi(t, \tau, \xi).$$

The transition property also follows from Theorem 4.26 since f clearly satisfies Assumption **A**. $\triangle$

Proof of Theorem 4.26

(1) Set $\eta := \psi(\sigma, \tau, \xi)$ and note that $\psi(\cdot, \tau, \xi)$ solves the initial-value problem

$$\dot{x}(t) = f(t, x(t)), \quad x(\sigma) = \eta.$$

The unique maximal solution of this initial-value problem is $\psi(\,\cdot\,,\sigma,\eta)$ and thus

$$I(\tau,\xi) \subset I(\sigma,\eta) \qquad \text{and} \qquad \psi(t,\tau,\xi) = \psi(t,\sigma,\eta) \quad \forall\, t \in I(\tau,\xi)\,.$$

The last identity shows in particular that $\psi(\tau,\sigma,\eta) = \psi(\tau,\tau,\xi) = \xi$ and we see that $\psi(\,\cdot\,,\sigma,\eta)$ solves the initial-value problem

$$\dot{x}(t) = f(t,x(t))\,, \quad x(\tau) = \xi\,.$$

Since $\psi(\,\cdot\,,\tau,\xi)$ is the unique maximal solution of this initial-value problem, it follows that

$$I(\sigma,\eta) \subset I(\tau,\xi)\,.$$

Consequently, $I(\sigma,\eta) = I(\tau,\xi)$ and $\psi(t,\tau,\xi) = \psi(t,\sigma,\eta)$ for all $t \in I(\tau,\xi)$.

(2) Assume that the system is autonomous, and so there exists locally Lipschitz $\tilde{f}\colon G \to G$ such that $f(t,z) = \tilde{f}(z)$ for all $(t,z) \in \mathbb{R} \times G$. The function $y\colon I(\sigma,\xi) - \sigma + \tau \to G,\ t \mapsto \psi(t + \sigma - \tau,\sigma,\xi)$ is a solution of the initial-value problem given by $\dot{x} = \tilde{f}(x)$ and $x(\tau) = \xi$. Since $\psi(\,\cdot\,,\tau,\xi)$ is the unique maximal solution of this initial-value problem, it follows that $I(\sigma,\xi) - \sigma + \tau \subset I(\tau,\xi)$ and

$$\psi(t + \sigma - \tau,\sigma,\xi) = y(t) = \psi(t,\tau,\xi) \quad \forall\, t \in I(\sigma,\xi) - \sigma + \tau\,.$$

To complete the proof, simply note that, on interchanging the roles of σ and τ in the above argument, we have $I(\tau,\xi) - \tau + \sigma \subset I(\sigma,\xi)$ and so $I(\tau,\xi) \subset I(\sigma,\xi) - \sigma + \tau \subset I(\tau,\xi)$. Therefore, $I(\tau,\xi) = I(\sigma,\xi) - \sigma + \tau$. $\qquad\square$

The next result follows in a straightforward way from part (1) of Theorem 4.26.

Corollary 4.28

Let $f\colon J \times G \to \mathbb{R}^N$ satisfy Assumption **A**. Then the relation on $J \times G$ defined by

$$(\tau,\xi) \sim (\sigma,\eta) \quad \text{if} \quad \sigma \in I(\tau,\xi) \ \text{and} \ \psi(\sigma,\tau,\xi) = \eta$$

is an equivalence relation, and, for every $(\tau,\xi) \in J \times G$, the set $\{(t,\psi(t,\tau,\xi))\colon t \in I(\tau,\xi)\}$ (the graph of the maximal solution of the initial-value problem (4.8)) is the corresponding equivalence class.

Recall that a *partition* of a non-empty set X is a family of pairwise disjoint subsets of X, the union of which is X. If an equivalence relation is defined on X, then every element of X is a member of an equivalence class and every two equivalence classes are either equal or disjoint. Therefore, the set of all equivalence classes of X forms a partition of X. By Corollary 4.28, it follows that, if $f\colon J \times G \to \mathbb{R}^N$ is such that Assumption **A** holds and $(\tau_1,\xi_1),(\tau_2,\xi_2) \in$

$J \times G$, then the graphs $\mathcal{G}_1$ and $\mathcal{G}_2$, of the maximal solutions $\psi(\cdot, \tau_1, \xi_1)$ and $\psi(\cdot, \tau_2, \xi_2)$ either coincide or do not intersect: $\mathcal{G}_1 = \mathcal{G}_2$ if $(\tau_1, \xi_1) \sim (\tau_2, \xi_2)$, or $\mathcal{G}_1 \cap \mathcal{G}_2 = \emptyset$ if $(\tau_1, \xi_1) \not\sim (\tau_2, \xi_2)$.

Exercise 4.13

Prove Corollary 4.28.

Finally, we establish that ψ is locally Lipschitz and its domain $\text{dom}(\psi)$ is relatively open in $J \times J \times G$.

Theorem 4.29

Let $f\colon J \times G \to \mathbb{R}^N$ satisfy Assumption **A** and let ψ be the transition map generated by f. Then $\text{dom}(\psi)$ is relatively open in $J \times J \times G$ and ψ is locally Lipschitz.

We preface the proof of this result with a lemma in which we assume $G = \mathbb{R}^N$ and identify a further assumption on f (in addition to Assumption **A**) under which we may conclude that $\text{dom}(\psi) = J \times J \times \mathbb{R}^N$ and that ψ is locally Lipschitz. This lemma plays an important role in the proof of Theorem 4.29.

Lemma 4.30

Let $f\colon J \times \mathbb{R}^N \to \mathbb{R}^N$ be such that Assumption **A** holds (with $G = \mathbb{R}^N$) and let ψ be the transition map generated by f. Assume further that, for every compact interval $K \subset J$, there exists $L > 0$ such that

$$\|f(t, z)\| \le L\big(1 + \|z\|\big) \quad \forall (t, z) \in K \times \mathbb{R}^N.$$

Then $\text{dom}(\psi) = J \times J \times \mathbb{R}^N$ and ψ is locally Lipschitz.

Proof

By Proposition 4.24, it follows that, for each $(\tau, \xi) \in J \times \mathbb{R}^N$, the unique maximal solution of the initial-value problem has interval of existence $I(\tau, \xi) = J$. Therefore, $\text{dom}(\psi) = J \times J \times \mathbb{R}^N$.

Next, we show that ψ is continuous. Seeking a contradiction, suppose otherwise. Then there exist $(t, \tau, \xi) \in J \times J \times \mathbb{R}^N$, $\varepsilon > 0$, and a sequence $\big((t_n, \tau_n, \xi_n)\big)$ in $J \times J \times \mathbb{R}^N$ converging to (t, τ, ξ) such that

$$\|\psi(t_n, \tau_n, \xi_n) - \psi(t, \tau, \xi)\| \ge \varepsilon \quad \forall n \in \mathbb{N}.$$

By Proposition 4.1, there exists a compact interval $I \subset J$ containing t and τ and such that, if t (respectively, τ) is an interior point of J, then t (respectively,

τ) is an interior point of I. We may assume that the sequences (t_n) and (τ_n) are contained in I. Denote the length of the interval I by δ. By hypothesis, there exists $L > 0$ such that

$$\|f(s,z)\| \leq L\big(1 + \|z\|\big) \quad \forall\, (s,z) \in I \times \mathbb{R}^N.$$

Let $M > 0$ be such that $\|\xi_n\| \leq M$ for all $n \in \mathbb{N}$. For each $n \in \mathbb{N}$, define

$$x_n : I \to \mathbb{R}^N, \ s \mapsto x_n(s) := \psi(s, \tau_n, \xi_n).$$

Since $x_n(s) = \xi_n + \int_{\tau_n}^s f(\sigma, x_n(\sigma))\mathrm{d}\sigma$, we have, for all $s \in I$,

$$\|x_n(s)\| \leq \|\xi_n\| + L\left|\int_{\tau_n}^s \big(1 + \|x_n(\sigma)\|\big)\mathrm{d}\sigma\right| \leq (M + \delta L) + L\left|\int_{\tau_n}^s \|x_n(\sigma)\|\mathrm{d}\sigma\right|.$$

An application of Gronwall's lemma (Lemma 2.4) now yields

$$\|x_n(s)\| \leq (M + \delta L)\, e^{\delta L} =: R \quad \forall\, s \in I, \ \forall\, n \in \mathbb{N}.$$

Furthermore, $x_n(s) - x_n(\sigma) = \int_\sigma^s f(\tau, x_n(\tau))\mathrm{d}\tau$ for all $s, \sigma \in I$, and hence,

$$\|x_n(s) - x_n(\sigma)\| \leq L(1 + R)|s - \sigma| \quad \forall\, s, \sigma \in I, \ \forall\, n \in \mathbb{N}.$$

Therefore, (x_n) is a bounded and equicontinuous sequence in $C(I, \mathbb{R}^N)$, and so, by the Arzelà-Ascoli theorem (Theorem A.24), has a subsequence – which we do not relabel – converging uniformly to $x \in C(I, \mathbb{R}^N)$. Writing $B := \overline{\mathbb{B}}(0, R)$ (the closed ball of radius R centred at 0), then $x(t), x_n(t) \in B$ for all $t \in I$ and all $n \in \mathbb{N}$. By Proposition 4.15, there exists $L_0 > 0$ such that

$$\|f(\sigma, x_n(\sigma)) - f(\sigma, x(\sigma))\| \leq L_0 \|x_n(\sigma) - x(\sigma)\| \quad \forall\, \sigma \in I.$$

and, by uniform convergence of (x_n) to x on I, we may conclude that

$$\lim_{n \to \infty} \int_\tau^s (f(\sigma, x_n(\sigma)) - f(\sigma, x(\sigma)))\mathrm{d}\sigma = 0 \quad \forall\, s \in I.$$

Since

$$x_n(s) - \xi_n = \int_{\tau_n}^\tau f(\sigma, x_n(\sigma))\mathrm{d}\sigma + \int_\tau^s f(\sigma, x_n(\sigma))\mathrm{d}\sigma \quad \forall\, s \in I, \ \forall\, n \in \mathbb{N},$$

it follows that

$$x(s) = \xi + \int_\tau^s f(\sigma, x(\sigma))\mathrm{d}\sigma \quad \forall\, s \in I.$$

Consequently, x solves (4.8) on I. By uniqueness of the (maximal) solution of the initial-value problem (4.8), we have

$$x(s) = \psi(s, \tau, \xi) \quad \forall\, s \in I.$$

Choosing $n \in \mathbb{N}$ sufficiently large so that

$$\|x_n(s) - x(s)\| < \varepsilon/2 \ \text{ for all } s \in I \quad \text{and} \quad \|x(t_n) - x(t)\| < \varepsilon/2,$$

we arrive at a contradiction

$$\varepsilon \leq \|\psi(t_n, \tau_n, \xi_n) - \psi(t, \tau, \xi)\| \leq \|x_n(t_n) - x(t_n)\| + \|x(t_n) - x(t)\| < \varepsilon.$$

Therefore, our supposition is false and so ψ is continuous.

It remains to prove that ψ is locally Lipschitz. To this end, let $(t, \tau, \xi) \in \mathrm{dom}(\psi) = J \times J \times \mathbb{R}^N$. Let $J_0 \subset J$ be an interval containing t and τ and such that properties (1) and (2) of Proposition 4.1 hold. Write $B_1 = \mathbb{B}(\xi, 1)$, the open ball of radius 1 centred at $\xi \in \mathbb{R}^N$, and observe that

$$\mathcal{R} := J_0 \times J_0 \times B_1$$

is relatively open in $J \times J \times \mathbb{R}^N$ and contains (t, τ, ξ). To complete the proof, it suffices to establish the existence of $\Gamma > 0$ such that

$$\|\psi(t_2, \tau_2, \xi_2) - \psi(t_1, \tau_1, \xi_1)\| \leq \Gamma\big(|t_2 - t_1| + |\tau_2 - \tau_1| + \|\xi_2 - \xi_1\|\big)$$
$$\forall \, (t_1, \tau_1, \xi_1), (t_2, \tau_2, \xi_2) \in \mathcal{R}. \quad (4.13)$$

Let $\bar{J}_0$ and $\bar{B}_1$ denote the closures of J_0 and B_1. Recall that $\bar{J}_0 \subset J$. Set

$$C := \bar{J}_0 \times \bar{J}_0 \times \bar{B}_1 \subset J \times J \times \mathbb{R}^N$$

and note that C is compact. Thus, by continuity, ψ attains its maximum on C and we define

$$P := \max\{\|\psi(s, \rho, \eta)\| : (s, \rho, \eta) \in C\} < \infty \quad \text{and} \quad B_2 := \mathbb{B}(0, P).$$

By the hypotheses on f there exists a constant $\Lambda > 0$ such that

$$\|f(s, z) - f(s, \zeta)\| \leq \Lambda \|z - \zeta\| \quad \forall \, s \in \bar{J}_0, \ \forall \, z, \zeta \in \bar{B}_2$$

and

$$\|f(s, z)\| \leq \Lambda(1 + \|z\|) \quad \forall \, s \in \bar{J}_0, \ \forall \, z \in \mathbb{R}^N.$$

As an immediate consequence of the latter estimate we have that

$$\sup\big\{\|f(s, \psi(s, \rho, \eta))\| : (s, \rho, \eta) \in C\big\} \leq \Lambda(1 + P) =: Q.$$

Observe that, for all $s, \tau_1, \tau_2 \in \bar{J}_0$ and all $\xi_1, \xi_2 \in \bar{B}_1$,

$$\|\psi(s, \tau_2, \xi_2) - \psi(s, \tau_1, \xi_1)\|$$
$$\leq \|\xi_2 - \xi_1\| + \left| \int_{\tau_2}^{s} \|f(\sigma, \psi(\sigma, \tau_2, \xi_2)) - f(\sigma, \psi(\sigma, \tau_1, \xi_1))\| d\sigma \right|$$
$$+ \left| \int_{\tau_1}^{\tau_2} \|f(\sigma, \psi(\sigma, \tau_1, \xi_1))\| d\sigma \right|$$
$$\leq \|\xi_2 - \xi_1\| + \Lambda \left| \int_{\tau_2}^{s} \|\psi(\sigma, \tau_2, \xi_2) - \psi(\sigma, \tau_1, \xi_1)\| d\sigma \right| + Q|\tau_2 - \tau_1|.$$

Writing $\bar{J}_0 = [a, b]$, an application of Gronwall's lemma (Lemma 2.4) gives

$$\|\psi(s, \tau_2, \xi_2) - \psi(s, \tau_1, \xi_1)\| \leq \left(\|\xi_2 - \xi_1\| + Q|\tau_2 - \tau_1| \right) e^{\Lambda(b-a)}$$
$$\forall s, \tau_1, \tau_2 \in \bar{J}_0, \ \forall \xi_1, \xi_2 \in \bar{B}_1.$$

Therefore, for all $(t_1, \tau_1, \xi_1), (t_2, \tau_2, \xi_2) \in \mathcal{R}$, we have

$$\|\psi(t_2, \tau_2, \xi_2) - \psi(t_1, \tau_1, \xi_1)\| \leq \|\psi(t_2, \tau_2, \xi_2) - \psi(t_2, \tau_1, \xi_1)\|$$
$$+ \|\psi(t_2, \tau_1, \xi_1) - \psi(t_1, \tau_1, \xi_1)\|$$
$$\leq \left(\|\xi_2 - \xi_1\| + Q|\tau_2 - \tau_1| \right) e^{\Lambda(b-a)}$$
$$+ \left| \int_{t_1}^{t_2} \|f(t, \psi(t, \tau_1, \xi_1))\| dt \right|$$
$$\leq \left(\|\xi_2 - \xi_1\| + Q|\tau_2 - \tau_1| \right) e^{\Lambda(b-a)} + Q|t_2 - t_1|$$

and so (4.13) holds with $\Gamma := (1 + Q)e^{\Lambda(b-a)}$. □

Proof of Theorem 4.29

Let $(\tau, \xi) \in J \times G$ and let $x: I(\tau, \xi) \to G$ denote the unique solution of the initial-value problem (4.8) given by $x(t) = \psi(t, \tau, \xi)$ for all $t \in I(\tau, \xi)$. Fixing $t \in I(\tau, \xi)$ arbitrarily, we have $(t, \tau, \xi) \in \mathrm{dom}(\psi)$. To conclude that $\mathrm{dom}(\psi)$ is relatively open in $J \times J \times G$, it suffices to prove the existence of a set $\mathcal{R}$ that is relatively open in $J \times J \times G$, is contained in $\mathrm{dom}(\psi)$ and contains (t, τ, ξ). Let J_0 be an interval, containing both t and τ, which is relatively open in $I(\tau, \xi)$ and has compact closure $\bar{J}_0 \subset I(\tau, \xi)$ (Proposition 4.1 ensures the existence of such an interval J_0). Since $x(s) \in G$ for all $s \in \bar{J}_0$ and G is open, we may choose $\varepsilon > 0$ sufficiently small so that the compact set $T_\varepsilon := \{(s, z) \in \bar{J}_0 \times \mathbb{R}^N: \|z - x(s)\| \leq 2\varepsilon\}$ is contained in $J \times G$ (see Figure 4.4). Observe that the set

$$K := \left\{ z \in \mathbb{R}^N: \ (s, z) \in T_\varepsilon \text{ for some } s \in \bar{J}_0 \right\}$$

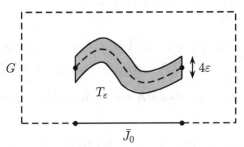

Figure 4.4 Compact "tubular" set T_ε depicted in $J \times G$. The dashed curve is the graph $\{(s, x(s)\colon s \in \bar{J}_0\}$.

is a compact subset of G. By Assumption **A**, Proposition 4.15 and Lemma 4.21, there exists $L_1 > 0$ such that

$$\|f(s, z_1) - f(s, z_2)\| \le L_1 \|z_1 - z_2\| \quad \forall\, (s, z_1), (s, z_2) \in \bar{J}_0 \times K. \tag{4.14}$$

With reference to Figure 4.5, let $g\colon \mathbb{R}_+ \to [0, 1]$ be a continuously differentiable function with the properties that

$$g(r) = 1 \quad \forall\, r \in [0, \varepsilon] \quad \text{and} \quad g(r) = 0 \quad \forall\, r \ge 2\varepsilon.$$

Observe that $\|z - x(s)\| g(\|z - x(s)\|) < 2\varepsilon$ for all $(s, z) \in \bar{J}_0 \times \mathbb{R}^N$ and so

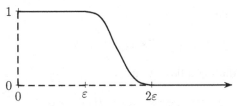

Figure 4.5 Graph of a typical function $g\colon \mathbb{R}_+ \to [0, 1]$

$$x(s) + (z - x(s))g(\|z - x(s)\|) \in K \subset G \quad \forall\, (s, z) \in \bar{J}_0 \times \mathbb{R}^N. \tag{4.15}$$

Therefore, the following is a well-defined map

$$\tilde{f}\colon \bar{J}_0 \times \mathbb{R}^N \to \mathbb{R}^N, \quad (s, z) \mapsto \tilde{f}(s, z) := f\big(s, x(s) + (z - x(s))g(\|z - x(s)\|)\big).$$

Moreover, $\tilde{f}$ satisfies Assumption **A** (with $G = \mathbb{R}^N$). Let $\tilde{\psi}$ be the transition map generated by $\tilde{f}$. The plan is to apply Lemma 4.30 to $\tilde{\psi}$. To this end, we need to show that $\|\tilde{f}(s,z)\| \leq L(1+\|z\|)$ for all $(s,z) \in \bar{J}_0 \times \mathbb{R}^N$. Invoking (4.14) and (4.15), we have

$$\|\tilde{f}(s,z) - \tilde{f}(s,0)\| \leq L_1\|(z-x(s))g(\|z-x(s)\|) + x(s)g(\|x(s)\|)\|$$
$$\leq L_1[\|z\| + \|x(s)\|] \quad \forall\, (s,z) \in \bar{J}_0 \times \mathbb{R}^N.$$

By compactness of $\bar{J}_0$, Assumption **A** and Lemma 4.21, there exists $L_2 > 0$ such that

$$\|f(s, x(s) - x(s)g(\|x(s)\|))\| + L_1\|x(s)\| \leq L_2 \quad \forall\, s \in \bar{J}_0$$

and so $\|\tilde{f}(s,0)\| + L_1\|x(s)\| \leq L_2$ for all $s \in \bar{J}_0$. Defining $L := \max\{L_1, L_2\}$, we have

$$\|\tilde{f}(s,z)\| \leq \|\tilde{f}(s,0)\| + \|\tilde{f}(s,z) - \tilde{f}(s,0)\| \leq L[1 + \|z\|] \quad \forall\, (s,z) \in \bar{J}_0 \times \mathbb{R}^N.$$

By Lemma 4.30, $\mathrm{dom}(\tilde{\psi}) = \bar{J}_0 \times \bar{J}_0 \times \mathbb{R}^N$ and $\tilde{\psi}$ is locally Lipschitz. Moreover, $\tilde{\psi}$ is uniformly continuous on the compact set $\bar{J}_0 \times \bar{J}_0 \times K$ and so there exists $\delta > 0$ such that, for all $(s_1, \sigma_1, \eta_1), (s_2, \sigma_2, \eta_2) \in \bar{J}_0 \times \bar{J}_0 \times K$,

$$\left.\begin{array}{l} |s_2 - s_1| + |\sigma_2 - \sigma_1| + \|\eta_2 - \eta_1\| \leq \delta \\ \qquad\qquad \implies \|\tilde{\psi}(s_2, \sigma_2, \eta_2) - \tilde{\psi}(s_1, \sigma_1, \eta_1)\| \leq \varepsilon. \end{array}\right\} \quad (4.16)$$

Introducing the "tubular" set

$$S_\varepsilon := \{(s,z) \in \bar{J}_0 \times \mathbb{R}^N : \|z - x(s)\| \leq \varepsilon\} \subset T_\varepsilon,$$

it is readily verified that

$$\tilde{f}(s,z) = f(s,z) \quad \forall\, (s,z) \in S_\varepsilon,$$

from which it immediately follows that

$$\tilde{\psi}(s, \tau, \xi) = \psi(s, \tau, \xi) = x(s) \quad \forall\, s \in J_0. \qquad (4.17)$$

Let $\sigma \in J_0$ and $\eta \in G$. Assume that $(s, \tilde{\psi}(s, \sigma, \eta)) \in S_\varepsilon$ for all $s \in J_0$ and write $\tilde{x}(s) := \tilde{\psi}(s, \sigma, \eta)$ for all $s \in J_0$. Then, since $\tilde{f}(s, \tilde{x}(s)) = f(s, \tilde{x}(s))$ for all $t \in J_0$, it follows that $\tilde{x}$ solves the initial-value problem $\dot{y}(s) = f(s, y(s))$, $y(\sigma) = \eta$, and so $\psi(s, \sigma, \eta) = \tilde{\psi}(s, \sigma, \eta)$ for all $s \in J_0$. In summary, we have

$$\left.\begin{array}{l} (s, \tilde{\psi}(s,\sigma,\eta)) \in S_\varepsilon \quad \forall\, s \in J_0 \\ \qquad \implies (s, \sigma, \eta) \in \mathrm{dom}(\psi) \text{ and } \psi(s,\sigma,\eta) = \tilde{\psi}(s,\sigma,\eta) \,\forall\, s \in J_0. \end{array}\right\} \quad (4.18)$$

Recalling that J_0 is relatively open in J and contains τ, there exists λ, with $0 < \lambda \leq \delta/2$, such that $J_1 := (\tau - \lambda, \tau + \lambda) \cap J$ is contained in J_0. Since ξ is an interior point of K, there exists μ, with $0 < \mu \le \delta/2$, such that $\mathbb{B}(\xi,\mu) \subset K$. Defining

$$\mathcal{R} := J_0 \times J_1 \times \mathbb{B}(\xi,\mu)$$

we see that $\mathcal{R}$ is relatively open in $J \times J \times G$ and contains (t,τ,ξ). To show that $\mathcal{R} \subset \mathrm{dom}(\psi)$, let $(\sigma,\eta) \in J_1 \times \mathbb{B}(\xi,\mu)$. Invoking (4.16) and (4.17), we have

$$\|\tilde\psi(s,\sigma,\eta) - x(s)\| = \|\tilde\psi(s,\sigma,\eta) - \tilde\psi(s,\tau,\xi)\| \le \varepsilon \ \ \forall\, s \in J_0.$$

Therefore, $(s,\tilde\psi(s,\sigma,\eta)) \in S_\varepsilon$ for all $s \in J_0$ and so, by (4.18), $(s,\sigma,\eta) \in \mathrm{dom}(\psi)$ for all $s \in J_0$. Since $(\sigma,\eta) \in J_1 \times \mathbb{B}(\xi,\mu)$ is arbitrary, we have $(s,\sigma,\eta) \in \mathrm{dom}(\psi)$ for all $(s,\sigma,\eta) \in J_0 \times J_1 \times \mathbb{B}(\xi,\mu) = \mathcal{R}$ and so $\mathcal{R} \subset \mathrm{dom}(\psi)$. We may now conclude that $\mathrm{dom}(\psi)$ is relatively open in $J \times J \times G$. Finally, by (4.18), we also have

$$\psi(s,\sigma,\eta) = \tilde\psi(s,\sigma,\eta) \ \ \forall\,(s,\sigma,\eta) \in \mathcal{R}$$

and so, since $\tilde\psi$ is locally Lipschitz, it follows that ψ is also locally Lipschitz. $\quad\square$

4.5 Periodic solutions

Here, we assume that $J = \mathbb{R}$ and $f\colon \mathbb{R} \times G \to \mathbb{R}^N$ satisfies Assumption **A**. We will investigate conditions under which the solution $\psi(\cdot,\tau,\xi)$ of the initial-value problem (4.8) is q-periodic for some $q > 0$, that is, $I(\tau,\xi) = \mathbb{R}$ and $\psi(t+q,\tau,\xi) = \psi(t,\tau,\xi)$ for all $t \in \mathbb{R}$. To this end, we introduce a family, parameterized by $q > 0$, of maps Σ_q given by

$$\Sigma_q(\zeta) := \psi(q,0,\zeta) \ \ \forall\, \zeta \in \mathrm{dom}(\Sigma_q) := \{z \in G\colon q \in I(0,z)\}.$$

Theorem 4.31

Assume that f is periodic in its first argument, with period $p > 0$, that is

$$f(t+p,z) = f(t,z) \ \ \forall\,(t,z) \in \mathbb{R} \times G.$$

Let $n \in \mathbb{N}$ and $(\tau,\xi) \in \mathbb{R} \times G$. The solution of (4.8) is np-periodic if, and only if, $0 \in I(\tau,\xi)$ and $\psi(0,\tau,\xi)$ is a fixed point of Σ_{np}.

Proof

Necessity. Assume that the solution $\psi(\cdot,\tau,\xi)$ of (4.8) is np-periodic. Then, in particular, $I(\tau,\xi) = \mathbb{R}$, implying that $0 \in I(\tau,\xi)$. Setting $\zeta := \psi(0,\tau,\xi)$ and invoking statement (1) of Theorem 4.26, we conclude that $I(0,\zeta) =$

$I(0, \psi(0, \tau, \xi)) = I(\tau, \xi) = \mathbb{R}$ and $\psi(t, 0, \zeta) = \psi(t, 0, \psi(0, \tau, \xi)) = \psi(t, \tau, \xi)$ for all $t \in \mathbb{R}$. It follows that

$$\Sigma_{np}(\zeta) = \psi(np, 0, \zeta) = \psi(np, \tau, \xi) = \psi(0, \tau, \xi) = \zeta.$$

Therefore, ζ is a fixed point of Σ_{np}.

Sufficiency. By hypothesis, $0 \in I(\tau, \xi)$ and $\zeta := \psi(0, \tau, \xi)$ is a fixed point of Σ_{np}. Then $np \in I(0, \zeta)$. By statement (1) of Theorem 4.26, $I(0, \zeta) = I(\tau, \xi)$. Write $Z := I(0, \zeta) - np = \{t - np \colon t \in I(0, \zeta)\}$ and define $z \colon Z \to G$ by $z(t) := \psi(t + np, 0, \zeta)$ for all $t \in Z$. Then $z(0) = \psi(np, 0, \zeta) = \Sigma_{np}(\zeta) = \zeta$. Moreover,

$$z(t) = \psi(t + np, 0, \zeta) = \psi(np, 0, \zeta) + \int_{np}^{t+np} f(s, \psi(s, 0, \zeta))\mathrm{d}s$$

$$= \zeta + \int_0^t f(s + np, \psi(s + np, 0, \zeta))\mathrm{d}s = \zeta + \int_0^t f(s, z(s))\mathrm{d}s \quad \forall t \in Z,$$

where in the last step we have used the periodicity property of f. Consequently, z solves the initial-value problem $\dot{x}(t) = f(t, x(t))$, $x(0) = \zeta$, the unique maximal solution of which is $\psi(\cdot, 0, \zeta)$. Therefore, $Z \subset I(0, \zeta)$ and

$$z(t) = \psi(t, 0, \zeta) \quad \forall t \in Z. \tag{4.19}$$

Since $I(0, \zeta) - np = Z \subset I(0, \zeta)$, it immediately follows that $\inf I(0, \zeta) = -\infty$. To show that $I(0, \zeta) = \mathbb{R}$, it suffices to prove that $\sup I(0, \zeta) = \infty$. Seeking a contradiction, suppose that $\sup I(0, \zeta) =: \omega < \infty$. Let (t_k) be a sequence in $I(0, \zeta)$ with $t_k \to \omega$ as $k \to \infty$. Setting $s_k := t_k - np$ for all $k \in \mathbb{N}$, we see that (s_k) is a sequence in Z with $s_k \to \omega - np$ as $k \to \infty$ and, invoking (4.19), we obtain

$$\psi(t_k, 0, \zeta) = \psi(s_k + np, 0, \zeta) = z(s_k) = \psi(s_k, 0, \zeta) \quad \forall k \in \mathbb{N}.$$

Therefore,

$$\lim_{k \to \infty} \psi(t_k, 0, \zeta) = \lim_{k \to \infty} \psi(s_k, 0, \zeta) = \psi(\omega - np, 0, \zeta) \in G. \tag{4.20}$$

Since $J = \mathbb{R}$, the interval $I(0, \zeta)$ is open and so $\omega \in J \backslash I(0, \zeta)$. Consequently, an application of Theorem 4.25 yields a contradiction to (4.20). We may now conclude that $I(\tau, \xi) = I(0, \zeta) = \mathbb{R}$ and $\psi(t + np, 0, \zeta) = \psi(t, 0, \zeta)$ for all $t \in \mathbb{R}$. Therefore, using statement (1) of Theorem 4.26, it follows that, for all $t \in \mathbb{R}$,

$$\psi(t + np, \tau, \xi) = \psi(t + np, 0, \psi(0, \tau, \xi)) = \psi(t + np, 0, \zeta)$$
$$= \psi(t, 0, \zeta) = \psi(t, 0, \psi(0, \tau, \xi)) = \psi(t, \tau, \xi),$$

showing that $\psi(\cdot, \tau, \xi)$ is np-periodic. $\qquad\qquad\qquad\qquad\qquad\qquad\square$

Example 4.32

Let $J = \mathbb{R}$, $G = \mathbb{R}^N$ and consider the initial-value problem for the linear system with piecewise continuous periodic $A : \mathbb{R} \to \mathbb{R}^{N \times N}$ of period $p > 0$:

$$\dot{x}(t) = A(t)x(t), \quad x(\tau) = \xi, \tag{4.21}$$

In this case, ψ is given by

$$\psi(t, \tau, \xi) := \Phi(t, \tau)\xi \quad \forall \, (t, \tau, \xi) \in \mathbb{R} \times \mathbb{R} \times \mathbb{R}^N,$$

where Φ is the transition matrix function generated by A. The family of maps Σ_q, $q > 0$, is given by $\Sigma_q(\zeta) = \Phi(q, 0)\zeta$ for all $\zeta \in \mathbb{R}^N$. Let $n \in \mathbb{N}$ and $(\tau, \xi) \in \mathbb{R} \times \mathbb{R}^N$. By Theorem 4.31, the solution of (4.21) is np-periodic if, and only if,

$$\Phi(np, 0)\Phi(0, \tau)\xi = \Phi(0, \tau)\xi.$$

Noting that $\Phi(np, 0) = \Phi^n(p, 0)$ (recall (2.32)), we see that the above condition is equivalent to

$$\Phi^n(p, 0) \text{ has eigenvalue 1 with associated eigenvector } \Phi(0, \tau)\xi. \tag{4.22}$$

If it is additionally assumed that

$$(\lambda/\mu)^n \neq 1 \quad \text{for all } \lambda, \mu \in \sigma(\Phi(p, 0)) \text{ such that } \lambda \neq \mu \tag{4.23}$$

(that is, for all eigenvalues λ and μ of $\Phi(p, 0)$, λ/μ is not a non-trivial n-th root of unity), then, on invoking the spectral mapping theorem (Theorem 2.19), (4.22) is equivalent to

$$\left. \begin{array}{l} \Phi(p, 0) \text{ has eigenvalue } \lambda \text{ such that } \lambda^n = 1 \text{ and } \Phi(0, \tau)\xi \text{ is} \\[4pt] \text{an associated eigenvector.} \end{array} \right\} \tag{4.24}$$

Consequently, if the additional assumption (4.23) is satisfied, then the solution of (4.21) is np-periodic if, and only if, (4.24) holds. $\qquad \triangle$

Exercise 4.14

Let $A : \mathbb{R} \to \mathbb{R}^{N \times N}$ be piecewise continuous and periodic of period $p > 0$. In the above Example 4.32 it was shown that the solution of (4.21) is np-periodic if, and only if, (4.22) holds. Deduce Proposition 2.20 from this equivalence.

4.6 Autonomous differential equations

In this section we study *autonomous* ordinary differential equations. These are differential equations, as in (4.1) or (4.8), but with the distinguishing feature that f does not depend on t. Of course, the theory developed so far in this chapter applies to autonomous equations (with $J = \mathbb{R}$). Let $G \subset \mathbb{R}^N$ be a non-empty open set. Throughout this section, we assume that $f \colon G \to \mathbb{R}^N$ is locally Lipschitz, that is, for every $z \in G$ there exists a neighbourhood $U \subset G$ of z and a number $L \geq 0$ such that

$$\|f(z_1) - f(z_2)\| \leq L\|z_1 - z_2\| \quad \forall z_1, z_2 \in U.$$

We consider the autonomous initial-value problem

$$\dot{x}(t) = f(x(t)), \quad x(0) = \xi \in G, \tag{4.25}$$

wherein, without loss of generality (see part (2) of Theorem 4.26), we have set $\tau = 0$. In this setting, a solution of (4.25) is a continuously differentiable function $x \colon I \to G$, on some interval I containing 0, such that (4.25) holds.

4.6.1 Flows and continuous dependence

The theory of existence, maximality and uniqueness of solutions, developed in the previous sections of this chapter (in particular, Theorems 4.18 and 4.22, with $J = \mathbb{R}$), implies the existence of a map

$$(t, \xi) \mapsto \varphi(t, \xi) \in G$$

such that $\varphi(\,\cdot\,, \xi)$ is the unique maximal solution of the initial-value problem (4.25). The domain $D := \mathrm{dom}(\varphi) \subset \mathbb{R} \times G$ of φ is given by

$$D = \{(t, \xi) \in \mathbb{R} \times G \colon t \in I_\xi\},$$

where I_ξ denotes the interval of existence of the unique maximal solution of (4.25). In the notation used in Theorem 4.26, we have that

$$I_\xi = I(0, \xi) \quad \text{and} \quad \varphi(\,\cdot\,, \xi) = \psi(\,\cdot\,, 0, \xi) \tag{4.26}$$

We refer to φ as the *local flow* or *local dynamical system* generated by the differential equation $\dot{x} = f(x)$ (or, more simply, generated by f). If ξ is such that $I_\xi = \mathbb{R}$, then we say that the solution $\varphi(\,\cdot\,, \xi)$ is *global*. If $I_\xi = \mathbb{R}$ for all $\xi \in G$, then φ is said to be a *flow* or *dynamical system*.

Example 4.33

Consider (4.25) with $G = \mathbb{R}$ and $f\colon G \to \mathbb{R}$, $x \mapsto x^2$. If $\xi = 0$, then the unique global solution is the zero function. If $\xi > 0$, then separation of variables and integration yields the unique maximal solution

$$x\colon I_\xi = (-\infty, 1/\xi) \to \mathbb{R}, \quad t \mapsto \frac{\xi}{1 - t\xi},$$

and, if $\xi < 0$, then the unique maximal solution is

$$x\colon I_\xi = (1/\xi, \infty) \to \mathbb{R}, \quad t \mapsto \frac{\xi}{1 - t\xi}.$$

Therefore, f generates a local flow

$$\varphi\colon (t, \xi) \mapsto \frac{\xi}{1 - t\xi} \quad \text{with domain } D := \{(t, \xi) \in \mathbb{R} \times \mathbb{R}\colon t\xi < 1\}.$$

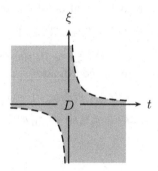

Figure 4.6 Example 4.33: domain D of the local flow φ

$\triangle$

Evidently, in the above example, the (local) flow φ is continuous and its domain D (the interior of the region lying between the two curves given by $\xi = 1/t$ in the first and third quadrants, see Figure 4.6) is an open set. Furthermore, we note that φ is not only continuous, but is locally Lipschitz (as follows, for example, from Proposition 4.14). Of course, these properties are to be expected, since the following general result is an immediate consequence of Theorem 4.29.

Theorem 4.34

$D = \operatorname{dom}(\varphi)$ is open and φ is locally Lipschitz.

Exercise 4.15

Compute the local flow $\varphi: D \subset \mathbb{R}^2 \to \mathbb{R}$ generated by $f: \mathbb{R} \to \mathbb{R}$, $z \mapsto z(1-z)$.

Exercise 4.16

Let $f: \mathbb{R}^2 \to \mathbb{R}^2$ be given by

$$f(z) = f(z_1, z_2) := \big(z_2 + z_1(1 - \|z\|^2),\ -z_1 + z_2(1 - \|z\|^2)\big).$$

Show that f generates a local flow $\varphi: D \to \mathbb{R}^2$ given by

$$\varphi(t, \xi) = \big(\|\xi\|^2 + (1 - \|\xi\|^2)e^{-2t}\big)^{-1/2} R(t)\xi,$$

where the function $R: \mathbb{R} \to \mathbb{R}^{2\times 2}$ is given by

$$R(t) := \begin{pmatrix} \cos t & \sin t \\ -\sin t & \cos t \end{pmatrix} \quad \forall t \in \mathbb{R}. \tag{4.27}$$

(and so $R(t)\xi$ is a clockwise rotation of ξ through t radians) and

$$D := \big\{(t, \xi) \in \mathbb{R} \times \mathbb{R}^2 : \|\xi\|^2 + (1 - \|\xi\|^2)e^{-2t} > 0\big\}.$$

(*Hint.* Show that, for $\xi = (\xi_1, \xi_2) \neq 0$, the initial-value problem (4.25) may be expressed – in polar coordinates – as

$$\dot{r}(t) = r(t)\big(1 - r^2(t)\big), \quad \dot{\theta}(t) = -1, \quad (r(0), \theta(0)) = (r^0, \theta^0),$$

where $r^0 = \|\xi\|$, $r^0 \cos\theta^0 = \xi_1$ and $r^0 \sin\theta^0 = \xi_2$.)

Theorem 4.35

The local flow $\varphi: D \to G$ has the following properties:

(1) $\varphi(0, \xi) = \xi$ for all $\xi \in G$;

(2) if $\xi \in G$ and $\tau \in I_\xi$, then $I_{\varphi(\tau,\xi)} = I_\xi - \tau$ and

$$\varphi(t + \tau, \xi) = \varphi(t, \varphi(\tau, \xi)), \quad \forall t \in I_\xi - \tau. \tag{4.28}$$

The property stated in (2) is referred to as the *group property* of the local flow φ. The following exercise explains the rationale behind the terminology.

Exercise 4.17

Assume that $I_\xi = \mathbb{R}$ for all $\xi \in G$ (and so φ is a flow). For every $t \in \mathbb{R}$, define a map

$$\Phi_t: G \to G, \quad z \mapsto \varphi(t, z).$$

Show that $\Phi_s \circ \Phi_t = \Phi_{s+t}$ for all $s, t \in \mathbb{R}$, where $\circ$ denotes composition. Conclude that, endowed with the operation $\circ$, the set $\{\Phi_t : t \in \mathbb{R}\}$ forms a commutative group.

Proof of Theorem 4.35

Part (1) follows immediately from the definition of the local flow φ. To prove part (2), we invoke (4.26) and Theorem 4.26 to obtain

$$I_{\varphi(\tau,\xi)} = I(0, \psi(\tau, 0, \xi)) = I(\tau, \psi(\tau, 0, \xi)) - \tau = I(0, \xi) - \tau = I_\xi - \tau,$$

and, moreover, for $t \in I_\xi - \tau$,

$$\varphi(t, \varphi(\tau, \xi)) = \psi(t, 0, \psi(\tau, 0, \xi)) = \psi(t + \tau, \tau, \psi(\tau, 0, \xi)) = \psi(t + \tau, 0, \xi),$$

showing that $\varphi(t, \varphi(\tau, \xi)) = \varphi(t + \tau, \xi)$ for all $t \in I_\xi - \tau$. $\square$

From Corollary 4.28, we know that, if $\eta \neq \psi(\sigma, \tau, \xi)$, then the graphs of the solutions $\psi(\cdot, \sigma, \eta)$ and $\psi(\cdot, \tau, \xi)$ of (4.8) do not intersect. In the autonomous case (4.25), more is true: defining the orbit $O(\xi)$ of $\xi \in G$ by

$$O(\xi) := \{\varphi(t, \xi) : t \in I_\xi\}, \tag{4.29}$$

the next result says, in particular, that, if $\eta \notin O(\xi)$, then the orbits $O(\xi)$ and $O(\eta)$ do not intersect. Note that graphs of solutions are subsets of $J \times G$, whilst orbits are subsets of G.

Corollary 4.36

The relation on G defined by

$$\xi \sim \eta \text{ if } \eta \in O(\xi),$$

is an equivalence relation and, for every $\xi \in G$, the orbit $O(\xi)$ is the corresponding equivalence class.

Exercise 4.18

Prove Corollary 4.36.

4.6.2 Limit sets

Consider once more (4.25), with f locally Lipschitz, and let φ denote the local flow generated by f. For $\xi \in G$, let $I_\xi = (\alpha_\xi, \omega_\xi)$ denote the interval of existence of the maximal solution of the initial-value problem (4.25). In addition to the concept of the orbit $O(\xi)$ of $\xi \in G$ (see (4.29)), we define the *positive semi-orbit* $O^+(\xi) := \{\varphi(t, \xi) : t \in [0, \omega_\xi)\}$ of ξ and the *negative semi-orbit* $O^-(\xi) := \{\varphi(t, \xi) : t \in (\alpha_\xi, 0]\}$ of ξ.

A point $z \in \mathbb{R}^N$ is an ω-*limit point* of ξ (or of its orbit $O(\xi)$, or of its semi-orbit $O^+(\xi)$) if there exists a sequence (t_n) in $[0, \omega_\xi)$ such that $t_n \to \omega_\xi$ and $\varphi(t_n, \xi) \to z$ as $n \to \infty$. The ω-*limit set* $\Omega(\xi)$ of ξ (or of its orbit $O(\xi)$, or of its semi-orbit $O^+(\xi)$) is the set of all ω-limit points of ξ. Similarly, a point $z \in \mathbb{R}^N$ is an α-*limit point* of ξ (or of its orbit $O(\xi)$, or of its semi-orbit $O^-(\xi)$) if there exists a sequence (t_n) in $(\alpha_\xi, 0]$ such that $t_n \to \alpha_\xi$ and $\varphi(t_n, \xi) \to z$ as $n \to \infty$. The α-*limit set* $A(\xi)$ of ξ (or of its orbit $O(\xi)$, or of its semi-orbit $O^-(\xi)$) is the set of all α-limit points of ξ.

A set $U \subset G$ is said to be *positively invariant* (respectively, *negatively invariant*) under the (local) flow if each $\xi \in U$ has positive semi-orbit $O^+(\xi)$ (respectively, negative semi-orbit $O^-(\xi)$) contained in U; the set U is said to be *invariant* under the (local) flow if it is both positively invariant and negatively invariant under the (local) flow, that is, $O(\xi) \subset U$ for all $\xi \in U$.

Example 4.37

Let $f : \mathbb{R}^2 \to \mathbb{R}^2$ be as in Exercise 4.16, the generator of the local flow

$$\varphi : D \to \mathbb{R}^2, \ (t, \xi) \mapsto \varphi(t, \xi) = \left(\|\xi\|^2 + (1 - \|\xi\|^2)e^{-2t} \right)^{-1/2} R(t)\xi,$$

where $R : \mathbb{R} \to \mathbb{R}^{2 \times 2}$ is defined by (4.27). The domain D of φ is

$$D = \left\{ (t, \xi) \in \mathbb{R} \times \mathbb{R}^2 : \|\xi\|^2 + (1 - \|\xi\|^2)e^{-2t} > 0 \right\}.$$

Let Δ be the open unit disc in $\mathbb{R}^2$, that is, $\Delta = \{(z_1, z_2) \in \mathbb{R}^2 : z_1^2 + z_2^2 < 1\}$.

The aims of this example are (i) to show that the sets $\{0\}$, Δ, $\partial\Delta$ and $\mathbb{R}^2 \backslash \bar{\Delta}$ are invariant, and (ii) to find, for every $\xi \in \mathbb{R}^2$, the associated ω and α-limit set.

Since, for each $t \in \mathbb{R}$, $R(t)$ is an orthogonal matrix (representing a rotation through t radians), we have that $\|R(t)\xi\| = \|\xi\|$ for each $\xi \in \mathbb{R}^2$, a fact which will be used freely in the arguments below.

Case 1. If $\xi = 0$, then $\varphi(t, \xi) = 0$ for all $t \in I_\xi = \mathbb{R}$ and thus the set $\{0\}$ is invariant and

$$A(\xi) = O(\xi) = \Omega(\xi) = \{0\}.$$

Case 2. If $\|\xi\| = 1$, then, for all $t \in I_\xi = \mathbb{R}$, $\varphi(t, \xi) = R(t)\xi$ and thus, in particular, $\|\varphi(t, \xi)\| = 1$. We conclude that $\partial\Delta$ is invariant and

$$A(\xi) = O(\xi) = \Omega(\xi) = \partial\Delta.$$

Case 3. Assume $0 < \|\xi\| < 1$. Then $\|\varphi(t, \xi)\| < 1$ for all $t \in I_\xi = \mathbb{R}$, showing that Δ is invariant. Furthermore, let $z \in \partial\Delta$ be arbitrary and let $T \geq 0$ be such that $R(T)\xi = \|\xi\|z$. Define the sequence (t_n) by $t_n := T + 2n\pi$. Then

$$\varphi(t_n, \xi) = \left(\|\xi\|^2 + (1 - \|\xi\|^2)e^{-2t_n} \right)^{-1/2} R(T)\xi \to z \text{ as } n \to \infty,$$

and so $z \in \Omega(\xi)$. Since $z \in \partial\Delta$ is arbitrary, we have $\partial\Delta \subset \Omega(\xi)$. Moreover, $\|\varphi(t,\xi)\| \to 1$ as $t \to \infty$, and thus it is clear that there are no other ω-limit points, showing that $\Omega(\xi) = \partial\Delta$. Since $\|\varphi(t,\xi)\| \to 0$ as $t \to -\infty$, it follows that $A(\xi) = \{0\}$.

Case 4. Finally, consider the remaining case wherein $\|\xi\| > 1$. Here (the only case in which $I_\xi \neq \mathbb{R}$), the maximal interval of existence I_ξ is given by $I_\xi = (\alpha_\xi, \infty)$, where

$$\alpha_\xi := \ln\left(\sqrt{\|\xi\|^2 - 1}/\|\xi\|\right) < 0.$$

Since $\|\varphi(t,\xi)\| > 1$ for all $t \in I_\xi$, we see that $\mathbb{R}^2\backslash\bar{\Delta}$ is invariant. Moreover, by the same argument as used in Case 3, we find that $\Omega(\xi) = \partial\Delta$. Finally, $\|\varphi(t_n,\xi)\| \to \infty$ as $t \to \alpha_\xi$, and so we may conclude that $A(\xi) = \emptyset$.   △

Exercise 4.19

Prove the following:

$$A(\xi) = \bigcap_{\tau \in I_\xi \cap (-\infty,0]} \overline{O^-(\varphi(\tau,\xi))}, \qquad \Omega(\xi) = \bigcap_{\tau \in I_\xi \cap [0,\infty)} \overline{O^+(\varphi(\tau,\xi))}.$$

We define the *distance* between two non-empty sets $U, V \subset \mathbb{R}^N$ by

$$\mathrm{dist}(U,V) = \inf\left\{\|u - v\|\colon u \in U, v \in V\right\}.$$

The distance $\mathrm{dist}(u,V)$ from a point $u \in \mathbb{R}^N$ to a set $V \subset \mathbb{R}^N$, defined in (4.4), can be written as

$$\mathrm{dist}(u,V) = \mathrm{dist}(\{u\}, V).$$

For a function $h : \mathbb{R}_+ \to \mathbb{R}^N$, we say that $h(t)$ approaches a non-empty set $U \subset \mathbb{R}^N$ as $t \to \infty$ if

$$\lim_{t\to\infty} \mathrm{dist}(h(t), U) = 0.$$

Note that, if $h(t)$ approaches the set $U \subset \mathbb{R}^N$ as $t \to \infty$, then, for every set $V \subset \mathbb{R}^N$ such that $U \subset V$, $h(t)$ approaches V as $t \to \infty$.

The basic properties of ω-limit sets are given in the following theorem. An analogous result holds for α-limit sets.

Theorem 4.38

Let $\xi \in G$. If the closure of the positive semi-orbit $O^+(\xi)$ is compact and contained in G, then $\omega_\xi = \infty$, the ω-limit set $\Omega(\xi) \subset G$ is non-empty, compact, connected and invariant (under the local flow generated by (4.25)), $I_\eta = \mathbb{R}$ for all $\eta \in \Omega(\xi)$ and

$$\lim_{t\to\infty} \mathrm{dist}(\varphi(t,\xi), \Omega(\xi)) = 0,$$

that is, $\varphi(t,\xi)$ approaches $\Omega(\xi)$ as $t \to \infty$.

Proof

Set $\gamma^+ := O^+(\xi)$. We proceed in several steps.

Step 1. First, we will prove that $\omega_\xi = \infty$, $\Omega(\xi) \subset G$ and $\Omega(\xi) \neq \emptyset$. By hypothesis, the closure of γ^+ is compact and contained in G. Therefore it follows immediately from part (1) of Corollary 4.10 that $\omega_\xi = \infty$. Since the closure $\text{cl}(\gamma^+)$ of γ^+ satisfies $\text{cl}(\gamma^+) \subset G$ and since $\Omega(\xi) \subset \text{cl}(\gamma^+)$, it follows that $\Omega(\xi) \subset G$. To show that $\Omega(\xi) \neq \emptyset$, let $(t_n) \subset [0, \infty)$ be a sequence with $t_n \to \infty$ as $n \to \infty$. By boundedness of γ^+, the sequence $(\varphi(t_n, \xi))$ is bounded and so, by the Bolzano[4]-Weierstrass theorem (Theorem A.16), contains a convergent subsequence, with limit z say. Thus, $z \in \Omega(\xi)$ and hence, $\Omega(\xi) \neq \emptyset$.

Step 2. Next, we establish compactness of $\Omega(\xi)$. By boundedness of γ^+, we see that $\Omega(\xi)$ is bounded. We show that $\Omega(\xi)$ is also closed (and so is compact): this we do by establishing that every convergent sequence in $\Omega(\xi)$ has its limit in $\Omega(\xi)$. Let (z_n) be a convergent sequence in $\Omega(\xi)$ with limit z. For each $k \in \mathbb{N}$, there exists $n_k \in \mathbb{N}$ such that

$$\|z_{n_k} - z\| \leq \frac{1}{2k}$$

and, since z_{n_k} is an ω-limit point of ξ, there exists $t_k \geq k$ such that

$$\|\varphi(t_k, \xi) - z_{n_k}\| \leq \frac{1}{2k}\,.$$

Therefore, for each $k \in \mathbb{N}$,

$$\|\varphi(t_k, \xi) - z\| \leq \|\varphi(t_k, \xi) - z_{n_k}\| + \|z_{n_k} - z\| \leq \frac{1}{k}.$$

The sequence (t_k) has the properties

$$t_k \to \infty \quad \text{and} \quad \varphi(t_k, \xi) \to z \quad \text{as } k \to \infty\,.$$

Therefore, $z \in \Omega(\xi)$, showing that $\Omega(\xi)$ is closed.

Step 3. We now establish connectedness of $\Omega(\xi)$. Seeking a contradiction, suppose that $\Omega(\xi)$ is not connected. Then there exist non-empty, disjoint and compact sets C_1 and C_2 with $\Omega(\xi) = C_1 \cup C_2$ and $\text{dist}(C_1, C_2) =: \delta > 0$. Let $d_i(t) := \text{dist}(\varphi(t, \xi), C_i)$, $i = 1, 2$. For each $k \in \mathbb{N}$ there exist $t_k^1, t_k^2 \geq k$ such that

$$d_1(t_k^1) < \delta/2 \quad \text{and} \quad d_2(t_k^2) < \delta/2\,.$$

Since $d_1(t) + d_2(t) \geq \delta$ for all $t \geq 0$ and since the functions d_1 and d_2 are continuous, there exists t_k between t_k^1 and t_k^2 such that

$$d_1(t_k) = \delta/2 \quad \text{and} \quad d_2(t_k) \geq \delta/2\,,$$

---

[4] Bernard Placidus Johann Bolzano (1781-1848), Bohemian.

showing that

$$\text{dist}(\varphi(t_k, \xi), \Omega(\xi)) \geq \delta/2 > 0, \quad \forall k \in \mathbb{N}. \tag{4.30}$$

On the other hand, the sequence $(\varphi(t_k, \xi))$ lies in the compact set $\text{cl}(\gamma^+)$. Therefore there exists a subsequence of $(\varphi(t_k, \xi))$ converging to a point $\eta \in \Omega(\xi)$, in contradiction to (4.30).

Step 4. Here, we prove invariance of $\Omega(\xi)$. Let $\eta \in \Omega(\xi)$ and let $t \in I_\eta$. We will show that $\varphi(t, \eta) \in \Omega(\xi)$. Since η is an ω-limit point, there exists a sequence (t_n) such that $t_n \to \infty$ and $\varphi(t_n, \xi) \to \eta$ as $n \to \infty$. For all n sufficiently large, $s_n := t + t_n \in I_\xi$. Furthermore, since the domain D of φ is open (by Proposition 4.34) and $(t, \eta) \in D$, it follows that $(t, \varphi(t_n, \xi)) \in D$ for all n sufficiently large. Thus $t \in I_{\varphi(t_n, \xi)}$ and

$$\varphi(s_n, \xi) = \varphi(t + t_n, \xi) = \varphi(t, \varphi(t_n, \xi)).$$

By continuity of φ,

$$\varphi(s_n, \xi) = \varphi(t, \varphi(t_n, \xi)) \to \varphi(t, \eta) \quad \text{as } n \to \infty.$$

Therefore, $\varphi(t, \eta) \in \Omega(\xi)$ for all $t \in I_\eta$. Moreover, since $\Omega(\xi)$ is a compact subset of G, we obtain from part (3) of Corollary 4.10 that $I_\eta = \mathbb{R}$.

Step 5. Finally, we show that $\varphi(t, \xi)$ approaches $\Omega(\xi)$ as $t \to \infty$. Seeking a contradiction, assume that the limit relation does not hold. Then there exists $\varepsilon > 0$ and a sequence (t_k) with $t_k \to \infty$ as $k \to \infty$ and such that

$$\text{dist}(\varphi(t_k, \xi), \Omega(\xi)) \geq \varepsilon \ \forall k \in \mathbb{N}. \tag{4.31}$$

Since $(\varphi(t_k, \xi))$ lies in the compact set $\text{cl}(\gamma^+)$, there exists a subsequence of $(\varphi(t_k, \xi))$ converging to a point η. Obviously, $\eta \in \Omega(\xi)$. On the other hand, by (4.31) and the continuity of the function $z \mapsto \text{dist}(z, \Omega(\xi))$, we conclude that $\text{dist}(\eta, \Omega(\xi)) \geq \varepsilon$, yielding the desired contradiction. $\qquad \square$

Exercise 4.20

Under the assumptions of Theorem 4.38, show that $\Omega(\xi)$ is the "smallest" closed set approached by $\varphi(t, \xi)$ as $t \to \infty$ (in the sense that if $S \subset \mathbb{R}^N$ is closed and is approached by $\varphi(t, \xi)$ as $t \to \infty$, then $\Omega(\xi) \subset S$).

4.6.3 Equilibria and periodic points

We say that $\xi \in G$ is an *equilibrium point* (or simply an *equilibrium*) of (4.25) if $f(\xi) = 0$. In the literature, equilibrium points are also called *critical points*. If ξ is an equilibrium of (4.25), then the constant function $\mathbb{R} \to G, t \mapsto \xi$

is a solution. Such solutions are said to be *equilibrium solutions.* If ξ is an equilibrium, then $O(\xi) = A(\xi) = \Omega(\xi) = \{\xi\}$. If $\zeta \in G$ is such that $\sup I_\zeta = \infty$ and $\xi := \lim_{t\to\infty} \varphi(t, \zeta)$ exists and is contained in G, then ξ is an equilibrium (see Exercise 4.4).

A point $\xi \in G$ is said to be a *periodic point* if $I_\xi = \mathbb{R}$ and there exists $p > 0$ such that

$$\varphi(t + p, \xi) = \varphi(t, \xi) \quad \forall t \in \mathbb{R}, \tag{4.32}$$

that is, the solution $\varphi(\cdot, \xi)$ is p-periodic. Any such $p > 0$ is called a *period* of ξ. Note that, if ξ is an equilibrium, then (4.32) holds for all $p > 0$ and so ξ is a periodic point. If $\xi \in G$ is a periodic point, then the associated orbit $O(\xi)$ is said to be a *periodic orbit.* A periodic point (respectively, orbit) that is not an equilibrium will be referred to as *a non-equilibrium periodic point* (respectively, *a non-equilibrium periodic orbit*).

Lemma 4.39

Let $\xi \in G$. If there exist $\sigma, \tau \in I_\xi$ such that $\sigma < \tau$, and $\varphi(\sigma, \xi) = \varphi(\tau, \xi)$ (that is, $\varphi(\cdot, \xi)$ is not injective), then $I_\xi = \mathbb{R}$ and

$$\varphi(t + \tau - \sigma, \xi) = \varphi(t, \xi) \quad \forall t \in \mathbb{R},$$

that is, ξ is a periodic point and $\tau - \sigma$ is a period of ξ.

As an immediate consequence of Lemma 4.39, we note that $\xi \in G$ is a periodic point if, and only if, there exists a positive $p \in I_\xi$ such that $\varphi(p, \xi) = \xi$.

If $\xi \in G$ is a periodic point with period $p > 0$, then $O(\xi) = \varphi(\mathbb{R}, \xi) = \varphi([0, p], \xi)$ and so ξ has compact orbit $O(\xi)$ (in Theorem 4.44 below we will show that the converse is also true: if $O(\xi)$ is compact, then ξ is a periodic point). Moreover, the α and ω-limit sets of a periodic point ξ coincide with its orbit: $A(\xi) = O(\xi) = \Omega(\xi)$.

Proof of Lemma 4.39

Assume $\sigma, \tau \in I_\xi$, $\sigma < \tau$ and $\varphi(\sigma, \xi) = \varphi(\tau, \xi)$. By part (b) of Theorem 4.35

$$I_\xi - \sigma = I_{\varphi(\sigma, \xi)} = I_{\varphi(\tau, \xi)} = I_\xi - \tau,$$

and so, since $\sigma \neq \tau$, $I_\xi = \mathbb{R}$. Again using Theorem 4.35 part (b), we have

$$\varphi(t + \tau - \sigma, \xi) = \varphi(t - \sigma, \varphi(\tau, \xi)) = \varphi(t - \sigma, \varphi(\sigma, \xi)) = \varphi(t, \xi) \quad \forall t \in \mathbb{R}.$$

Consequently, ξ is a periodic point and $\tau - \sigma$ is a period of ξ. $\qquad \square$

Implicit in the following proposition is the result that, if ξ is a periodic point that is not an equilibrium, then it has a smallest positive period, which we refer to as the *minimal period* of ξ (or of $\varphi(\cdot, \xi)$).

Proposition 4.40

Let ξ be a periodic point. Denote the set of periods of ξ by

$$P := \{p \in \mathbb{R} : p > 0, \ \varphi(p, \xi) = \xi\}$$

and write $\tilde{p} := \inf P$.

(1) $\xi \in G$ is an equilibrium if, and only if, $P = (0, \infty)$.

(2) $\xi \in G$ is not an equilibrium if, and only if, $P = \{n\tilde{p} : n \in \mathbb{N}\}$.

Proof

Since ξ is a periodic point, we have $I_\xi = \mathbb{R}$.

(1) Assume that $\xi \in G$ is an equilibrium. Then, for each $p > 0$, $\varphi(p, \xi) = \xi$ for all $t \in \mathbb{R}$ and so $P = (0, \infty)$. Now assume that $P = (0, \infty)$. Then $\varphi(p, \xi) = \xi$ for all $p > 0$. Furthermore, $\varphi(0, \xi) = \xi$ and, for each $t < 0$, on setting $p = -t \in (0, \infty)$, we have $\varphi(t, \xi) = \varphi(t + p, \xi) = \varphi(0, \xi) = \xi$. Therefore, ξ is an equilibrium.

(2) Assume that ξ is not an equilibrium. We first show that $\tilde{p} > 0$. Seeking a contradiction, suppose that $\tilde{p} = 0$. Let $\varepsilon > 0$ be arbitrary. By continuity of $\varphi(\cdot, \xi)$, there exists $\delta > 0$ such that

$$0 \leq \tau \leq \delta \implies \|\varphi(\tau, \xi) - \xi\| \leq \varepsilon.$$

By supposition, $\tilde{p} = 0$ and so there exists $p \in P$ with $0 < p < \delta$. Let $t \in \mathbb{R}$ be arbitrary and let let $m \in \mathbb{Z}$ be the unique integer such that $mp \leq t < (m+1)p$. Then $t = mp + qp$ for some $q \in [0, 1)$. Therefore,

$$\varphi(t, \xi) = \varphi(mp + qp, \xi) = \varphi(qp, \varphi(mp, \xi)) = \varphi(qp, \xi).$$

and $0 \leq qp < \delta$, whence

$$\|\varphi(t, \xi) - \xi\| = \|\varphi(qp, \xi) - \xi\| \leq \varepsilon.$$

Since the latter holds for all $t \in \mathbb{R}$ and all $\varepsilon > 0$, it follows that $\varphi(t, \xi) = \xi$ for all $t \in \mathbb{R}$, which contradicts the hypothesis that ξ is not an equilibrium. Thus, our supposition is false and so $\tilde{p} > 0$. Next, we show that $\tilde{p} \in P$. Let (p_n) be a sequence in P with $p_n \to \tilde{p}$ as $n \to \infty$. Then $\xi = \varphi(p_n, \xi)$ for all $n \in \mathbb{N}$ and so, by continuity of φ,

$$\xi = \lim_{n \to \infty} \varphi(p_n, \xi) = \varphi(\tilde{p}, \xi).$$

Therefore, $\tilde{p} \in P$. We proceed to show that $P = \{n\tilde{p} \colon n \in \mathbb{N}\}$. Suppose otherwise. Then there exist $p \in P$ and $n \in \mathbb{N}$ such that $n\tilde{p} < p < (n+1)\tilde{p}$ and so $0 < p - n\tilde{p} < \tilde{p}$. Since $p, \tilde{p} \in P$, we have $\varphi(p - n\tilde{p}, \xi) = \varphi(p, \xi) = \xi$ and so $p - n\tilde{p} \in P$, whence the contradiction $\inf P \le p - n\tilde{p} < \tilde{p} = \inf P$. Thus, we have shown that, if ξ is not an equilibrium, then necessarily $P = \{n\tilde{p} \colon n \in \mathbb{N}\}$.

Finally, we prove sufficiency by contraposition. Assume ξ is an equilibrium. Then, by the result in part (1), $P = (0, \infty) \ne \{n\tilde{p} \colon n \in \mathbb{N}\}$. Therefore, if $P = \{n\tilde{p} \colon n \in \mathbb{N}\}$, then ξ is not an equilibrium. $\qquad\square$

Corollary 4.41

Assume that ξ is a non-equilibrium periodic point. Let $\tilde{p} > 0$ be the minimal period of ξ. Then $\varphi(\cdot, \xi)$ is injective on $[0, \tilde{p})$.

Proof

Suppose that $\varphi(\cdot, \xi)$ is not injective on $[0, \tilde{p})$. Then there exist $\sigma, \tau \in [0, \tilde{p})$ such that $\sigma < \tau$ and $\varphi(\sigma, \xi) = \varphi(\tau, \xi)$ and so, by Lemma 4.39, $\tau - \sigma < \tilde{p}$ is a period of ξ, contradicting the fact that $\tilde{p}$ is the minimal period of ξ. $\qquad\square$

Recall that a *closed Jordan curve* in $\mathbb{R}^N$ is subset $\Gamma \subset \mathbb{R}^N$ such that there exists a continuous surjective function $\gamma \colon [a, b] \to \Gamma$ which is injective on $[a, b)$ and satisfies $\gamma(a) = \gamma(b)$. As an immediate consequence of Corollary 4.41, we have the following.

Corollary 4.42

If ξ is a periodic point that is not an equilibrium point, then its orbit $O(\xi)$ is a closed Jordan curve.

Example 4.43

Consider again the function f and associated local flow φ as in Exercise 4.16. If $\|\xi\| = 1$, then the initial-value problem (4.25) has periodic solution

$$\mathbb{R} \to \mathbb{R}^2, \quad t \mapsto x(t) := \varphi(t, \xi) = R(t)\xi$$

with minimal period 2π. Therefore, each ξ with $\|\xi\| = 1$ is a periodic point with orbit $O(\xi) = \{z \in \mathbb{R}^2 \colon \|z\| = 1\}$ (the unit circle in the plane centred at zero). $\qquad\triangle$

We have already observed that, if $\xi \in G$ is a periodic point, then its orbit $O(\xi)$

is compact. The next result shows that the converse is also true: if $\xi \in G$ has compact orbit $O(\xi)$, then ξ is a periodic point.

Theorem 4.44

A point $\xi \in G$ is periodic if, and only if, $O(\xi)$ is compact.

Proof

Let $\xi \in G$. We have previously shown that, if ξ is a periodic point, then $O(\xi)$ is compact. Conversely, assume that $O(\xi)$ is compact. Then, by Corollary 4.10, $I_\xi = \mathbb{R}$ and, by Theorem 4.38, $\Omega(\xi)$ is non-empty. Moreover, it is clear that $\Omega(\xi) \subset O(\xi)$. For $\eta \in \Omega(\xi)$, the invariance of $\Omega(\xi)$ (guaranteed by Theorem 4.38) implies that $O(\eta) \subset \Omega(\xi)$. Since $\eta \in O(\xi)$, we obtain from Corollary 4.36 that $O(\eta) = O(\xi)$ and thus $O(\xi) = \Omega(\xi)$. Setting

$$O_n := \{\varphi(t, \xi): \ t \in [-n, n]\} \quad \forall n \in \mathbb{N},$$

we have

$$\Omega(\xi) = O(\xi) = \bigcup_{n \in \mathbb{N}} O_n.$$

By Lemma 4.39, it is sufficient to show that $\varphi(\cdot, \xi)$ is not injective. Seeking a contradiction, suppose that $\varphi(\cdot, \xi)$ is injective. Then, $\varphi(t, \xi) \notin O_1$ for all $t > 1$. Consequently, since $\xi \in O(\xi) = \Omega(\xi)$, there exists $t_1 > 1$ such that $\varphi(t_1, \xi) \in \mathbb{B}(\xi, 1)$ and $\varphi(t_1, \xi) \notin O_1$. It follows that there exists $\varepsilon_1 \in (0, 1)$ such that

$$\overline{\mathbb{B}}(\varphi(t_1, \xi), \varepsilon_1) \subset \mathbb{B}(\xi, 1) \qquad \text{and} \qquad \overline{\mathbb{B}}(\varphi(t_1, \xi), \varepsilon_1) \cap O_1 = \emptyset.$$

Moreover, since $\varphi(t_1, \xi) \in O(\xi) = \Omega(\xi)$, there exist $t_2 > 2$ and $\varepsilon_2 \in (0, 1/2)$ such that

$$\overline{\mathbb{B}}(\varphi(t_2, \xi), \varepsilon_2) \subset \mathbb{B}(\varphi(t_1, \xi), \varepsilon_1) \qquad \text{and} \qquad \overline{\mathbb{B}}(\varphi(t_2, \xi), \varepsilon_2) \cap O_2 = \emptyset.$$

Continuing this process, we see that there exist sequences $(t_n)_{n \in \mathbb{N}}$ and $(\varepsilon_n)_{n \in \mathbb{N}}$ such that, for all $n \in \mathbb{N}$, $t_n > n$, $\varepsilon_n \in (0, 1/n)$,

$$\overline{\mathbb{B}}(\varphi(t_n, \xi), \varepsilon_n) \subset \mathbb{B}(\varphi(t_{n-1}, \xi), \varepsilon_{n-1}) \quad \text{for all integers } n \geq 2$$

and

$$\overline{\mathbb{B}}(\varphi(t_n, \xi), \varepsilon_n) \cap O_n = \emptyset \quad \forall n \in \mathbb{N}.$$

Now choose, for every $n \in \mathbb{N}$, $\eta_n \in \overline{\mathbb{B}}(\varphi(t_n, \xi), \varepsilon_n)$. Then $(\eta_n)_{n \in \mathbb{N}}$ is a Cauchy sequence and therefore has a limit η. It follows from the construction that

$$\eta \in \overline{\mathbb{B}}(\varphi(t_n, \xi), \varepsilon_n) \quad \text{and} \quad \eta \notin O_n; \quad \forall n \in \mathbb{N}.$$

Consequently, $\varphi(t_n, \xi) \to \eta$ as $n \to \infty$ and $\eta \notin \Omega(\xi) = \bigcup_{n \in \mathbb{N}} O_n$, yielding the desired contradiction. $\qquad \square$

We conclude this section with the observation that every non-empty compact and convex set in G that is positively invariant under the (local) flow φ must contain an equilibrium.

Theorem 4.45

Let $C \subset G$ be non-empty, convex and compact. If C is positively invariant under the local flow φ, then C contains an equilibrium.

Proof

Let (p_n) be a sequence in $(0, \infty)$ with $p_n \to 0$ as $n \to \infty$. By positive invariance of C, $\varphi(p_n, \xi) \in C$ for all $\xi \in C$ and all $n \in \mathbb{N}$. Therefore, for each $n \in \mathbb{N}$, we may define a continuous map $f_n \colon C \to C$ by $f_n(\xi) := \varphi(p_n, \xi)$ for all $\xi \in C$. By the Brouwer[5] fixed point theorem (Theorem A.26), it follows that there exists a sequence (ξ_n) in C with $\xi_n = f_n(\xi_n) = \varphi(p_n, \xi_n)$ for all $n \in \mathbb{N}$. Thus, for each $n \in \mathbb{N}$, ξ_n is a periodic point and so, in particular, $I_{\xi_n} = \mathbb{R}$. By compactness of C and passing to a subsequence if necessary, we may assume that (ξ_n) is convergent with limit $\xi \in C$.

We claim that ξ is an equilibrium. It is sufficient to show that $\varphi(t, \xi) = \xi$ for all $t \in I_\xi$. To this end, let $t \in I_\xi$ be arbitrary and note that

$$\|\varphi(t, \xi) - \xi\|$$
$$\leq \|\varphi(t, \xi - \varphi(t, \xi_n)\| + \|\varphi(t, \xi_n) - \varphi(p_n, \xi_n)\| + \|\varphi(p_n, \xi_n) - \xi\|. \quad (4.33)$$

By convergence of ξ_n to ξ and continuity of φ we have that

$$\|\varphi(t, \xi) - \varphi(t, \xi_n)\| \to 0 \quad \text{and} \quad \|\varphi(p_n, \xi_n) - \xi\| = \|\xi_n - \xi\| \to 0 \quad \text{as } n \to \infty.$$

It remains to show that the second term on the right of (4.33) also converges to 0 as $n \to \infty$. By p_n-periodicity of ξ_n, we have

$$\varphi(mp_n, \xi_n) = \xi_n \ \ \forall m \in \mathbb{Z}, \ \ \forall n \in \mathbb{N}.$$

For each $n \in \mathbb{N}$, there exists $m_n \in \mathbb{Z}$ such that $m_n p_n \leq t < (m_n + 1)p_n$ and so there exists $q_n \in [0, 1)$ such that $t = m_n p_n + q_n p_n$. Therefore,

$$\varphi(t, \xi_n) = \varphi(m_n p_n + q_n p_n, \xi_n) = \varphi(q_n p_n, \xi_n) \ \ \forall n \in \mathbb{N}$$

and so, as $t \to \infty$,

$$\|\varphi(t, \xi_n) - \varphi(p_n, \xi_n)\| = \|\varphi(q_n p_n, \xi_n) - \xi_n\| \to \|\varphi(0, \xi) - \xi\| = \|\xi - \xi\| = 0,$$

completing the proof. $\qquad \square$

---

[5] Luitzen Egbertus Jan Brouwer (1881-1966), Dutch.

4.7 Planar systems

Here, we restrict attention to two-dimensional systems of the form (4.25), that is, we will consider (4.25) with $f\colon G \to \mathbb{R}^2$, where $G \subset \mathbb{R}^2$ is a non-empty open set. Our main concern will be the question of existence of periodic solutions of (4.25), addressed in the spirit of Poincaré[6] and Bendixson[7]. Throughout this section, it is assumed that f is locally Lipschitz.

4.7.1 The Poincaré-Bendixson theorem

The following theorem provides a sufficient condition for the existence of a non-constant periodic solution.

Theorem 4.46 (Poincaré-Bendixson theorem)

If $\xi \in G$ is such that the closure of $O^+(\xi)$ is compact and contained in G and $\Omega(\xi)$ contains no equilibrium points, then $\Omega(\xi)$ is the orbit of a periodic point.

The proof of this theorem requires a number of preliminaries. We begin by introducing the notion of a transversal. A *transversal* of f through $\zeta \in G$ is any compact line segment L in G, containing ζ as an inner point, such that $f(z)$ is not zero and not parallel to L for all $z \in L$.

With reference to Figure 4.7, by a *line segment* in G, we mean a compact set of the form

Figure 4.7

$$L := \{(1-\mu)a + \mu b \colon 0 \le \mu \le 1\}$$

such that $L \subset G$. The points a and b are called the *endpoints* of L. In the following, we shall denote such a line segment by $[\![a, b]\!]$. By an *inner point* of $[\![a, b]\!]$ we mean a point $c = (1-\mu)a + \mu b$ with $0 < \mu < 1$, that is, a point c in $[\![a, b]\!]$ which is not equal to a or b.[8] A transversal of f

---

[6] Jules Henri Poincaré (1854-1912), French.

[7] Ivar Otto Bendixson (1861-1935), Swedish.

[8] The concept of an inner point should not be confused with that of an interior point from topology: whilst every point in $[\![a, b]\!]$ which is not an endpoint is an inner point, the line segment $[\![a, b]\!]$ (viewed as a subset of $\mathbb{R}^2$) does not have any interior points.

through $\zeta \in G$ is therefore a line segment $[\![a, b]\!]$ in G with the properties:

(i) $\zeta \in [\![a, b]\!]$ and $\zeta \neq a, b$;

(ii) $f(z) \neq 0$ for all $z \in [\![a, b]\!]$;

(iii) $|\langle f(z), b - a \rangle| < \|f(z)\| \|b - a\|$ for all $z \in [\![a, b]\!]$.

We now give an interpretation of the concept of a transversal in terms of the local flow generated by f. Assume that L is a transversal of f, let $\xi \in G$ and set $x(t) := \varphi(t, \xi)$ for all $t \in I := I_\xi$. If there exists $\tau \in I$ such that $x(\tau) \in L$, then $f(x(\tau))$ is not zero and not parallel to L, implying that the orbit $O(\xi)$ must cross L at time $x(\tau)$. Moreover, if v is a unit vector orthogonal to L and if there exist $\sigma, \tau \in I$ such that $x(\sigma), x(\tau) \in L$, then

$$\langle \dot{x}(\sigma), v \rangle \langle \dot{x}(\tau), v \rangle = \langle f(x(\sigma)), v \rangle \langle f(x(\tau)), v \rangle > 0,$$

showing that if the orbit $O(\xi)$ crosses L at $x(\sigma)$ and $x(\tau)$, then these crossings are in the same directional sense, see Figure 4.8(a); in particular, the behaviour illustrated in Figure 4.8(b) cannot occur.

(a)   (b)

Figure 4.8 Feasible (a) and infeasible (b) intersections with a transversal L

The following lemma shows that, for every point $\zeta \in G$ which is not an equilibrium, there exists a transversal through ζ.

Lemma 4.47

Let $\zeta \in G$. If $f(\zeta) \neq 0$, then there exists a transversal of f through ζ.

Proof

Let $\zeta \in G$ and assume that $f(\zeta) \neq 0$. By continuity of f and openness of G, there exists a neighbourhood $U \subset G$ of ζ such that $f(z) \neq 0$ for all $z \in U$. Choose $c, d \in U$ such that $c \neq d$, $[\![c, d]\!] \subset U$, $\zeta = (c + d)/2$ and $\langle f(\zeta), d - c \rangle = 0$ (in words: choose a line segment $[\![c, d]\!]$ in U such that ζ is at the centre of $[\![c, d]\!]$ and $f(\zeta)$ is orthogonal to $[\![c, d]\!]$). Let

$$k := \min_{z \in [\![c, d]\!]} \|f(z)\| > 0.$$

By continuity of f and since $\langle f(\zeta), d - c \rangle = 0$, there exist inner points a and b of $[\![c, \zeta]\!]$ and $[\![\zeta, d]\!]$, respectively, such that

$$|\langle f(z), d - c \rangle| < k\|d - c\| \quad \forall z \in [\![a, b]\!].$$

Obviously, $[\![a, b]\!] \subset [\![c, d]\!]$, ζ is an inner point of $[\![a, b]\!]$, $f(z) \neq 0$ for all $z \in [\![a, b]\!]$ and there exists $\varepsilon \in (0, 1)$ such that $b - a = \varepsilon(d - c)$. Multiplying both sides of the above inequality by ε leads to

$$|\langle f(z), b - a \rangle| < k\|b - a\| \leq \|f(z)\|\|b - a\| \quad \forall z \in [\![a, b]\!].$$

Therefore, $[\![a, b]\!]$ is a transversal of f through ζ.   □

A *compact arc* on $O(\xi)$ is a set of the form $\{\varphi(t, \xi) \colon t_1 \leq t \leq t_2\}$, where $t_1, t_2 \in I_\xi$ with $t_1 < t_2$. The following lemma describes properties of the intersection of an orbit and a transversal.

Lemma 4.48

Let $\xi \in G$, let A be a compact arc on $O(\xi)$ and let L be a transversal of f. The following statements hold.

(1) The intersection $A \cap L$ is finite.

(2) If ξ is not a periodic point, then the intersection $A \cap L$ has the following monotonicity property: if $t_1, t_2, t_3 \in I_\xi$ are such that $t_1 < t_2 < t_3$ and $x_i := \varphi(t_i, \xi) \in A \cap L$ for $i = 1, 2, 3$, then x_2 is an inner point of $[\![x_1, x_3]\!]$.

(3) If ξ is a periodic point, then the intersection $O(\xi) \cap L$ contains at most one point.

The proof of this lemma relies on the following famous result.

Theorem 4.49 (Jordan curve theorem)

A closed Jordan curve Γ in $\mathbb{R}^2$ separates $\mathbb{R}^2$ into two connected sets in the sense that the open set $\mathbb{R}^2 \backslash \Gamma$ is the union of two non-empty, open, connected and disjoint sets: a bounded set, the interior of Γ, and an unbounded set, the exterior of Γ, and Γ is the boundary of each of the two sets.

Whilst the Jordan curve theorem appears obvious, it is difficult to prove.[9]

---

[9] See, for example, the book by R.B. Burckel, *An Introduction to Classical Complex Analysis, Volume 1*, Academic Press, New York, 1979, and the paper by R. Maehara, "The Jordan curve theorem via the Brouwer fixed point theorem", *The American Mathematical Monthly*, vol. 91, 1984, pp. 641-643.

Proof of Lemma 4.48

Set $x(t) := \varphi(t, \xi)$ for all $t \in I := I_\xi$. Then A can be written in form $A = \{x(t) \colon \sigma \le t \le \tau\}$ for some $\sigma, \tau \in I$ with $\sigma < \tau$.

We first prove statement (1). Seeking a contradiction, suppose that $A \cap L$ is infinite. Then there exist numbers s_n such that $\sigma \le s_n \le \tau$ and $x(s_n) \in A \cap L$ for all $n \in \mathbb{N}$ and $x(s_n) \ne x(s_m)$ if $n \ne m$. By the Bolzano-Weierstrass theorem (Theorem A.16), it follows that the sequence (s_n) has a subsequence – which we do not relabel – converging to a limit $s \in [\sigma, \tau]$. Consequently, $x(s_n) \to x(s)$ as $n \to \infty$ and thus,

$$\lim_{n \to \infty} \frac{1}{s_n - s} \big(x(s_n) - x(s) \big) = \dot{x}(s) = f(x(s)). \tag{4.34}$$

Since $x(s_n) \in L$ for all $n \in \mathbb{N}$ and L is compact, we have $x(s) \in L$. It follows that, $x(s_n) - x(s)$ is parallel to L for all $n \in \mathbb{N}$ and hence $f(x(s))$ is parallel to L, yielding the desired contradiction.

We proceed to prove statement (2). Assume that $\xi \in G$ is not periodic and let $t_1 < t_2 < t_3$ be points of I such that $x^i = x(t_i) \in A \cap L$ for $i = 1, 2, 3$. We consider two scenarios.

Scenario 1. We first assume $x(t) \notin L$ for all $t \in (t_1, t_2)$. Setting $\Gamma_0 := \{x(t) \colon t_1 \le t \le t_2\}$ (the arc on $O(\xi)$ from x^1 to x^2), it is then clear that $\Gamma := \Gamma_0 \cup [\![x^1, x^2]\!]$ is a closed Jordan curve. By the Jordan curve theorem (Theorem 4.49), Γ separates $\mathbb{R}^2$ into two connected parts: the interior and the exterior of Γ. For $t_0 < t_1$ sufficiently close to t_1 and $t_4 > t_2$ sufficiently close to t_2, the points $x^0 := x(t_0)$ and $x^4 := x(t_4)$ are on opposite sides of Γ. With reference to Figure 4.9, there are two possible cases: x^4 is in the interior of Γ or x^4 is in the exterior of Γ. Assume that the first case holds (the other can be treated

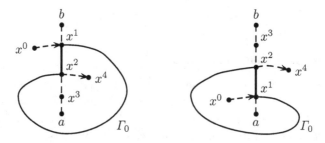

Figure 4.9 Two possible cases of the Jordan curve $\Gamma = \Gamma_0 \cup [\![x^1, x^2]\!]$

similarly). Then, for any $t \in I \cap (t_2, \infty)$, $x(t)$ cannot cross Γ_0, by uniqueness, and it cannot cross $[\![x^1, x^2]\!]$, since any crossing of $[\![x^1, x^2]\!]$ must be from the exterior to the interior of Γ. Consequently, $x(t)$ remains in the interior of Γ for all $t \in I \cap (t_2, \infty)$, implying that x^3 is in the interior of Γ and thus, $x^2 \in [\![x^1, x^3]\!]$. Since, by hypothesis, x is not periodic, it follows that x^2 is an inner point of $[\![x^1, x^3]\!]$.

Scenario 2. Now assume that $x(t) \in L$ for some $t \in (t_1, t_2)$. By the result in part (1), there are only finitely many such points, which we label and order as follows: $t_1 < \tau_k < \cdots < \tau_1 < t_2$. Write $z^i := \varphi(\tau_i, \xi)$, $i = 1, \ldots, k$. For notational convenience, we also write $\tau_{k+1} := t_1$, $\tau_0 := t_2$, $z^{k+1} := x^1$ and $z^0 := x^2$. Applied in the context of the triple $\tau_1 < t_2 < t_3$, the result in Scenario 1 implies that $x^2 = z^0$ is an inner point of $[\![z^1, x^3]\!]$. For $i = 1, \ldots, k$, the same result applied in the context of the triple $\tau_{i+1} < \tau_i < \tau_{i-1}$ implies that z^i is an inner point of $[\![z^{i+1}, z^{i-1}]\!]$. It immediately follows that x^2 is an inner point of $[\![z^{i+1}, x^3]\!]$ for $i = 0, \ldots, k$. In particular, x^2 is an inner point of $[\![z^{k+1}, x^3]\!] = [\![x^1, x^3]\!]$.

Finally, to prove statement (3), assume that ξ is a periodic point. Seeking a contradiction, suppose that $O(\xi)$ intersects L at more than one point. Then there exist $t_1, t_2 \in I = \mathbb{R}$ such that $t_1 < t_2$, $x^1 := x(t_1) \in L$, $x^2 := x(t_2) \in L$, $x(t) \notin L$ for all $t \in (t_1, t_2)$ and $x^1 \neq x^2$. Let Γ, Γ_0, t_0, t_4, x^0 and x^4 be as before and consider Figure 4.9. The argument used in the proof of statement (2) shows that $x(t)$ cannot cross Γ for any $t > t_2$. On the other hand, by periodicity, there exists $s > t_2$ such that $x(s) = x^0$ and consequently, $x(t)$ must cross Γ for some $t > t_2$, yielding a contradiction.   □

The next lemma is not surprising: it shows that if η is "sufficiently close" to a transversal L, then there exists t with "small" modulus $|t|$ such that $\varphi(t, \eta) \in L$.

Lemma 4.50

If L is a transversal of f through $\zeta \in G$, then, for each $\varepsilon > 0$, there exists an open neighbourhood $U \subset G$ of ζ such that, for all $\eta \in U$, there exists $t \in (-\varepsilon, \varepsilon)$ such that $\varphi(t, \eta) \in L$.

Proof

Let $L := [\![a, b]\!]$ be a transversal through $\zeta \in G$, let v be a unit vector orthogonal to L and define $k := \langle \zeta, v \rangle$. Defining the line

$$L^e := \{z \in \mathbb{R}^2 : \langle z, v \rangle - k = 0\} = \{a + \lambda(b - a) : \lambda \in \mathbb{R}\},$$

it is clear that $L \subset L^e$. Obviously, by choosing a sufficiently small open neighbourhood W of ζ, we have that

$$W \cap L = W \cap L^e. \qquad (4.35)$$

Let $\varepsilon > 0$ be given. Since $\varphi(0,\zeta) = \zeta$ and φ is continuous (by Theorem 4.34), there exist an open interval $T \subset (-\varepsilon,\varepsilon)$ containing 0 and an open neighbourhood $V \subset \mathbb{R}^2$ of ζ such that $T \times V$ is contained in the domain of φ and

$$\varphi(t,\eta) \in W \quad \forall (t,\eta) \in T \times V. \tag{4.36}$$

The function $\theta \colon T \times V \to \mathbb{R}$, $(t,\eta) \mapsto \langle \varphi(t,\eta), v \rangle - k$ is continuous and the partial derivative $\partial_1 \theta$ with respect to its first argument exists and is continuous. Moreover, $\theta(0,\zeta) = 0$ and $\partial_1 \theta(0,\zeta) = \langle f(\zeta), v \rangle \neq 0$. By the implicit function theorem (Theorem A.35), there exist an open neighbourhood $U \subset V$ of ζ and a continuous function $\tau \colon U \to T$ such that $\tau(\zeta) = 0$ and $\theta(\tau(\eta),\eta) = 0$ for all $\eta \in U$. Therefore, for each $\eta \in U$, there exists $t = \tau(\eta) \in T$ such that $\varphi(t,\eta) \in L^e$. Combining this with (4.35) and (4.36), shows that $\varphi(t,\eta) \in L$, completing the proof. $\qquad\square$

Before we come to the proof of the Poincaré-Bendixson theorem, we establish the following result on the intersection of a transversal and an ω-limit set.

Lemma 4.51

Assume $\xi \in G$ is such that the closure of $O^+(\xi)$ is compact and contained in G. If L is a transversal of f, then $\Omega(\xi) \cap L$ has at most one element.

Proof

If ξ is a periodic point, then $O(\xi) = \Omega(\xi)$ and the claim follows from statement (3) of Lemma 4.48. Assume now that ξ is not a periodic point. By the assumption on $O^+(\xi)$, it follows that $[0,\infty) \subset I_\xi$ (by Corollary 4.10). Set $x(t) := \varphi(t,\xi)$ for all $t \geq 0$. Assume that $\zeta \in G$ is such that $f(\zeta) \neq 0$ and let L be a transversal through ζ (L exists by virtue of Lemma 4.47). Seeking a contradiction, suppose that there exist two distinct points $z_1, z_2 \in \Omega(\xi) \cap L$. Then $f(z_1) \neq 0$ and $f(z_2) \neq 0$, and so (by extending L if necessary), we may assume that z_1 and z_2 are inner points of L. Obviously, there exist line segments $L_1, L_2 \subset L$ such that $L_1 \cap L_2 = \emptyset$ and z_1 and z_2 are inner points of L_1 and L_2, respectively. In particular, L_1 and L_2 are transversals of f through z_1 and z_2, respectively. By Lemma 4.50 there exist open neighbourhoods U_1 and U_2 of z_1 and z_2, respectively, such that for every $\eta_1 \in U_1$ and every $\eta_2 \in U_2$, there exist $\tau_1, \tau_2 \in (-1,1)$ such that $\varphi(\tau_1,\eta_1) \in L_1$ and $\varphi(\tau_2,\eta_2) \in L_2$.

Since z_1 and z_2 are in $\Omega(\xi)$, there exist $s_1 < s_2 < s_3$ such that $s_1 > 1$, $s_2 - s_1 > 2$, $s_3 - s_2 > 2$, $x(s_1) \in U_1$, $x(s_3) \in U_1$ and $x(s_2) \in U_2$. Consequently, there exist $\tau_1, \tau_2, \tau_3 \in (-1,1)$ such that

$$\varphi(\tau_1, x(s_1)),\ \varphi(\tau_3, x(s_3)) \in L_1 \qquad \text{and} \qquad \varphi(\tau_2, x(s_2)) \in L_2.$$

Setting $t_i := s_i + \tau_i$, we have $x(t_i) = \varphi(s_i + \tau_i, \xi) = \varphi(\tau_i, x(s_i))$ for $i = 1, 2, 3$. Therefore, $x(t_1), x(t_3) \in L_1$ and $x(t_2) \in L_2$. Now L_1 and L_2 are disjoint, and thus $x(t_2) \notin [\![x(t_1), x(t_3)]\!]$. Since, by construction, $t_1 < t_2 < t_3$, it follows that the monotonicity property of statement (2) of Lemma 4.48 is violated, yielding the desired contradiction. Therefore, our supposition is false and so $\Omega(\xi) \cap L$ has at most one element. $\qquad\square$

We are now in a position to prove the Poincaré-Bendixson theorem.

Proof of Theorem 4.46

Invoking Theorem 4.38, it follows in particular that $\Omega(\xi)$ is non-empty, compact, connected and invariant. Let $\eta \in \Omega(\xi)$. Then $x(t) := \varphi(t, \eta) \in \Omega(\xi)$ for all $t \in \mathbb{R}$ and $\Omega(\eta) \subset \Omega(\xi)$. Since $\Omega(\xi)$ contains no equilibrium points, $\Omega(\eta)$ contains no equilibrium points. Let $\zeta \in \Omega(\eta)$ be arbitrary. Then $f(\zeta) \neq 0$ and so, by Lemma 4.47, there exists a transversal L through ζ. By Lemma 4.50, there exists an open neighbourhood U of ζ such that, for all $z \in U$, there exists $\tau \in (-1, 1)$ such that $\varphi(\tau, z) \in L$. Since $\zeta \in \Omega(\eta)$, there exist positive s_1 and s_2 such that $s_2 > s_1 + 2$ and $x(s_1), x(s_2) \in U$. Consequently, there exist $\tau_1, \tau_2 \in (-1, 1)$ such that $\varphi(\tau_1, x(s_1)), \varphi(\tau_2, x(s_2)) \in L$ and thus,

$$x(s_i + \tau_i) = \varphi(s_i + \tau_i, \eta) = \varphi(\tau_i, \varphi(s_i, \eta)) = \varphi(\tau_i, x(s_i)) \in \Omega(\xi) \cap L, \quad i = 1, 2.$$

By Lemma 4.51, it follows that $x(s_1 + \tau_1) = x(s_2 + \tau_2)$. By construction, $s_1 + \tau_1 \neq s_2 + \tau_2$, implying that x is periodic. Thus, $O(\eta)$ is a periodic orbit contained in $\Omega(\xi)$.

It remains only to show that $\Omega(\xi) = O(\eta)$. Seeking a contradiction, suppose that $\Omega(\xi) \backslash O(\eta) \neq \emptyset$. By connectedness of $\Omega(\xi)$ and closedness of $O(\eta)$, it follows that $\Omega(\xi) \backslash O(\eta)$ is not closed. Consequently, there exists a sequence (z_n) in $\Omega(\xi) \backslash O(\eta)$ which converges to a limit $z \in O(\eta)$ as $n \to \infty$. Let Z be a transversal through z. By Lemma 4.50, there exist $k \in \mathbb{N}$ (sufficiently large) and $t \in \mathbb{R}$ such that $\varphi(t, z_k) \in Z$. Now z and $\varphi(t, z_k)$ are in $\Omega(\xi) \cap Z$ and thus, by Lemma 4.51, $z = \varphi(t, z_k)$. Hence, $z_k = \varphi(-t, z) \in O(\eta)$, contradicting the fact that $z_k \in \Omega(\xi) \backslash O(\eta)$. $\qquad\square$

In applications, the Poincaré-Bendixson theorem is frequently used to prove existence of non-equilibrium periodic points via the following argument: assume that $\mathcal{A} \subset G$ is a compact domain (typically annular, see Figure 1.2) with the property that the vector $f(z)$ is not directed outward at any point z of the boundary of $\mathcal{A}$. Then, every $\xi \in \mathcal{A}$ has semi-orbit $O^+(\xi)$ in $\mathcal{A}$. By compactness of $\mathcal{A}$, it follows that $O^+(\xi)$ is bounded and $\Omega(\xi)$ is a non-empty subset of $\mathcal{A}$.

If $\mathcal{A}$ contains no equilibrium points, then $\Omega(\xi)$ contains no equilibrium points and so, by Theorem 4.46, $\Omega(\xi)$ must be the orbit of a periodic point.

Example 4.52

The above methodology applies in the case of the "twin-tunnel diode" example from circuit theory given in Section 1.1.1, thereby confirming the existence of at least one non-constant periodic solution. $\triangle$

Exercise 4.21

Consider the system $\dot{x}(t) = f(x(t))$ with $f\colon \mathbb{R}^2 \to \mathbb{R}^2$ given by

$$f(z) = f(z_1, z_2) := \big(z_1 g(z_1, z_2) + z_2 \,,\, z_2 g(z_1, z_2) - z_1\big),$$

where $g(z_1, z_2) := 3 + 2z_1 - z_1^2 - z_2^2$. Prove that there exists at least one non-constant periodic solution.

Exercise 4.22

A second-order scalar differential equation of the form

$$\ddot{y}(t) + d(y(t))\dot{y}(t) + k(y(t)) = 0 \tag{4.37}$$

is generally referred to as Liénard's[10] equation. In a mechanical context, $d(y)\dot{y}$ represents a friction term that is linear in the velocity and $k(y)$ models a restoring force.

Consider the particular case wherein k is the identity function, that is $k(u) = u$ for all $u \in \mathbb{R}$. Furthermore, assume that the function $d\colon \mathbb{R} \to \mathbb{R}$ is locally Lipschitz and even $(d(u) = d(-u)$ for all $u \in \mathbb{R})$, and has the properties $d(0) < 0$ and $D(u) := \int_0^u d(v)\mathrm{d}v \to \infty$ as $v \to \infty$.

(a) Deduce that the function $D\colon \mathbb{R} \to \mathbb{R}$ is odd $(D(u) = -D(-u)$ for all $u \in \mathbb{R})$ and there exist constants $b \geq a > 0$ such that $D(a) = 0 = D(b)$, $D(u) < 0$ for all $0 < u < a$, and $D(u) > 0$ for all $u > b$ (see Figure 4.10).

(b) Writing $x_1(t) = y(t)$, $x_2(t) = \dot{y}(t) + D(y(t))$ and $x(t) = \big(x_1(t), x_2(t)\big)$, show that Liénard's equation (with $k(u) = u$ for all $u \in \mathbb{R}$) may be rewritten as

$$\dot{x}_1(t) = x_2(t) - D(x_1(t)), \quad \dot{x}_2(t) = -x_1(t). \tag{4.38}$$

(c) The system (4.38) is of the form $\dot{x}(t) = f(x(t))$ with $f\colon \mathbb{R}^2 \to \mathbb{R}^2$ given by

$$f(z) = f(z_1, z_2) := \big(z_2 - D(z_1), -z_1\big) \quad \forall\, (z_1, z_2) \in \mathbb{R}^2.$$

---

[10] Alfred-Marie Liénard (1869-1958), French.

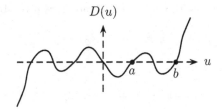

Figure 4.10 Graph of typical function D

Let φ denote the local flow generated by f. By investigating the nature of the function f on the circle of radius a centred at $0 \in \mathbb{R}^2$, deduce that, if $\xi \in \mathbb{R}^2$ is such that $\|\xi\| \geq a$, then $\|\varphi(t, \xi)\| \geq a$ for all $t \in I_\xi \cap \mathbb{R}_+$. In other words, the exterior of the open disc of radius a centred at 0 is positively invariant under the local flow.

(d) Define $m := \max\{|D(u)| : 0 \leq u \leq b\}$. Since $D(u) \to \infty$ as $u \to \infty$

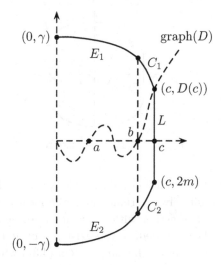

Figure 4.11 The curve $\Gamma = E_1 \cup C_1 \cup L \cup C_2 \cup E_2$

and $D(b) = 0$, there exists $c > 0$ such that $D(c) = \sqrt{4m^2 + 3b^2}/2$. Consider the curve Γ, shown in Figure 4.11, comprising the union of the line segment L joining $(c, D(c))$ and $(c, 2m)$, the circular arc (with centre 0 and radius r_1 given by $r_1^2 = c^2 + D^2(c)$)

$$C_1 := \{(z_1, z_2) \in \mathbb{R}^2 : z_1^2 + z_2^2 = r_1^2, \ b \leq z_1 \leq c, \ z_2 > 0\},$$

the circular arc (with centre 0 and radius r_2 given by $r_2^2 = c^2 + 4m^2$)

$$C_2 := \{(z_1, z_2) \in \mathbb{R}^2 : z_1^2 + z_2^2 = r_2^2, \ b \le z_1 \le c, \ z_2 < 0\},$$

the ellipsoidal arc

$$E_1 := \{(z_1, z_2) \in \mathbb{R}^2 : z_1^2 + 2z_2^2 = 2r_1^2 - b^2, \ 0 \le z_1 \le b, \ z_2 > 0\},$$

and the ellipsoidal arc

$$E_2 := \{(z_1, z_2) \in \mathbb{R}^2 : 2z_1^2 + z_2^2 = r_2^2 + b^2, \ 0 \le z_1 \le b, \ z_2 < 0\}.$$

Defining $\gamma := \sqrt{b^2 + c^2 + 4m^2}$, verify that $(0, \gamma)$ and $(0, -\gamma)$ are the end points of the curve Γ, as indicated in Figure 4.11.

(e) Now consider the closed Jordan curve Γ^* comprising Γ and its rotation through $180°$ about the origin. The curve Γ^* forms the outer boundary of the annular region $\mathcal{A}$ shown in Figure 4.12, wherein the inner boundary is the circle of radius a centred at 0. By investigating

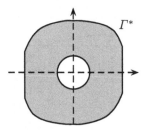

Figure 4.12   Annular region $\mathcal{A}$

the nature of the function f on Γ^*, deduce that the union of Γ^* and its interior is positively invariant under the local flow φ. Furthermore, by using this fact, together with the result in (c) above and the Poincaré-Bendixson theorem, deduce the existence of at least one non-constant periodic solution of the Liénard system (4.38) with orbit in $\mathcal{A}$.

(f) Show that there exists $v \in [-\gamma, -a] \cup [a, \gamma]$ such that the solution of (4.37) satisfying $y(0) = 0$ and $\dot{y}(0) = v$ is periodic.

Example 4.53

Consider again the "twin-tunnel diode" example from circuit theory given in Section 1.1.1, that is, the system

$$\dot{x}_1(t) = x_2(t), \quad \dot{x}_2(t) = -x_1(t) + x_2(t) - x_2^3(t).$$

Writing $y(t) = -x_2(t)$, we have

$$\ddot{y}(t) = -\ddot{x}_2(t) = \dot{x}_1(t) - (1 - 3x_2^2(t))\dot{x}_2(t) = x_2(t) - (1 - 3x_2^2(t))\dot{x}_2(t)$$
$$= -y(t) - (3y^2(t) - 1)\dot{y}(t).$$

Therefore, the example is, in fact, a particular case (a so-called van der Pol[11] equation) of Liénard's equation, to which the result in Exercise 4.22 applies. Thus, we may conclude the existence of a non-equilibrium periodic orbit (a conclusion also arrived at - via a different construction - in Example 4.52). △

4.7.2 First integrals and periodic orbits

Recall that a first integral for the planar system under consideration, namely,

$$\dot{x}(t) = f(x(t)), \quad f \colon G \subset \mathbb{R}^2 \to \mathbb{R}^2 \text{ locally Lipschitz} \qquad (4.39)$$

is a non-constant continuously differentiable function $E \colon G \to \mathbb{R}$ such that $\langle (\nabla E)(z), f(z) \rangle = 0$ for all $z \in G$ (see Exercise 1.2).

Proposition 4.54

Assume that (4.39) has precisely one critical point $\zeta \in G$ and let E be a first integral. If $E(\zeta) < E(\xi)$ for all $\xi \in G \backslash \{\zeta\}$ and every non-empty level set $E^{-1}(\alpha) := \{z \in G \colon E(z) = \alpha\}$ with $\alpha > E(\zeta)$ is a closed Jordan curve, then, for every $\xi \in G \backslash \{\zeta\}$, $O(\xi) = E^{-1}(E(\xi))$ and $O(\xi)$ is a periodic orbit.

Proof

Let $\xi \in G \backslash \{\zeta\}$. Set $\alpha := E(\xi)$. Then, by hypothesis, $E^{-1}(\alpha)$ is a closed Jordan curve. Since E is a first integral, we have $\varphi(t, \xi) \in E^{-1}(\alpha)$ for all $t \in I_\xi$. Since $E^{-1}(\alpha)$ is compact, it follows that $I_\xi = \mathbb{R}$ (by Corollary 4.10) and $\Omega(\xi) \subset \overline{O(\xi)} \subset E^{-1}(\alpha)$. Since $E^{-1}(\alpha)$ contains no critical points, it follows from the Poincaré-Bendixson theorem (Theorem 4.46) that $\Omega(\xi)$ is the orbit of a non-constant periodic solution and so is a closed Jordan curve. Therefore, both $E^{-1}(\alpha)$ and $\Omega(\xi)$ are closed Jordan curves, and so, since $\Omega(\xi) \subset E^{-1}(\alpha)$, we may conclude[12] that $\Omega(\xi) = E^{-1}(\alpha)$, whence $\overline{O(\xi)} = \Omega(\xi)$. Consequently, $O(\xi) \subset \Omega(\xi)$ and, since $\Omega(\xi)$ is an orbit, it follows from Corollary 4.36 that $O(\xi) = \Omega(\xi)$. Therefore, $O(\xi)$ is a periodic orbit. □

---

[11] Balthasar van der Pol (1889-1959), Dutch.

[12] Here, we are using the fact that, if Γ_1 and Γ_2 are closed Jordan curves with $\Gamma_1 \subset \Gamma_2$, then $\Gamma_1 = \Gamma_2$. From an intuitive viewpoint, this fact is unsurprising. For this reason and for brevity, we do not give a proof (which is not entirely straightforward).

Example 4.55

Consider the Lotka–Volterra predator-prey system (as given in Section 1.1.5)

$$\dot{p}(t) = p(t)\big(- a + bq(t)\big), \quad \dot{q}(t) = q(t)\big(c - dp(t)\big)$$

where a, b, c, d are positive constants, $p(t) \in (0, \infty)$ denotes the prey population at time t and $q(t) \in (0, \infty)$ denotes the predator population at time t. Setting $G = (0, \infty) \times (0, \infty)$, introducing the function

$$f \colon G \to \mathbb{R}^2, \quad z = (z_1, z_2) \mapsto f(z) := \big(- az_1 + bz_1 z_2 \, , \, cz_2 - dz_1 z_2 \big)$$

and writing $x(t) = (p(t), q(t))$, the system takes the form $\dot{x}(t) = f(x(t))$. This system has precisely one critical point $\zeta = (\zeta_1, \zeta_2) := (c/d, a/b)$.

We will show, via Proposition 4.54, that every non-equilibrium orbit is periodic. It is readily verified that $E \colon G \to \mathbb{R}$ given by $E(z) = E(z_1, z_2) := d\, z_1 - c \ln z_1 + b\, z_2 - a \ln z_2$ is a first integral (see Exercise 1.2). Introduce functions $E_1, E_2 \colon (0, \infty) \to \mathbb{R}$ given by $E_1(s) := d\, s - c \ln s$ and $E_2 := bs - a \ln s$. Observe that

$$E(z) = E(z_1, z_2) = E_1(z_1) + E_2(z_2) \ \forall \, (z_1, z_2) = z \in G,$$
$$E_i'(z_i) < 0 \ \forall \, z_i \in (0, \zeta_i), \quad E_i'(\zeta_i) = 0, \quad E_i'(z_i) > 0 \ \forall \, z_i \in (\zeta_i, \infty), \quad i = 1, 2,$$
$$E_i(z_i) \to \infty \text{ as } z_i \to 0 \quad \text{and} \quad E_i(z_i) \to \infty \text{ as } z_i \to \infty, \quad i = 1, 2.$$

Note that $\min_{z_i > 0} E_i(z_i) = E_i(\zeta_i) =: \lambda_i, \ i = 1, 2$. Thus, the first integral E has the following properties:

$$E(z) > E(\zeta) = \lambda_1 + \lambda_2 =: \mu \ \forall z \in G \backslash \{\zeta\} \quad \text{and} \quad \sup_{z \in G} E(z) = \infty.$$

Thus, $E^{-1}(\alpha) \neq \emptyset$ if, and only if, $\alpha \in [\mu, \infty)$. We proceed to show that $E^{-1}(\alpha)$ is a closed Jordan curve for all $\alpha > \mu$. Let $\alpha \in (\mu, \infty)$. We will construct a continuous function $\gamma \colon [0, 1] \to \mathbb{R}^2$ such that $\gamma(0) = \gamma(1)$, γ is injective on $[0, 1)$ and $E^{-1}(\alpha) = \{\gamma(t) \colon t \in [0, 1]\}$. To this end, consider the equation $\alpha = E(z) = E(z_1, z_2) = E_1(z_1) + E_2(z_2)$ or, equivalently,

$$E_2(z_2) = \alpha - E_1(z_1) =: F(z_1).$$

By properties of E_1, we have $F'(\zeta_1) = 0$, $F'(z_1) > 0$ for all $z_1 \in (0, \zeta_1)$, $F'(z_1) < 0$ for all $z_1 \in (\zeta_1, \infty)$, $F(z_1) \to -\infty$ as $z_1 \to 0$, and $F(z_1) \to -\infty$ as $z_1 \to \infty$. In particular,

$$\max_{z_1 > 0} F(z_1) = F(\zeta_1) > \mu - E_1(\zeta_1) = E(\zeta) - E_1(\zeta_1) = E_2(\zeta_2) = \lambda_2 \, .$$

We may now infer the existence of $u, v \in (0, \infty)$, with $u < v$, such that

$$F(u) = \lambda_2 = F(v), \ F(z_1) > \lambda_2 \ \forall \, z_1 \in (u, v), \ F(z_1) < \lambda_2 \ \forall \, z_1 \in (0, u) \cup (v, \infty).$$

By properties of E_2 we see that E_2 maps the two intervals $(0, \zeta_2]$ and $[\zeta_2, \infty)$ bijectively to $[\lambda_2, \infty)$. We denote the respective inverses by

$$P\colon [\lambda_2, \infty) \to (0, \zeta_2] \quad \text{and} \quad Q\colon [\lambda_2, \infty) \to [\zeta_2, \infty).$$

Since $(z_1, z_2) \in E^{-1}(\alpha)$ if, and only if, $F(z_1) = E_2(z_2)$ and with reference to the schematic in Figure 4.13, it now follows that

$$E^{-1}(\alpha) = \{(z_1, z_2) \in G \colon z_1 \in [u, v] \text{ and } z_2 = P(F(z_1)) \text{ or } z_2 = Q(F(z_1))\}.$$

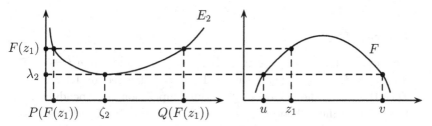

Figure 4.13 Solving the equation $F(z_1) = E_2(z_2)$.

Defining the continuous functions

$$\gamma_1 \colon [u, v] \to (0, \zeta_2], \ z_1 \mapsto P(F(z_1)), \quad \gamma_2 \colon [u, v] \to [\zeta_2, \infty), \ z_1 \mapsto Q(F(z_1)),$$

we have $E^{-1}(\alpha) = \{(z_1, \gamma_1(z_1))\colon u \le z_1 \le v\} \cup \{(z_1, \gamma_2(z_1))\colon u \le z_1 \le v\}$. Note that $\gamma_1(u) = \gamma_1(v) = \gamma_2(u) = \gamma_2(v) = \zeta_2$. Introducing

$$\gamma_0 \colon [0, 1] \to [u, v], \ t \mapsto \gamma_0(t) := \begin{cases} u + 2(v - u)t, & 0 \le t \le 1/2 \\ u + 2(v - u)(1 - t), & 1/2 < t \le 1 \end{cases}$$

and defining the continuous function $\gamma \colon [0, 1] \to \mathbb{R}^2$ by

$$\gamma(t) := \begin{cases} (\gamma_0(t), \gamma_1(\gamma_0(t))), & 0 \le t \le 1/2 \\ (\gamma_0(t), \gamma_2(\gamma_0(t))), & 1/2 < t \le 1 \end{cases}$$

we may conclude that $E^{-1}(\alpha) = \{\gamma(t)\colon t \in [0, 1]\}$. Moreover,

$$\gamma(0) = (u, \gamma_1(u)) = (u, \gamma_2(u)) = \gamma(1)$$

and, since γ_0 is evidently injective on $[0, 1)$, it follows that γ is injective on $[0, 1)$. Therefore, $E^{-1}(\alpha)$ is a closed Jordan curve. By Proposition 4.54, it follows that, for each $\xi \in G \backslash \{\zeta\}$, $O(\xi) = E^{-1}(E(\xi))$ and $O(\xi)$ is a periodic orbit. For illustrative parameter values $a = b = c = d = 1$, Figure 4.14 depicts the equilibrium at $(1, 1)$ and typical periodic orbits. $\triangle$

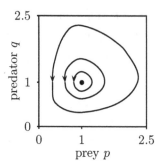

Figure 4.14 The equilibrium and periodic orbits of the predator-prey system.

Exercise 4.23

Define $G := \{(z = (z_1, z_2) \in \mathbb{R}^2 : z_2^2 - 2\cos z_1 < 2\}$ and define $f\colon G \to \mathbb{R}^2$
by $f(z) = f(z_1, z_2) := (z_2, -\sin z_1)$. Then, recalling Section 1.1.2, $\dot{x}(t) =$
$f(x(t))$ governs the behaviour of an undamped nonlinear pendulum on
the open neighbourhood G of the equilibrium $0 \in \mathbb{R}^2$. Use Proposition
4.54 to deduce that, for each $\xi \in G\backslash\{0\}$, $O(\xi)$ is a periodic orbit.

4.7.3 Limit cycles

For the planar system (4.39) under consideration (with associated local flow
φ), an ω-*limit cycle* is a non-equilibrium periodic orbit γ with the property
that, for some $\xi \in G$ with $\xi \notin \gamma$, we have

$$\mathbb{R}_+ \subset I_\xi \quad \text{and} \quad \operatorname{dist}(\varphi(t, \xi), \gamma) \to 0 \text{ as } t \to \infty. \tag{4.40}$$

In words, an ω-limit cycle is a non-equilibrium periodic solution of (4.39) that
is approached, as $t \to \infty$, by some other solution of (4.39). By compactness of γ
we see that, if $\xi \in G\backslash\gamma$ and (4.40) holds, then the positive semi-orbit $O^+(\xi)$ has
compact closure contained in G and so, by Theorem 4.38 and Exercise 4.20, the
ω-limit set $\Omega(\xi)$ is non-empty, compact, invariant under φ and is the smallest
closed set approached by $\varphi(t, \xi)$ as $t \to \infty$. It follows that $\Omega(\xi) \subset \gamma$. Moreover,
since $\gamma = O(\zeta)$ for every $\zeta \in \gamma$, choosing $\zeta \in \Omega(\xi)$ and invoking the invariance
of $\Omega(\xi)$, we may infer that $\gamma \subset \Omega(\xi)$. Therefore, $\gamma = \Omega(\xi)$ and so we arrive at
an equivalent characterization of an ω-limit cycle as a non-equilibrium periodic
orbit γ that coincides with the ω-limit set $\Omega(\xi)$ of some $\xi \in G\backslash\gamma$ with $\mathbb{R}_+ \subset I_\xi$.

Whilst we will consider only ω-limit cycles there is, of course, a comple-
mentary notion of an α-*limit cycle*, namely, a non-equilibrium periodic or-
bit γ with the property that, for some $\xi \in G\backslash\gamma$, we have $(-\infty, 0] \subset I_\xi$ and

$\text{dist}(\varphi(t, \xi), \gamma) \to 0$ as $t \to -\infty$. This notion has an equivalent characterization: an α-limit cycle is a non-equilibrium periodic orbit γ that coincides with the α-limit set $A(\xi)$ of some $\xi \in G \backslash \gamma$. We also remark that the above concepts of α and ω-limit cycles – presented in the context of planar systems – extend to higher-dimensional systems.

With reference to Example 4.55, we know that every non-equilibrium solution of the Lotka-Volterra predator-prey system is periodic. Note, however, this system has no ω-limit cycles. The next result identifies conditions on (4.39) which guarantee the existence of an ω-limit cycle.

Proposition 4.56

Let $C \subset G$ be a compact set that is positively invariant under the local flow φ generated by (4.39). Assume that 0 is an interior point of C and is the only equilibrium in C. Assume further that f is differentiable at 0. Write $A := (Df)(0)$ with spectrum $\sigma(A) = \{\lambda_1, \lambda_2\}$. If $\text{Re}\,\lambda_i > 0$ for $i = 1, 2$, then there exists at least one ω-limit cycle in C.

Proof

By positive invariance of C, the positive semi-orbit $O^+(\xi)$ of every $\xi \in C$ is contained in C. Therefore, the closure of $O^+(\xi)$ is a compact subset of G and so $\mathbb{R}_+ \subset I_\xi$ for every $\xi \in C$. The hypotheses, together with Theorem 5.33 (a result to be established in the next chapter), ensure that the equilibrium 0 is repelling in the sense it has a neighbourhood U such that U is contained in C and, for each $\xi \in U \backslash \{0\}$, there exists $\tau \in \mathbb{R}_+$ such that $\varphi(t, \xi) \notin U$ for all $t \geq \tau$. Let $\xi \in U$ be arbitrary. Then $\Omega(\xi) \subset C \backslash U$ and so contains no equilibrium points. By the Poincaré-Bendixson theorem (Theorem 4.46), it follows that $\Omega(\xi)$ is the orbit of a periodic point. Finally, since ξ is in U, we have $O^+(\xi) \neq \Omega(\xi)$ and so we may conclude that $\Omega(\xi)$ is an ω-limit cycle. $\square$

Example 4.57

Consider again the system given in Exercise 4.16, with $G = \mathbb{R}^2$ and $f: \mathbb{R}^2 \to \mathbb{R}^2$ given by $f(z) = f(z_1, z_2) := (z_2 + z_1(1 - \|z\|^2), -z_1 + z_2(1 - \|z\|^2))$. Let C be the closed unit disc $\{z \in \mathbb{R}^2 : \|z\| \leq 1\}$. Then

$$\langle z, f(z) \rangle = \|z\|^2 (1 - \|z\|^2) = 0 \quad \forall z \in \partial C,$$

and so solutions starting in C cannot exit C in forwards time. Thus, the compact set C is positively invariant. Moreover, 0 is the unique equilibrium in C

and

$$A = (Df)(0) = \begin{pmatrix} 1 & 1 \\ -1 & 1 \end{pmatrix}$$

with spectrum $\sigma(A) = \{1 + i, 1 - i\}$. Therefore, by Proposition 4.56, we may conclude the existence of a limit cycle in C. This, of course, is entirely consistent with Exercise 4.16 and Example 4.37, the conjunction of which shows (by explicit computation of the local flow) that the unit circle $\gamma = \partial C$ is a periodic orbit and coincides with the ω-limit set $\Omega(\xi)$ of every ξ with $0 < \|\xi\| < 1$. $\triangle$

5

Stability and asymptotic behaviour

The focus of this chapter is threefold in theme. Firstly, the topic of *stability of equilibria* will be investigated in the context of an autonomous differential equation $\dot{x} = f(x)$ with an equilibrium at 0 (i.e. $f(0) = 0$). Loosely speaking, this topic addresses the following question: in forwards time, do solutions which start "close" to 0 stay close to 0? A related and more specific question is: in forwards time, do solutions which start close to 0 converge to 0? The latter issue is that of *attractivity* of the equilibrium. The notions of stability and attractivity lead to the concept of *asymptotic stability*: the equilibrium 0 is said be asymptotically stable if it is stable and attractive.

Attractivity captures a very specific feature of the long-term behaviour of solutions, namely, convergence to an equilibrium (a singleton set). Less specific questions regarding long-term behaviour of solutions can be asked. For example, can one identify or characterize general sets (not necessarily equilibria) which are approached by solutions in forwards time? Such questions form the second theme of the chapter, which pertains to *asymptotic behaviour* of solutions.

The third theme considers systems with inputs (which, at one extreme, may be extraneous disturbances or perturbations or, at the other extreme, may be control functions open to choice) of the form $\dot{x} = f(x, u)$ with initial condition $x(0) = \xi$. Assuming that f is sufficiently regular to ensure that, for each ξ and piecewise-continuous input u, the latter initial-value problem has a unique solution x, *input-to-state stability* is a concept that relates to certain boundedness properties of the map $(\xi, u) \mapsto x$.

The methodology underpinning all three themes has its origins in the

H. Logemann and E. P. Ryan, *Ordinary Differential Equations*,
Springer Undergraduate Mathematics Series,
DOI: 10.1007/978-1-4471-6398-5_5, © Springer-Verlag London 2014

pioneering work of Lyapunov[1] which, at its simplest, is predicated on the intuitively-appealing notion that, if, in forwards time, some measure of "energy" is conserved (respectively, dissipated) along a solution, then the solution is bounded (respectively, asymptotically comes to rest). Although Lyapunov's seminal memoire on the stability of motion, published in 1892 in Russian, was translated into French in 1907 (reprinted in the USA in 1949), it was only at the end of the 1950s that scientists in the West began to appreciate, use and further develop Lyapunov's pioneering ideas[2]. This contrasted with the pre-eminence that Lyapunov's methodology had achieved in the former Soviet Union as a major mathematical tool in the context of linear and nonlinear stability problems.

We first consider systems without input and proceed to develop a compendium of results pertaining to stability (or lack thereof) of equilibria and asymptotic behaviour of solutions in the context of the autonomous differential equation

$$\dot{x} = f(x)\,, \tag{5.1}$$

where $G \subset \mathbb{R}^N$ is a non-empty open set and $f \colon G \to \mathbb{R}^N$ is continuous. The chapter then closes with a presentation of basic concepts and results pertaining to input-to-state stability of systems with input.

5.1 Lyapunov stability theory

Throughout this section, $f \colon G \to \mathbb{R}^N$ is a continuous function. We assume the existence of at least one *equilibrium* or *equilibrium point* for (5.1), that is, a point $\zeta \in G$ such that $f(\zeta) = 0$; the corresponding constant function $\mathbb{R} \to G$, $t \mapsto \zeta$ is then a solution of (5.1), a so-called *equilibrium solution*. If $\zeta \in G$ is an equilibrium for (5.1), then, defining $\tilde{G} := G - \zeta = \{z - \zeta \colon z \in G\}$ and $\tilde{f} \colon \tilde{G} \to \mathbb{R}^N$ by $\tilde{f}(z) = f(z + \zeta)$, we have that $0 \in \tilde{G}$ and $\tilde{f}(0) = 0$. Furthermore, $x \colon I \to G$ satisfies (5.1) if, and only if, the function $y \colon I \to \tilde{G}$, $t \mapsto x(t) - \zeta$ satisfies $\dot{y} = \tilde{f}(y)$. Therefore, throughout this section and without loss of generality, we assume that $0 \in G$ and $f(0) = 0$.

The notion of *stability* of the equilibrium 0 encapsulates the following property: a solution remains close (quantified by $\varepsilon > 0$) to 0 in forwards time provided that it starts sufficiently close (quantified by $\delta > 0$) to 0, see Figure 5.1.

Formalizing this notion, we state the following definition.

---

[1] Aleksandr Mikhailovich Lyapunov (1857-1918), Russian.

[2] Lyapunov's memoire was eventually translated into English by A.T. Fuller in 1992: *International Journal of Control*, vol. 55, March 1992.

Definition 5.1

The equilibrium 0 is said to be *stable* (in the sense of Lyapunov) if, for each $\varepsilon > 0$, there exists $\delta > 0$ such that, for every maximal solution $x\colon I \to G$ of (5.1) with $0 \in I$ and $\|x(0)\| \leq \delta$, we have

$$\|x(t)\| \leq \varepsilon \quad \forall\, t \in I \cap \mathbb{R}_+. \tag{5.2}$$

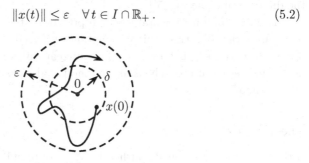

Figure 5.1 Stable equilibrium

If, in Definition 5.1, ε is such that $\overline{\mathbb{B}}(0,\varepsilon)$, the closed ball with centre 0 and radius ε, is contained in G (this holds for all sufficiently small $\varepsilon > 0$), then (5.2) together with Corollary 4.10 implies that $\mathbb{R}_+ \subset I$. Consequently, if the equilibrium 0 is stable, then there exists $\delta > 0$ such that, for every maximal solution $x\colon I \to G$ of (5.1) with $0 \in I$ and $\|x(0)\| \leq \delta$, $x(t)$ is defined for all $t \in \mathbb{R}_+$, or, equivalently, $\mathbb{R}_+ \subset I$.

Let $U \subset G$ be open. For a continuously differentiable function $V\colon U \to \mathbb{R}$, we define a continuous function $V_f\colon U \to \mathbb{R}$ by

$$V_f(z) = \langle (\nabla V)(z), f(z)\rangle = \sum_{j=1}^{N}(\partial_j V)(z)f_j(z) \quad \forall\, z \in U,$$

where ∇V denotes the gradient of V, $\partial_j V$ denotes the partial derivative of V with respect to the j-th coordinate and f_j denotes the j-th component of f. For each $z \in U$, $V_f(z)$ is the directional derivative of V at z in the direction $f(z)$ (not normalized). To see this, fix $z \in U$ arbitrarily and let $g\colon \mathbb{R} \to \mathbb{R}^N$ be given by $g(\eta) := z + \eta f(z)$. By Proposition A.34 (chain rule), applied in the context of the composition $V \circ g$, we have

$$\lim_{\eta \to 0}\frac{1}{\eta}\big(V(z+\eta f(z)) - V(z)\big) = (V \circ g)'(0) = \langle \nabla V(g(0)), g'(0)\rangle$$

$$= \langle \nabla V(z), f(z)\rangle = V_f(z).$$

Furthermore, if $x\colon I \to G$ is a solution of (5.1) with $x(t) \in U$ for all $t \in I$, then the chain rule (Proposition A.34) yields

$$(V \circ x)'(t) = \langle (\nabla V)(x(t)), \dot{x}(t)\rangle = \langle (\nabla V)(x(t)), f(x(t))\rangle \quad \forall\, t \in I,$$

and so

$$(V \circ x)'(t) = V_f(x(t)) \quad \forall t \in I. \tag{5.3}$$

In view of (5.3), V_f may be termed the *derivative of V along solutions of* (5.1). Note that computation of V_f does not require knowledge of solutions of the differential equation (5.1).

The following fundamental theorem provides a test for stability of an equilibrium. In the literature, this is sometimes referred to as "Lyapunov's direct method" (because it is based on the nonlinear differential equation itself and not on its linearization). The direct method is also known as "Lyapunov's second method".

Theorem 5.2

If there exist an open neighbourhood $U \subset G$ of 0 and a continuously differentiable function $V: U \to \mathbb{R}$ such that

$$V(0) = 0, \quad V(z) > 0 \ \forall z \in U \setminus \{0\}, \quad V_f(z) \le 0 \ \forall z \in U,$$

then 0 is a stable equilibrium for (5.1).

Proof

Let the hypotheses hold and, seeking a contradiction, suppose that 0 is not a stable equilibrium. Then there exist $\varepsilon > 0$ and, for each $n \in \mathbb{N}$, a maximal solution $x_n: I_n \to G$ and $t_n \in I_n \cap \mathbb{R}_+$ such that the closed ball $B := \overline{\mathbb{B}}(0, \varepsilon)$ is contained in U, $0 \in I_n$, $\|x_n(0)\| < \varepsilon/n$, and $\|x_n(t_n)\| > \varepsilon$. Define τ_n by

$$\tau_n := \min \{ t \in [0, t_n] : \|x_n(t)\| = \varepsilon \}.$$

Integration of (5.3) (with $x = x_n$) from 0 to τ_n gives

$$V(x_n(\tau_n)) = V(x_n(0)) + \int_0^{\tau_n} V_f(x_n(t)) \, dt \le V(x_n(0)) \quad \forall n \in \mathbb{N}, \tag{5.4}$$

where we have used the facts that $x_n(t) \in B \subset U$ for all $t \in [0, \tau_n]$ and $V_f(z) \le 0$ for all $z \in U$. By continuity and positivity of V on the compact set ∂B (the boundary of B), there exists $\mu > 0$ such that $V(z) \ge \mu$ for all $z \in \partial B$. Since $x_n(\tau_n) \in \partial B$, the left-hand side of (5.4) is bounded below by $\mu > 0$ for all $n \in \mathbb{N}$, whilst the right-hand side converges to $V(0) = 0$ as $n \to \infty$, yielding a contradiction. $\qquad \square$

A function V satisfying the hypotheses of Theorem 5.2 is termed a *Lyapunov function* for (5.1). Reiterating earlier comments, Theorem 5.2 can be applied without solving the differential equation (5.1). However, there is not a general method for finding Lyapunov functions: it is a matter of experience, ingenuity, or trial and error. Sometimes there are natural "candidates": for electrical and mechanical systems, energy considerations frequently provide a Lyapunov function (see Example 5.3). Therefore, Lyapunov functions are sometimes also described as "energy-like" functions.

Example 5.3

In this example, we will analyze stability properties of the nonlinear pendulum

$$\ddot{\theta} + a\dot{\theta} + b\sin\theta = 0$$

which was considered in Section 1.1.2. Here θ is the angular deviation from vertical, and $a = c/m \geq 0$ and $b = g/l > 0$ are constants, where $c \geq 0$ is the coefficient of friction, m is the mass, g is the gravitation constant and l is the length of the pendulum (see also Figure 1.3). Writing $x = (x_1, x_2) = (\theta, \dot{\theta})$ and introducing the function $f: \mathbb{R}^2 \to \mathbb{R}^2$ given by $f(z) = f(z_1, z_2) := (z_2, -b\sin z_1 - az_2)$, the above equation may be expressed in the form $\dot{x} = f(x)$. Obviously, for every $n \in \mathbb{Z}$, the point $(n\pi, 0)$ is an equilibrium point. We claim that the equilibrium 0 is stable. To this end, note that the sum of kinetic and potential energies at time t is given by

$$\frac{1}{2}ml^2 x_2^2(t) + mgl\big(1 - \cos(x_1(t))\big) = \frac{ml^2}{2}\left[x_2^2(t) + 2b\big(1 - \cos(x_1(t))\big)\right],$$

which motivates us to consider the function V given by

$$V(z) = V(z_1, z_2) := z_2^2 + 2b(1 - \cos z_1)$$

as a candidate Lyapunov function. We note that $V(2n\pi, 0) = 0$ for every $n \in \mathbb{Z}$. Defining $U := (-2\pi, 2\pi) \times \mathbb{R}$, it is clear that $V(z) > 0$ for every $z \in U\backslash\{0\}$. Moreover, for all $(z_1, z_2) \in U$,

$$V_f(z_1, z_2) = 2z_2 b\sin z_1 - 2z_2(b\sin z_1 + az_2) = -2az_2^2 \leq 0.$$

This shows that the function $V : U \to \mathbb{R}$ is a Lyapunov function for the above system. Consequently, by Theorem 5.2, the equilibrium 0 is stable. $\triangle$

Example 5.4

The system

$$\dot{x}_1 = x_2, \quad \dot{x}_2 = -x_1^3 - x_2 + x_2|x_2|; \quad x(t) = (x_1(t), x_2(t)) \in \mathbb{R}^2$$

evidently has a unique equilibrium at 0. In this case,

$$f\colon \mathbb{R}^2 \to \mathbb{R}^2, \ z = (z_1, z_2) \mapsto \left(z_2, \ -z_1^3 - z_2 + z_2|z_2|\right).$$

To investigate the stability of this equilibrium, consider the function

$$V\colon \mathbb{R}^2 \to \mathbb{R}, \ z = (z_1, z_1) \mapsto z_1^4 + 2z_2^2.$$

Clearly, $V(z) > 0$ for all $z \neq 0$ and $V(0) = 0$. Moreover,

$$V_f(z) = 4z_1^3 z_2 + 4z_2(-z_1^3 - z_2 + z_2|z_2|) = -4z_2^2(1 - |z_2|).$$

Therefore, if we define $U := \{z \in \mathbb{R}^2 : \|z\|^2 = z_1^2 + z_2^2 < 1\}$ (that is, the open unit disc centred at 0), we have

$$V_f(z) \leq 0 \ \ \forall z \in U.$$

Thus, by Theorem 5.2, the equilibrium 0 is stable.   △

In the case where $G = \mathbb{R}^N$ and V is a Lyapunov function with bounds of class $\mathcal{K}_\infty$, one can conclude a property stronger than stability (sometimes referred to as *stability in-the-large*). First, we make precise what we mean by the class $\mathcal{K}_\infty$. A function $a\colon \mathbb{R}_+ \to \mathbb{R}_+$ is said to be a $\mathcal{K}$ function or a function of class $\mathcal{K}$ if it is continuous, strictly increasing, and $a(0) = 0$. The class of unbounded $\mathcal{K}$ functions is denoted by $\mathcal{K}_\infty$. Obviously, if $a \in \mathcal{K}_\infty$, then $a(s) \to \infty$ as $s \to \infty$. Observe that, if $a \in \mathcal{K}_\infty$, then a is bijective and hence has an inverse function a^{-1}; it is readily seen that $a^{-1} \in \mathcal{K}_\infty$. We also note that $\mathcal{K}_\infty$ is closed under composition (that is, $a_1, a_2 \in \mathcal{K}_\infty$ implies $a_1 \circ a_2 \in \mathcal{K}_\infty$).

Proposition 5.5

Let $G = \mathbb{R}^N$. If there exist a continuously differentiable function $V\colon \mathbb{R}^N \to \mathbb{R}$ and functions $a_1, a_2 \in \mathcal{K}_\infty$ such that

$$a_1(\|z\|) \leq V(z) \leq a_2(\|z\|) \ \text{ and } \ V_f(z) \leq 0 \ \text{ for all } z \in \mathbb{R}^N,$$

then, for each $\xi \in \mathbb{R}^N$, every maximal solution $x\colon I \to \mathbb{R}^N$ of the initial-value problem $\dot{x} = f(x)$, $x(0) = \xi$, is such that $\mathbb{R}_+ \subset I$ and

$$\|x(t)\| \leq a_3(\|\xi\|) \ \ \forall t \in \mathbb{R}_+, \ \text{ where } \ a_3 := a_1^{-1} \circ a_2 \in \mathcal{K}_\infty.$$

Proof

Let $\xi \in \mathbb{R}^N$ be arbitrary, and let $x\colon I \to \mathbb{R}^N$ be a maximal solution of the initial-value problem $\dot{x} = f(x)$, $x(0) = \xi$. Then

$$V(x(t)) - V(\xi) = \int_0^t V_f(x(s)) \, \mathrm{d}s \leq 0 \ \ \forall t \in I \cap \mathbb{R}_+.$$

Therefore, $a_1(\|x(t)\|) \leq V(x(t)) \leq V(\xi) \leq a_2(\|\xi\|)$ for all $t \in I \cap \mathbb{R}_+$ and so $\|x(t)\| \leq (a_1^{-1} \circ a_2)(\|\xi\|)$ for all $t \in I \cap \mathbb{R}_+$. By Corollary 4.10, it follows that $\mathbb{R}_+ \subset I$, whence the assertion of the proposition. □

Note that, as a particular consequence, the hypotheses of Proposition 5.5 ensure that 0 is a stable equilibrium.

Example 5.6

Let $g \colon \mathbb{R} \to \mathbb{R}$ be continuous and such that, for some constants $0 < \rho \leq \sigma$,

$$\rho y^2 \leq yg(y) \leq \sigma y^2 \quad \forall y \in \mathbb{R}. \tag{5.5}$$

The system

$$\dot{x}_1 = -x_2, \quad \dot{x}_2 = g(x_1); \quad (x_1(t), x_2(t)) \in \mathbb{R}^2$$

evidently has a unique equilibrium at 0. This system can be written in the form $\dot{x} = f(x)$ with

$$f \colon \mathbb{R}^2 \to \mathbb{R}^2, \quad z = (z_1, z_2) \mapsto \big(-z_2 \,, g(z_1) \big).$$

Introducing the function

$$V \colon \mathbb{R}^2 \to \mathbb{R}, \quad z = (z_1, z_2) \mapsto 2 \int_0^{z_1} g(y)\mathrm{d}y + z_2^2,$$

we have

$$V_f(z) = -2g(z_1)z_2 + 2z_2g(z_1) = 0 \quad \forall z = (z_1, z_2) \in \mathbb{R}^2.$$

Moreover, invoking (5.5) and defining $\alpha := \min\{1, \rho\}$ and $\beta := \max\{1, \sigma\}$, gives

$$\alpha\|z\|^2 \leq \rho z_1^2 + z_2^2 \leq V(z) \leq \sigma z_1^2 + z_2^2 \leq \beta\|z\|^2 \quad \forall z = (z_1, z_2) \in \mathbb{R}^2.$$

Therefore, the hypotheses of Proposition 5.5 hold (with $a_1(s) = \alpha s^2$ and $a_2(s) = \beta s^2$ for all $s \geq 0$) and so, for each $\xi \in \mathbb{R}^2$, every maximal solution $x \colon I \to \mathbb{R}^2$ of the initial-value problem $\dot{x} = f(x)$, $x(0) = \xi$, is such that $\mathbb{R}_+ \subset I$ and $\|x(t)\| \leq \gamma\|\xi\|$ for all $t \geq 0$, where $\gamma := \sqrt{\beta/\alpha}$. △

Exercise 5.1

Consider the predator-prey model of Lotka-Volterra introduced in Section 1.1.5 with all four parameters a, b, c and d equal to 1, that is,

$$\dot{p} = p(q-1), \quad \dot{q} = q(1-p).$$

The "natural domain" for this system is the open quadrant $(0, \infty) \times (0, \infty)$. Setting $x_1 := p - 1$ and $x_2 := q - 1$, the above equations can be reformulated as a system on $G := (-1, \infty) \times (-1, \infty)$:

$$\dot{x}_1 = (x_1 + 1)x_2 \,, \quad \dot{x}_2 = -x_1(x_2 + 1) \,.$$

The origin $(0, 0)$ is the unique equilibrium of this system in G (corresponding to the equilibrium $(1, 1)$ of the original system). Show that $(0, 0)$ is stable.

(*Hint.* See part (d) of Exercise 1.2 and make use of the first integral.)

Exercise 5.2

Consider the following planar system on $\mathbb{R}^2$:

$$\dot{x}_1 = x_2 \,, \quad \dot{x}_2 = -x_1 - x_2 + g(x_2) \,,$$

where $g \colon \mathbb{R} \to \mathbb{R}$ is a continuous function satisfying $\lim_{w \to 0}(g(w)/w) = 0$. Prove that the equilibrium 0 is stable.

(*Hint.* You may find the function $V \colon (z_1, z_2) \mapsto z_1^2 + z_2^2$ useful.)

The equilibrium 0 of (5.1) is said to be *unstable* if it is not stable. The next result identifies conditions under which the equilibrium is unstable: results of this nature are referred to as *instability theorems* (and are frequently associated with the name Chetaev[3]).

Theorem 5.7

Assume that there exist an open neighbourhood $U \subset G$ of 0 and a continuously differentiable function $V \colon U \to \mathbb{R}$ satisfying the following hypotheses.

(1) For every $z \in U$, if $V(z) > 0$, then $V_f(z) > 0$.

(2) For every $\delta > 0$, there exists $\xi \in U$ with $\|\xi\| < \delta$ and $V(\xi) > 0$.

Then 0 is an unstable equilibrium for (5.1).

Proof

Let hypotheses (1) and (2) hold. Seeking a contradiction, suppose that 0 is a stable equilibrium. Choose $\varepsilon > 0$ sufficiently small so that the closed ball $B := \overline{\mathbb{B}}(0, 2\varepsilon)$ is contained in U. By stability of 0, there exists $\delta \in (0, \varepsilon)$ such that every maximal solution $x \colon I \to \mathbb{R}$ of (5.1), with $0 \in I$ and $\|x(0)\| \leq \delta$, is

---

[3] Nikolai Guryevich Chetaev (1902-1959), Russian.

such that $\|x(t)\| \leq \varepsilon$ for all $t \geq 0$. By hypothesis (2), there exists $\xi \in U$ with $\|\xi\| < \delta$ and $V(\xi) =: \alpha > 0$. Define

$$W := \{z \in B : V(z) \geq \alpha\} \qquad \text{and} \qquad \beta := \min_{z \in W} V_f(z).$$

Since W is compact and V_f is continuous, and, by hypothesis (1), V_f is positive on the set W, we may infer that $\beta > 0$. Let x be a maximal solution of (5.1) with $x(0) = \xi$. Then $\|x(t)\| \leq \varepsilon$ for all $t \geq 0$ and, consequently, $V \circ x$ is bounded on $\mathbb{R}_+$. If $x(t) \in W$ for all $t \geq 0$, then

$$V(x(t)) = V(x(0)) + \int_0^t V_f(x(s))\,\mathrm{d}s \geq \alpha + \beta t \ \ \forall t \geq 0$$

which is impossible since $V \circ x$ is bounded on $\mathbb{R}_+$. We may now conclude that the set $T := \{t \geq 0 : x(t) \notin W\}$ is not empty. Moreover, since $x(0) \in W$, it follows from hypothesis (1) that $(V \circ x)'(0) = V_f(x(0)) > 0$, so that $x(t) \in W$ for all sufficiently small $t > 0$, implying that $t^* := \inf T > 0$. Clearly, $x(t^*)$ is in the boundary of W and, since $V(x(t^*)) \geq \alpha + \beta t^* > \alpha$, it follows that $x(t^*)$ is in the boundary of B. Thus, we arrive at the contradiction $2\varepsilon = \|x(t^*)\| \leq \varepsilon$ and so 0 is not a stable equilibrium of (5.1). $\qquad\qquad\qquad\square$

Example 5.8

Let $g, h \colon \mathbb{R}^2 \to \mathbb{R}$ be continuous functions such that

$$\lim_{z \to 0} \frac{g(z)}{\|z\|^3} = \lim_{z \to 0} \frac{h(z)}{\|z\|^3} = 0.$$

Consider system (5.1) with $G = \mathbb{R}^2$ and

$$f \colon \mathbb{R}^2 \to \mathbb{R}^2, \quad z = (z_1, z_2) \mapsto f(z) := \left(z_1^3 + g(z),\, -z_2^3 + h(z)\right).$$

Choose $r > 0$ sufficiently small so that

$$\|z\| < r \quad \Longrightarrow \quad \|z\|^3 - 2\big(|g(z)| + |h(z)|\big) > 0. \tag{5.6}$$

Let U be the open disc of radius r centred at 0 and define $V \colon U \to \mathbb{R}$ by

$$V(z) = V(z_1, z_2) := z_1^2 - z_2^2 \ \ \forall z = (z_1, z_2) \in U.$$

Then, $V(0) = 0$ and, for all $z = (z_1, z_2) \in U \backslash \{0\}$,

$$V_f(z) = 2z_1^4 + 2z_2^4 + 2z_1 g(z) - 2z_2 h(z) \geq \|z\|^4 - 2\|z\|\big(|g(z)| + |h(z)|\big),$$

where we have used the estimate $\|z\|^4 = z_1^4 + 2z_1^2 z_2^2 + z_2^4 \leq 2z_1^4 + 2z_2^4$. Combining this with (5.6) yields that $V_f(z) > 0$ for all $z \in U \backslash \{0\}$. Therefore, hypothesis (1) of Theorem 5.7 holds; moreover, for $0 \neq z = (z_1, 0)$, we have $V(z) = z_1^2 > 0$ and so hypothesis (2) of Theorem 5.7 also holds. Therefore, 0 is an unstable equilibrium. $\qquad\qquad\qquad\triangle$

Exercise 5.3

Consider the frictionless ($a = 0$) nonlinear pendulum equation in first-order form, that is,

$$\dot{x}_1 = x_2, \quad \dot{x}_2 = -b\sin x_1,$$

where $b > 0$ is a constant (see also Section 1.1.2 and Example 5.3). The point $(\pi, 0)$ is an equilibrium of this system (pendulum at rest in the vertically upright position). Physical intuition suggests that this equilibrium is unstable. The aim of this exercise is to prove that this is indeed the case. To this end set, $y_1 := x_1 - \pi$ and $y_2 = x_2$, so that

$$\dot{y}_1 = y_2, \quad \dot{y}_2 = -b\sin(y_1 + \pi) = b\sin y_1.$$

Obviously, $(0, 0)$ is an equilibrium of this system (corresponding to the equilibrium $(\pi, 0)$ of the original equation). Show that $(0, 0)$ is unstable. (*Hint*. You may find the function $V \colon (z_1, z_2) \mapsto z_1 z_2$ useful.)

The study of dynamical processes frequently involves a stronger concept of stability: namely, that solutions "near" a stable equilibrium tend to the equilibrium as $t \to \infty$. We will return to this concept of *asymptotic stability* in due course: we first prove some general results pertaining to asymptotic behaviour of solutions.

5.2 Invariance principles

Here, we assemble some results on asymptotic behaviour of solutions of (5.1), under the assumption that f is locally Lipschitz (and so (5.1) generates a local flow φ): the terminology "invariance principles" stems from the fact that invariant sets play a pivotal role in these results[4], the proofs of which are based on the invariance property of ω-limit sets (recall Theorem 4.38). In this section, the previous assumptions that $0 \in G$ and $f(0) = 0$ are not required, and hence will not be imposed here.

We start with a simple, but very useful, observation, frequently associated with the name Barbălat[5].

---

[4] The interested reader may also find the following 'tutorial-style' article useful:
H Logemann & E P Ryan, "Asymptotic Behaviour of Nonlinear Systems", *The American Mathematical Monthly*, vol.111, 2004, pp. 864-889.
[5] Ioan Barbălat (1907-1988), Romanian.

Lemma 5.9 (Barbălat's lemma)

If $h: \mathbb{R}_+ \to \mathbb{R}$ is uniformly continuous and the limit $\lim_{t\to\infty} \int_0^t h(s)\,ds$ exists and is finite, then $h(t) \to 0$ as $t \to \infty$.

Proof

Suppose to the contrary that $h(t)$ does not converge to 0 as $t \to \infty$. Then there exist $\varepsilon > 0$ and a sequence (t_n) in $\mathbb{R}_+$ such that $t_n \to \infty$ as $n \to \infty$ and $|h(t_n)| \geq \varepsilon$ for all $n \in \mathbb{N}$. By uniform continuity of h, there exists $\delta > 0$ such that, for all $n \in \mathbb{N}$ and all $t \in \mathbb{R}_+$,

$$|t_n - t| \leq \delta \quad \Longrightarrow \quad |h(t_n) - h(t)| \leq \varepsilon/2.$$

Therefore, for all $t \in [t_n, t_n + \delta]$ and all $n \in \mathbb{N}$,

$$|h(t)| = |h(t_n) - (h(t_n) - h(t))| \geq |h(t_n)| - |h(t_n) - h(t)| \geq \varepsilon/2,$$

from which it follows that, for each $n \in \mathbb{N}$,

$$\left| \int_0^{t_n+\delta} h(s)\,ds - \int_0^{t_n} h(s)\,ds \right| = \left| \int_{t_n}^{t_n+\delta} h(s)\,ds \right| = \int_{t_n}^{t_n+\delta} |h(s)|\,ds \geq \frac{\varepsilon\delta}{2} > 0,$$

where the second equation follows from the fact that h does not change sign on the interval $[t_n, t_n+\delta]$. However, by hypothesis, the left-hand side of the above inequality tends to 0 as $n \to \infty$, yielding a contradiction. $\square$

In Lemma 5.9, the assumption that $\lim_{t\to\infty} \int_0^t h(s)\,ds$ exists and is finite is of course the same as the condition that the improper (Riemann) integral $\int_0^\infty h(s)\,ds$ converges.

Exercise 5.4

The aim of this exercise is to show that, in Lemma 5.9, the assumption of uniform continuity is essential in the sense that if, in the statement of Lemma 5.9, uniform continuity is relaxed to continuity, then the conclusion of Lemma 5.9 is not true in general.

(a) Defining

$$h(t) = \begin{cases} 2\cos t^2 - \dfrac{\sin t^2}{t^2}, & t > 0, \\ 1, & t = 0, \end{cases}$$

it is clear that h is continuous on $\mathbb{R}_+$ and that $h(t)$ does not converge to 0 as $t \to \infty$. Show that $\int_0^t h(s)ds \to 0$ as $t \to \infty$.

(b) By Lemma 5.9, the function h defined in part (a) cannot be uniformly continuous. Show directly, without appealing to Lemma 5.9, that h is not uniformly continuous.

(c) Find another example of a continuous function $h\colon \mathbb{R}_+ \to \mathbb{R}$ such that $\lim_{t\to\infty} \int_0^t h(s)ds$ exists and is finite, but $h(t)$ does not converge to 0 as $t \to \infty$. Show directly, without appealing to Lemma 5.9, that the example found is not uniformly continuous.

Barbălat's lemma plays an important role in the proof of our next result. As usual, for locally Lipschitz f, I_ξ denotes the interval of existence of the unique maximal solution x of (5.1) with $x(0) = \xi$.

Theorem 5.10 (Integral invariance principle)

Assume that $f\colon G \to \mathbb{R}^N$ is locally Lipschitz. Let φ denote the local flow generated by (5.1). Let $U \subset G$ be non-empty and open, and let $g\colon U \to \mathbb{R}$ be continuous. Assume that $\xi \in U$ is such that the closure of the positive semi-orbit $O^+(\xi)$ is compact and contained in U. Then $\mathbb{R}_+ \subset I_\xi$. If, in addition, the limit $\lim_{t\to\infty} \int_0^t g(\varphi(s,\xi))ds$ exists and is finite, $\varphi(t,\xi)$ approaches the largest invariant set in $g^{-1}(0) = \{z \in U : g(z) = 0\}$ as $t \to \infty$.

We remark that, if the closure of $O^+(\xi)$ is compact and contained in G, then we already know that $\varphi(t,\xi)$ approaches the ω-limit set $\Omega(\xi)$ as $t \to \infty$ (by Theorem 4.38). Moreover, $\Omega(\xi)$ is the smallest set approached by $\varphi(t,\xi)$ as $t \to \infty$ (recall Exercise 4.20). Therefore, the reader may ask "what is the point of Theorem 5.10?". To answer this question, we note that, if $\Omega(\xi)$ is known, then Theorem 5.10 indeed does not provide any additional information. However, in general, $\Omega(\xi)$ is not known (with the exception of simple one or two-dimensional examples) and, in such cases, Theorem 5.10 provides valuable information on the behaviour of $\varphi(t,\xi)$ as $t \to \infty$. We emphasize that the largest invariant set in $g^{-1}(0)$ can frequently be determined by exploiting the structure of the function f in (5.1) (see Example 5.11 for the basic idea in this context).

Proof of Theorem 5.10

By Corollary 4.10, $\mathbb{R}_+ \subset I_\xi$. Moreover, by Theorem 4.38, $O^+(\xi)$ has non-empty ω-limit set $\Omega(\xi)$, the solution $\varphi(t,\xi)$ approaches $\Omega(\xi)$ as $t \to \infty$ and $\Omega(\xi)$ is an invariant set. Hence, it suffices to prove that $\Omega(\xi) \subset g^{-1}(0)$. By hypothesis, the semi-orbit $O^+(\xi)$ has compact closure $C := \mathrm{cl}(O^+(\xi)) \subset U$. By continuity of f, there exists $M > 0$ such that $\|f(z)\| \leq M$ for all $z \in C$. For $\varepsilon > 0$, define

$\delta := \varepsilon/M$. Then, for all $s, t \in \mathbb{R}_+$,

$$|s - t| \leq \delta \quad \Longrightarrow \quad \|\varphi(s, \xi) - \varphi(t, \xi)\| \leq \left| \int_s^t \|f(\varphi(\sigma, \xi))\| d\sigma \right| \leq M|s - t| \leq \varepsilon,$$

and so $\varphi(\cdot, \xi)$ is uniformly continuous. Moreover, by continuity, g is uniformly continuous on the compact set C. It follows that $h(\cdot) := g(\varphi(\cdot, \xi))$ is uniformly continuous. By hypothesis, $\lim_{t \to \infty} \int_0^t h(s) ds$ exists and is finite, and so, using Lemma 5.9, we conclude that $h(t) = g(\varphi(t, \xi)) \to 0$ as $t \to \infty$.

Let $z \in \Omega(\xi)$ be arbitrary. Then there exists a sequence (t_n) in $[0, \infty)$ such that $t_n \to \infty$ and $\varphi(t_n, \xi) \to z$ as $n \to \infty$. Therefore, by continuity of g,

$$g(z) = \lim_{n \to \infty} g(\varphi(t_n, \xi)) = 0,$$

and so $z \in g^{-1}(0)$. Consequently, $\Omega(\xi) \subset g^{-1}(0)$. $\qquad\square$

Example 5.11

Consider the following three-dimensional system with $G = \mathbb{R}^3$:

$$\dot{x}_1 = x_2, \quad \dot{x}_2 = -x_1 - x_2 x_3, \quad \dot{x}_3 = x_2^2.$$

Let φ denote the associated local flow. The set E of all equilibrium points is given by $E := \{(0, 0, z_3) \colon z_3 \in \mathbb{R}\}$. Define $E_- := \{(0, 0, z_3) \colon z_3 \leq 0\}$.

Claim. For every $\xi \in \mathbb{R}^3$, $\mathbb{R}_+ \subset I_\xi$. Furthermore, $\lim_{t \to \infty} \varphi(t, \xi) = (0, 0, \|\xi\|)$ for $\xi \in \mathbb{R}^3 \setminus E_-$ and $\lim_{t \to \infty} \varphi(t, \xi) = (0, 0, -\|\xi\|)$ for $\xi \in E_-$.

Let $\xi = (\xi_1, \xi_2, \xi_3) \in \mathbb{R}^3$, set $x(\cdot) := \varphi(\cdot, \xi)$, write $x = (x_1, x_2, x_3)$ and note that

$$\frac{d}{dt} \|x(t)\|^2 = 2x_1(t)x_2(t) - 2x_2(t)x_1(t) - 2x_2^2(t)x_3(t) + 2x_3(t)x_2^2(t) = 0 \quad \forall t \in I_\xi.$$

Consequently, $\|x(t)\| = \|\xi\|$ for all $t \in I_\xi$ and so, in particular, the semi-orbit $O^+(\xi)$ is bounded, implying that its closure is compact. Thus, by Corollary 4.10, $\mathbb{R}_+ \subset I_\xi$. Furthermore,

$$\int_0^t x_2^2(s) \, ds = x_3(t) - \xi_3 \leq |x_3(t)| + |\xi_3| \leq \|x(t)\| + \|\xi\| = 2\|\xi\| \quad \forall t \geq 0,$$

showing that the function $t \mapsto \int_0^t x_2^2(s) ds$ is bounded. Since this function is also non-decreasing, we conclude that $\int_0^t x_2^2(s) ds$ converges to a finite limit as $t \to \infty$. By Theorem 5.10 (with $U = \mathbb{R}^3$ and $g \colon U \to \mathbb{R}$ given by $g(z) = g(z_1, z_2, z_3) = z_2^2$), $x(t)$ approaches the largest invariant subset in $g^{-1}(0) = \{(z_1, 0, z_3) \colon z_1, z_3 \in \mathbb{R}\}$. Assume that $y = (y_1, y_2, y_3)$ is a solution such that $y(t) \in g^{-1}(0)$ for all $t \in \mathbb{R}$. Then $y_2 = 0$ and it follows from the differential equation that $y_1 = 0$ and $\dot{y}_3 = 0$. Consequently, E is the largest invariant set in $g^{-1}(0)$ and $x(t)$ approaches E as $t \to \infty$. Since $\|x(t)\| = \|\xi\|$ for all $t \in \mathbb{R}$, we obtain that $x(t) \to (0, 0, \|\xi\|)$ or $x(t) \to (0, 0, -\|\xi\|)$ as $t \to \infty$. If $\xi \in \mathbb{R}^3 \setminus E_-$, then $-\|\xi\| < \xi_3$ and so, since $\dot{x}_3(t) \geq 0$ for all $t \in \mathbb{R}$, $x(t) \to (0, 0, \|\xi\|)$ as $t \to \infty$. Finally, if $\xi \in E_-$, then $x(t) = (0, 0, \xi_3)$ for all $t \in \mathbb{R}$, where $\xi_3 \leq 0$, and hence $x(t) = (0, 0, -\|\xi\|)$ for all $t \in \mathbb{R}$. This establishes the claim. $\triangle$

Exercise 5.5

Consider the following system with $G = \mathbb{R}^2$:

$$\dot{x} = x^2 \tanh(x)(1 - y), \quad \dot{y} = x^3 \tanh(x), \quad (x(0), y(0)) = (\xi_1, \xi_2) = \xi \in \mathbb{R}^2.$$

Show that the semi-orbit $O^+(\xi)$ is bounded and $\lim_{t \to \infty} x(t) = 0$.
(*Hint*. To show boundedness of $O^+(\xi)$, multiply both sides of the first differential equation by x and integrate. To prove $\lim_{t \to \infty} x(t) = 0$, use the integral invariance principle with $U = \mathbb{R}^2$ and $g \colon U \to \mathbb{R}$ given by $g(z_1, z_2) = z_1^3 \tanh(z_1)$.)

The following result, first proved by LaSalle[6], is a consequence of Theorem 5.10.

Theorem 5.12 (LaSalle's invariance principle)

Assume that f is locally Lipschitz and let φ denote the local flow generated by (5.1). Let $U \subset G$ be non-empty and open. Let $V \colon U \to \mathbb{R}$ be continuously differentiable and such that $V_f(z) \leq 0$ for all $z \in U$. If $\xi \in U$ is such that the closure of the semi-orbit $O^+(\xi)$ is compact and contained in U, then $\mathbb{R}_+ \subset I_\xi$ and $\varphi(t, \xi)$ approaches the largest invariant set in $V_f^{-1}(0)$ as $t \to \infty$.

Proof

By Corollary 4.10, $\mathbb{R}_+ \subset I_\xi$. Set $x(t) := \varphi(t, \xi)$ for all $t \geq 0$. By continuity of V and compactness of $\mathrm{cl}(O^+(\xi))$, the function V is bounded on $\mathrm{cl}(O^+(\xi))$ and so $V \circ x$ is bounded. Moreover, by (5.3) and the non-positivity of V_f,

$$(V \circ x)'(t) = V_f(x(t)) \leq 0 \quad \forall t \in \mathbb{R}_+,$$

and thus, $V \circ x$ is non-increasing. It follows that $V(x(t)) \to l$ as $t \to \infty$ for some $l \in \mathbb{R}$. Therefore,

$$\lim_{t \to \infty} \int_0^t V_f(x(s)) ds = \lim_{t \to \infty} V(x(t)) - V(x(0)) = l - V(\xi),$$

---

[6] Joseph Pierre LaSalle (1916-1983), US American.

and so, by Theorem 5.10 (with $g = V_f$), $x(t) = \varphi(t, \xi)$ approaches the largest invariant set in $V_f^{-1}(0)$ as $t \to \infty$. $\square$

Exercise 5.6

Show that the conclusions of the above theorem remain valid if the hypothesis "$V_f(z) \leq 0$ for all $z \in U$" is replaced by "$V_f(z) \geq 0$ for all $z \in U$". Explain this apparent anomaly.

Example 5.13

Consider the following planar system with $G = \mathbb{R}^2$:

$$\dot{x}_1 = -x_2(1 + x_1 x_2), \quad \dot{x}_2 = 2x_1.$$

Defining $f \colon \mathbb{R}^2 \to \mathbb{R}^2$ by

$$f(z_1, z_2) := (-z_2(1 + z_1 z_2), 2z_1) \quad \forall (z_1, z_2) \in \mathbb{R}^2,$$

we see that the above system can be written in the form $\dot{x} = f(x)$. Let φ denote the associated local flow. We claim that, for every $\xi \in \mathbb{R}^2$, $\mathbb{R}_+ \subset I_\xi$ and $\varphi(t, \xi) \to 0$ as $t \to \infty$.
With $V \colon \mathbb{R}^2 \to \mathbb{R}$ given by

$$V(z_1, z_2) := 2z_1^2 + z_2^2 \quad \forall (z_1, z_2) \in \mathbb{R}^2,$$

we have

$$V_f(z_1, z_2) = -4z_1 z_2 (1 + z_1 z_2) + 4z_1 z_2 = -4z_1^2 z_2^2 \leq 0 \quad \forall (z_1, z_2) \in \mathbb{R}^2.$$

Hence,

$$\frac{\mathrm{d}}{\mathrm{d}t} V(\varphi(t, \xi)) = V_f(\varphi(t, \xi)) \leq 0 \quad \forall t \in [0, \omega_\xi),$$

where $\omega_\xi := \sup I_\xi$. Consequently,

$$\|\varphi(t, \xi)\|^2 \leq V(\varphi(t, \xi)) \leq V(\xi) \quad \forall t \in [0, \omega_\xi).$$

Hence, the positive semi-orbit $O^+(\xi)$ is bounded, implying that its closure is compact. Invoking Theorem 5.12, we conclude that $\mathbb{R}_+ \subset I_\xi$ and $\varphi(t, \xi)$ approaches the largest invariant set M in $V_f^{-1}(0) = \{(z_1, z_2) \in \mathbb{R}^2 : z_1 z_2 = 0\}$ as $t \to \infty$. Thus, to conclude that $\varphi(t, \xi) \to 0$ as $t \to \infty$, it suffices to show that $M = \{0\}$. Let $\zeta = (\zeta_1, \zeta_2) \in M$ and write $\varphi(t, \zeta) = (x_1(t), x_2(t))$. By invariance of M, we have $x_1(t)x_2(t) = 0$ for all $t \in \mathbb{R}_+$. Differentiating, we obtain $0 = \dot{x}_1(t)x_2(t) + x_1(t)\dot{x}_2(t) = -x_2^2(t) + 2x_1^2(t)$ for all $t \in \mathbb{R}_+$. Evaluation of $x_1 x_2$ and its derivative at $t = 0$ gives $\zeta_1 \zeta_2 = 0 = -\zeta_2^2 + 2\zeta_1^2$, whence $\zeta_1 = 0 = \zeta_2$. Therefore, $M = \{0\}$. $\triangle$

Exercise 5.7

Consider again the nonlinear pendulum, which, in first order form, is described by

$$\dot{x}_1 = x_2, \quad \dot{x}_2 = -ax_2 - b\sin x_1,$$

where $a \geq 0$ and $b > 0$ are constants (see also Section 1.1.2 and Example 5.3). Show that if $a > 0$ (that is, the system is subject to friction), then there exists a neighbourhood U of 0 such that every solution (x_1, x_2) with $(x_1(0), x_2(0)) \in U$ satisfies $(x_1(t), x_2(t)) \to (0, 0)$ as $t \to \infty$.

Exercise 5.8

Consider the following planar system on $\mathbb{R}^2$:

$$\dot{x}_1 = x_2 - x_1^3(a_1 + b_1 x_1^2), \quad \dot{x}_2 = -x_1 - x_2^3(a_2 + b_2 x_2^2),$$

where a_1, a_2, b_1 and b_2 are positive constants. Show that for every $\xi \in \mathbb{R}^2$, $\mathbb{R}_+ \subset I_\xi$ and $\varphi(t, \xi) \to 0$ as $t \to \infty$.

Exercise 5.9

Prove LaSalle's invariance principle without using the integral invariance principle or Barbălat's lemma.
(*Hint.* Show that V is constant on $\Omega(\xi)$. Combine this with the invariance of $\Omega(\xi)$ to show that $V_f(z) = 0$ for all $z \in \Omega(\xi)$. Whilst it may be tempting to conclude from $V = \text{const}$ on $\Omega(\xi)$ that ∇V (and hence V_f) is equal to 0 on $\Omega(\xi)$, this is not a valid conclusion. Why not?)

5.3 Asymptotic stability

Throughout this section we continue to assume that f in (5.1) is locally Lipschitz and so (5.1) generates a local flow φ. Furthermore, we assume that $0 \in G$ and $f(0) = 0$ (i.e., 0 is an equilibrium of (5.1)).

Definition 5.14

The equilibrium 0 of (5.1) is said to be *attractive* if there exists $\delta > 0$ such that, for every $\xi \in G$ with $\|\xi\| \leq \delta$, the following properties hold: $\mathbb{R}_+ \subset I_\xi$ and $\varphi(t, \xi) \to 0$ as $t \to \infty$. We say that the equilibrium 0 is *asymptotically stable* (in the sense of Lyapunov) if it is stable and attractive.

The following exercise shows that attractivity does *not* imply stability in general.

Exercise 5.10

Consider the following system of differential equations in polar coordinates

$$\dot{r} = r(1 - r), \quad \dot{\theta} = \sin^2(\theta/2). \tag{5.7}$$

Let ψ denote the (local) flow generated by this system.

(a) Show that

    (i) $\lim\limits_{t\to\infty} \psi(t, (r^0, \theta^0)) = (1, 0) \quad \forall (r^0, \theta^0) \in (0, \infty) \times \{0\}$,

    (ii) $\lim\limits_{t\to\infty} \psi(t, (r^0, \theta^0)) = (1, 2\pi) \quad \forall (r^0, \theta^0) \in (0, \infty) \times (0, 2\pi)$.

(b) Verify that, in Cartesian coordinates, (5.7) leads to the following system on $\mathbb{R}^2 \backslash \{0\}$

$$\dot{x} = g(x, y)x - h(x, y)y, \quad \dot{y} = g(x, y)y + h(x, y)x,$$

where $g(x, y) := 1 - (x^2 + y^2)^{1/2}$ and $h(x, y) := \left(1 - x(x^2 + y^2)^{-1/2}\right)/2$.

(c) With g and h as defined in (b), consider the system

$$\dot{x} = g(x+1, y)(x+1) - h(x+1, y)y, \quad \dot{y} = g(x+1, y)y + h(x+1, y)(x+1)$$

on $\mathbb{R}^2 \backslash \{(-1, 0)\}$. Show that 0 is an attractive, but *unstable* equilibrium.

The following theorem provides a sufficient condition for asymptotic stability in terms of a Lyapunov function V and the set $V_f^{-1}(0)$.

Theorem 5.15

If there exist an open neighbourhood $U \subset G$ of 0 and a continuously differentiable function $V : U \to \mathbb{R}$ such that

$$V(0) = 0, \quad V(z) > 0 \ \ \forall z \in U \backslash \{0\}, \quad V_f(z) \le 0 \ \ \forall z \in U,$$

and $\{0\}$ is the only invariant set contained in $V_f^{-1}(0)$, then 0 is an asymptotically stable equilibrium of (5.1).

Proof

It is an immediate consequence of Theorem 5.2 that the equilibrium 0 is stable. It remains only to prove attractivity of 0. Let $\varepsilon > 0$ be such that $\overline{\mathbb{B}}(0, \varepsilon) \subset U \subset G$. By stability of 0, there exists $\delta \in (0, \varepsilon)$ such that, if $\|\xi\| \le \delta$, then $\mathbb{R}_+ \subset I_\xi$ and $\|\varphi(t, \xi)\| \le \varepsilon$ for all $t \ge 0$. Thus, for $\xi \in G$ with $\|\xi\| \le \delta$, $\mathrm{cl}(O^+(\xi))$ is compact and contained in $U \subset G$. By Theorem 5.12, it follows that $\varphi(t, \xi)$

approaches the largest invariant set in $V_f^{-1}(0)$ as $t \to \infty$. But, by hypothesis, $\{0\}$ is the only invariant subset of $V_f^{-1}(0)$. Therefore, $\varphi(t, \xi) \to 0$ as $t \to \infty$ for all $\xi \in G$ with $\|\xi\| \le \delta$. $\qquad\qquad\qquad\qquad\qquad\qquad\qquad\qquad\qquad\qquad\square$

Example 5.16

In this example, we describe a typical application of Theorem 5.15 in the context of a general class of nonlinear second-order systems. Consider

$$\ddot{y}(t) + g(y(t), \dot{y}(t)) = 0, \tag{5.8}$$

where $g \colon \mathbb{R}^2 \to \mathbb{R}$ is locally Lipschitz and continuously differentiable with respect to the second variable. Furthermore, we assume that $g(0, 0) = 0$. Setting $x = (x_1, x_2) = (y, \dot{y})$, the second-order system (5.8) can be expressed in the equivalent form

$$\dot{x} = f(x), \quad \text{where} \quad f \colon \mathbb{R}^2 \to \mathbb{R}^2, \ (z_1, z_2) \mapsto (z_2, -g(z_1, z_2)).$$

Let $\varepsilon > 0$, set $U = (-\varepsilon, \varepsilon) \times (-\varepsilon, \varepsilon)$, and define

$$V \colon U \to \mathbb{R}, \ (z_1, z_2) \mapsto \int_0^{z_1} g(s, 0)\, ds + \frac{z_2^2}{2}.$$

It follows from the mean-value theorem of differentiation that, for each $(z_1, z_2) \in U$, there exists a number $\theta = \theta(z_1, z_2) \in (0, 1)$ such that

$$V_f(z_1, z_2) = -z_2(g(z_1, z_2) - g(z_1, 0)) = -z_2^2 \partial_2 g(z_1, \theta z_2).$$

Claim. If $z_1 g(z_1, 0) > 0$ for all $z_1 \in (-\varepsilon, \varepsilon) \setminus \{0\}$ and $\partial_2 g(z_1, z_2) > 0$ for all $(z_1, z_2) \in U$ satisfying $z_1 z_2 \neq 0$, then the equilibrium 0 is asymptotically stable.

To establish this claim, we note first that $V(0, 0) = 0$, $V(z_1, z_2) > 0$ for all $(z_1, z_2) \in U \setminus \{0\}$ and $V_f(z_1, z_2) \le 0$ for all $(z_1, z_2) \in U$. Observe that $V_f^{-1}(0) \subset \{(z_1, z_2) \in U \colon z_1 z_2 = 0\}$. Writing $\varphi(t, \xi) = (x_1(t), x_2(t))$, we see that, for $\xi = (\xi_1, 0) \in U$ with $\xi_1 \neq 0$, $\dot{x}_2(0) = -g(\xi_1, 0) \neq 0$. Similarly, for $\xi = (0, \xi_2) \in U$ with $\xi_2 \neq 0$, $\dot{x}_1(0) = \xi_2 \neq 0$. We conclude that, in both cases, there exists $\tau > 0$ such that $x_1(t)x_2(t) \neq 0$ for all $t \in (0, \tau)$. Consequently, solutions with initial condition $\xi = (\xi_1, \xi_2) \in U \setminus \{0\}$ satisfying $\xi_1 \xi_2 = 0$ do not remain in $V_f^{-1}(0)$, showing that $\{0\}$ is the only invariant subset of $V_f^{-1}(0)$. The claim now follows from Theorem 5.15. $\qquad\qquad\qquad\qquad\qquad\qquad\triangle$

Exercise 5.11

Reconsider the planar system in Example 5.4. Use the result in Example 5.16 to deduce that 0 is an asymptotically stable equilibrium.

Exercise 5.12

As a special case of (5.8), consider the Liénard equation

$$\ddot{y}(t) + d(y(t))\dot{y}(t) + k(y(t)) = 0, \tag{5.9}$$

previously considered in a different context in Exercise 4.22. We assume that the functions $d \colon \mathbb{R} \to \mathbb{R}$ and $k \colon \mathbb{R} \to \mathbb{R}$ are locally Lipschitz and $k(0) = 0$.

(a) From the result in Example 5.16, deduce that 0 is an asymptotically stable equilibrium of the Liénard equation, provided that there exists $\varepsilon > 0$ such that $z_1 k(z_1) > 0$ and $d(z_1) > 0$ for all $z_1 \in (-\varepsilon, \varepsilon)$ with $z_1 \neq 0$.

(b) Assume that

(i) there exists $\varepsilon > 0$ such that $z_1 k(z_1) \geq 0$ and $d(z_1) \geq 0$ for all $z_1 \in (-\varepsilon, \varepsilon)$ (note that, in contrast to part (a) above, these inequalities are not strict);

(ii) for every $\delta > 0$, there exist $z_1^+ \in (0, \delta)$ and $z_1^- \in (-\delta, 0)$ such that $k(z_1^+)k(z_1^-) \neq 0$.

Show that 0 is a stable equilibrium. Give an example of a locally Lipschitz functions d and k (satisfying the assumptions (i) and (ii)) for which the stable equilibrium 0 fails to be asymptotically stable.

Observe that, in Theorem 5.15, if the hypothesis of non-positivity of V_f on U is strengthened to that of negativity of V_f on $U \backslash \{0\}$, then $V_f^{-1}(0) = \{0\}$, in which case the final hypothesis of the theorem is trivially satisfied. We may therefore conclude the following.

Corollary 5.17

If there exist an open neighbourhood $U \subset G$ of 0 and a continuously differentiable function $V \colon U \to \mathbb{R}$ such that

$$V(0) = 0, \quad V(z) > 0 \ \ \forall z \in U \backslash \{0\}, \quad V_f(z) < 0 \ \ \forall z \in U \backslash \{0\},$$

then 0 is an asymptotically stable equilibrium of (5.1).

Exercise 5.13

Note that Corollary 5.17 is underpinned by the assumption that f is locally Lipschitz (and so generates a local flow). The goal of this exercise is to show that the result is valid under the weaker assumption that f is merely continuous (and so may not generate a local flow). Of course, in this context, we must first define the concept of attractivity. Let $f \colon G \to$

$\mathbb{R}^N$ be continuous. The equilibrium 0 of (5.1) is said to be *attractive* if there exists $\delta > 0$ such that, for each $\xi \in G$ with $\|\xi\| \leq \delta$, every maximal solution $x \colon I \to \mathbb{R}^N$ of the initial-value problem $\dot{x} = f(x)$, $x(0) = \xi$, has the following properties: $\mathbb{R}_+ \subset I$ and $x(t) \to 0$ as $t \to \infty$. The equilibrium 0 is *asymptotically stable* (in the sense of Lyapunov) if it is stable and attractive. Let U and V be as in Corollary 5.17. Prove that 0 is an asymptotically stable equilibrium.

(*Hint.* Stability is a consequence of Theorem 5.2. The remaining issue is to establish attractivity. To this end, show that, for suitably small $\delta > 0$, every maximal solution x of the initial-value problem with $\|\xi\| \leq \delta$ is such that $V \circ x$ is uniformly continuous, then make use of Barbălat's lemma.)

We record (without proof[7], which is outside the scope of this book) an important result, namely, that Corollary 5.17 has a converse in the sense that, if 0 is an asymptotically stable equilibrium, then there exists a continuously differentiable function V with the requisite properties.

Theorem 5.18

If 0 is an asymptotically stable equilibrium of (5.1), then there exist an open neighbourhood $U \subset G$ of 0 and a continuously differentiable function $V \colon U \to \mathbb{R}$ such that

$$V(0) = 0, \quad V(z) > 0 \quad \forall z \in U \backslash \{0\}, \quad V_f(z) < 0 \quad \forall z \in U \backslash \{0\}.$$

Definition 5.19

Let 0 be an asymptotically stable equilibrium of (5.1). The *domain of attraction* of 0 is the set $\mathcal{A} := \{\xi \in G \colon \mathbb{R}_+ \subset I_\xi, \ \varphi(t, \xi) \to 0 \text{ as } t \to 0\}$.

In words, $\mathcal{A}$ is the set of all states that are attracted to the asymptotically stable equilibrium of (5.1).

Exercise 5.14

Assume that $0 \in G$ is an asymptotically stable equilibrium of (5.1). Prove that the domain of attraction $\mathcal{A}$ is an open set.

The following proposition asserts that, if 0 is an asymptotically stable equilibrium, then the property of attractivity of the equilibrium is uniform with respect to initial states in any compact subset of $\mathcal{A}$.

---

[7] The interested reader might consult Theorem 49.4 in W. Hahn, *Stability of Motion*, Springer-Verlag, Berlin, 1967.

Proposition 5.20

Assume that $0 \in G$ is an asymptotically stable equilibrium of (5.1). Let C be a non-empty compact subset of the domain of attraction $\mathcal{A}$. Then

$$\lim_{t \to \infty} \max_{\xi \in C} \|\varphi(t, \xi)\| = 0. \tag{5.10}$$

Proof

Seeking a contradiction, suppose that (5.10) fails to hold. Then there exist $\varepsilon > 0$ and sequences (ξ_n) in C and (t_n) in $\mathbb{R}_+$ such that $t_n \to \infty$ as $n \to \infty$ and

$$\|\varphi(t_n, \xi_n)\| > \varepsilon \quad \forall n \in \mathbb{N}. \tag{5.11}$$

Noting that (ξ_n) is a sequence in the compact set C and passing to a subsequence if necessary, we may assume, without loss of generality, that (ξ_n) is convergent with limit $\xi \in C$. Let (s_m) be a sequence in $\mathbb{R}_+$ with $s_m \to \infty$ as $m \to \infty$. For each $m \in \mathbb{N}$, there exists $n_m \in \mathbb{N}$ such that

$$s_m \leq t_{n_m} \quad \text{and} \quad \|\varphi(s_m, \xi_{n_m}) - \varphi(s_m, \xi)\| \leq \frac{1}{m},$$

where we have used continuity of φ (Theorem 4.34). Since $\varphi(s_m, \xi) \to 0$ as $m \to \infty$, it follows that $\varphi(s_m, \xi_{n_m}) \to 0$ as $m \to \infty$. By stability of 0, there exists $\mu > 0$ such that, for all $\zeta \in G$ with $\|\zeta\| \leq \mu$, we have $\|\varphi(t, \zeta)\| \leq \varepsilon$ for all $t \geq 0$. Now choose $m \in \mathbb{N}$ sufficiently large so that $\|\varphi(s_m, \xi_{n_m})\| \leq \mu$. Then, since $t_{n_m} - s_m \geq 0$, we may infer that

$$\|\varphi(t_{n_m}, \xi_{n_m})\| = \|\varphi(t_{n_m} - s_m, \varphi(s_m, \xi_{n_m}))\| \leq \varepsilon$$

which contradicts (5.11). This completes the proof. $\qquad \square$

With the exception of Proposition 5.5, the stability concepts described so far are entirely local, that is, they relate to solutions with initial conditions in a sufficiently small neighbourhood of the equilibrium point. We now introduce the concept of global asymptotic stability which combines local and global aspects.

Definition 5.21

Assume that $G = \mathbb{R}^N$. The equilibrium 0 of (5.1) is said to be *globally attractive* if, for every $\xi \in \mathbb{R}^N$, $\mathbb{R}_+ \subset I_\xi$ and $\varphi(t, \xi) \to 0$ as $t \to \infty$. The equilibrium 0 is said to be *globally asymptotically stable* if it is stable and globally attractive.

Clearly, the domain of attraction $\mathcal{A}$ of a globally asymptotically stable equilibrium is the whole space, that is, $\mathcal{A} = \mathbb{R}^N$.

Theorem 5.22

Assume that $G = \mathbb{R}^N$. Let the hypotheses of Theorem 5.15 hold with $U = G = \mathbb{R}^N$. If, in addition, V is radially unbounded , that is,

$$V(z) \to \infty \quad \text{as } \|z\| \to \infty,$$

then 0 is a globally asymptotically stable equilibrium of (5.1).

Proof

Stability of 0 is clear. Let $\xi \in \mathbb{R}^N$. Set $\omega_\xi := \sup I_\xi$ and $x(t) := \varphi(t, \xi)$ for all $t \in [0, \omega_\xi)$. If we can show that x is bounded, then by Corollary 4.10, $\omega_\xi = \infty$ and the claim follows from Theorem 5.12. By the hypotheses of Theorem 5.15, $(d/dt)V(x(t)) = V_f(x(t)) \leq 0$ for all $t \in [0, \omega_\xi)$ and so

$$0 \leq V(x(t)) \leq V(\xi) \quad \forall t \in [0, \omega_\xi),$$

showing that the function $V \circ x$ is bounded. The radial unboundedness of V now implies that x is bounded, completing the proof. □

Exercise 5.15

Show that a continuous function $V \colon \mathbb{R}^N \to \mathbb{R}_+$ is radially unbounded if, and only if, for each $c \in \mathbb{R}_+$, the sublevel set $\{z \in \mathbb{R}^N \colon V(z) \leq c\}$ is compact.

Exercise 5.16

Consider the Lorenz system

$$\dot{x}_1 = \sigma(x_2 - x_1), \quad \dot{x}_2 = rx_1 - x_2 - x_1 x_3, \quad \dot{x}_3 = x_1 x_2 - bx_3,$$

with three positive parameters b, r and σ. Prove that if $0 < r < 1$, then the origin is globally asymptotically stable.
(*Hint.* Consider the function $V \colon \mathbb{R}^3 \to \mathbb{R}$ given by $V(z_1, z_2, z_3) = rz_1^2 + \sigma z_2^2 + \sigma z_3^2$.)

In 1963, the meteorologist Lorenz[8] introduced the above system of differential equations to explain some of the unpredictable behaviour of the weather. If $r > 1$, then it can be shown that the origin is unstable (see Exercise 5.23). As the parameter r increases, the dynamics of the Lorenz system become very complicated (chaotic).[9]

---

[8] Edward Norton Lorenz (1917-2008), US American.
[9] See, for example, C. Sparrow, *The Lorentz Equations: Bifurcations, Chaos and Strange Attractors*, Springer-Verlag, New York, 1982.

Exercise 5.17

The aim of this exercise is to show that the condition of radial unboundedness in Theorem 5.22 is essential.

Let $f\colon \mathbb{R}^2 \to \mathbb{R}^2$ be given by

$$f(z) = f(z_1, z_2) = \begin{cases} (-z_1, z_2) & \text{if } z_1^2 z_2^2 \geq 1 \\ (-z_1, 2z_1^2 z_2^3 - z_2) & \text{if } z_1^2 z_2^2 < 1. \end{cases}$$

Define $V\colon \mathbb{R}^2 \to \mathbb{R}$ by

$$V(z) = V(z_1, z_2) = z_1^2 + \frac{z_2^2}{1 + z_2^2}.$$

(a) Show that the equilibrium 0 of (5.1) is asymptotically stable.

(b) Show that the equilibrium 0 is *not* globally asymptotically stable.

(c) Show that V is not radially unbounded.

5.4 Stability of linear systems

Consider a linear system with $G = \mathbb{R}^N$:

$$\dot{x} = Ax, \qquad A \in \mathbb{R}^{N \times N}. \tag{5.12}$$

Note that (5.12) is a special case of (5.1) with $f\colon \mathbb{R}^N \to \mathbb{R}^N$, $z \mapsto Az$. Clearly, 0 is an equilibrium and A generates the flow

$$\mathbb{R} \times \mathbb{R}^N \to \mathbb{R}^N, \ (t, \xi) \mapsto \varphi(t, \xi) := \exp(At)\xi.$$

The following proposition shows that, for linear systems, stability can be characterized in terms of the eigenvalues of A. Recall that an eigenvalue λ of A is said to be *semisimple* if its geometric and algebraic multiplicities coincide, or, equivalently, if the corresponding generalized eigenspace coincides with the eigenspace.

Proposition 5.23

Let $A \in \mathbb{R}^{N \times N}$. The following statements are equivalent.

(1) 0 is a stable equilibrium of (5.12).

(2) $\operatorname{Re} \lambda \leq 0$ for all $\lambda \in \sigma(A)$ and λ is semisimple for all $\lambda \in \sigma(A)$ with $\operatorname{Re} \lambda = 0$.

Proof

Assume that statement (1) holds and let $\varepsilon > 0$. Then there exists $\delta > 0$ such that, for all $\eta \in \mathbb{R}^N$ with $\|\eta\| \le \delta$,

$$\| \exp(At)\eta \| \le \frac{\varepsilon}{2} \quad \forall t \ge 0.$$

For all complex $\eta \in \mathbb{C}^N$ with $\|\eta\| \le \delta$, we have $\|\operatorname{Re}\eta\| \le \delta$ and $\|\operatorname{Im}\eta\| \le \delta$, and hence

$$\| \exp(At)\eta \| \le \| \exp(At)\operatorname{Re}\eta \| + \| \exp(At)\operatorname{Im}\eta \| \le \varepsilon \quad \forall t \ge 0.$$

(In the above, $\operatorname{Re}\eta$ and $\operatorname{Im}\eta$ should be interpreted in the componentwise sense, that is, as real vectors the components of which are, respectively, the real and imaginary parts of the components of η.) Now let $\xi \in \mathbb{C}^N$, $\xi \ne 0$, and set $\eta := (\delta/\|\xi\|)\xi$. Then $\|\eta\| \le \delta$. Hence,

$$\| \exp(At)\xi \| = \frac{\|\xi\|}{\delta} \| \exp(At)\eta \| \le \frac{\varepsilon}{\delta}\|\xi\| \quad \forall t \ge 0,$$

showing that $\| \exp(At) \| \le \varepsilon/\delta$ for all $t \ge 0$. Consequently, by statement (1) of Theorem 2.12, $\operatorname{Re}\lambda \le 0$ for all $\lambda \in \sigma(A)$. Invoking Corollary 2.13 shows that, if $\lambda \in \sigma(A)$ and $\operatorname{Re}\lambda = 0$, then λ is semisimple.

Finally, the implication (2)$\Rightarrow$(1) follows immediately from Corollary 2.13. $\qquad\square$

Definition 5.24

The equilibrium 0 of (5.12) is said to be *exponentially stable* if there exist constants $M \ge 1$ and $\alpha > 0$ such that

$$\| \exp(At)\xi \| \le M e^{-\alpha t}\|\xi\| \quad \forall t \ge 0, \ \forall \xi \in \mathbb{R}^N.$$

The following proposition shows in particular that, for linear systems, asymptotic, global asymptotic and exponential stability are equivalent concepts and that they can be characterized in terms of $\sigma(A)$. We say that A is a *Hurwitz matrix* or that A is *Hurwitz*[10] if $\sigma(A) \subset \{\lambda \in \mathbb{C} \colon \operatorname{Re}\lambda < 0\} =: \mathbb{C}_-$.

Proposition 5.25

Let $A \in \mathbb{R}^{N \times N}$. The following statements are equivalent.

(1) A is Hurwitz.

---

[10] Adolf Hurwitz (1859-1919), German.

(2) 0 is an exponentially stable equilibrium of (5.12).

(3) 0 is a globally asymptotically stable equilibrium of (5.12).

(4) 0 is an asymptotically stable equilibrium of (5.12).

(5) 0 is an attractive equilibrium of (5.12).

Proof

The implication (1)⇒(2) is an immediate consequence of Corollary 2.13 and the implications (2)⇒(3)⇒(4)⇒(5) hold trivially. It remains to show that (5)⇒(1). To this end assume that (5) holds. Then there exists $\delta > 0$ such that $\lim_{t\to\infty} \exp(At)\xi = 0$ for all real $\xi \in \mathbb{R}^N$ with $\|\xi\| \le \delta$. For all complex $\xi \in \mathbb{C}^N$ with $\|\xi\| \le \delta$, we have $\|\mathrm{Re}\,\xi\| \le \delta$ and $\|\mathrm{Im}\,\xi\| \le \delta$, and thus

$$\exp(At)\xi = \exp(At)\mathrm{Re}\,\xi + i\exp(At)\mathrm{Im}\,\xi \to 0 \ \text{ as } t \to \infty.$$

Now let $\xi \in \mathbb{C}^N$ with $\|\xi\| > \delta$ and set $\eta := (\delta/\|\xi\|)\xi$. Then $\|\eta\| \le \delta$ and thus

$$\exp(At)\xi = \frac{\|\xi\|}{\delta}\exp(At)\eta \to 0 \ \text{ as } t \to \infty.$$

Hence, $\lim_{t\to\infty} \exp(At)\xi = 0$ for every $\xi \in \mathbb{C}^N$, and it follows from Corollary 2.13 that (1) holds.   □

We are now in a position to deduce a necessary and sufficient condition for exponential stability. We preface this result with an exercise, recording a basic fact frequently used in this and the next chapter.

Exercise 5.18

Let $M \in \mathbb{R}^{N \times N}$ and consider the *quadratic form*

$$q\colon \mathbb{R}^N \to \mathbb{R}, \quad z \mapsto \langle z, Mz \rangle.$$

Show that $(\nabla q)(z) = (M + M^*)z$ for all $z \in \mathbb{R}^N$.

Theorem 5.26

The matrix A is Hurwitz if, only if, for each symmetric positive-definite matrix $Q \in \mathbb{R}^{N \times N}$, the equation

$$PA + A^*P + Q = 0 \tag{5.13}$$

has a symmetric positive-definite solution $P \in \mathbb{R}^{N \times N}$.

Equation (5.13) is called the *Lyapunov matrix equation*. We note that, in the scalar case (that is, $N = 1$), Theorem 5.26 is trivially true. It is remarkable that it generalizes to higher dimensions in a straightforward way. The proof given below shows that, if P is a symmetric positive-definite solution of (5.13) for some positive-definite matrix $Q \in \mathbb{R}^{N \times N}$, then the function $V \colon \mathbb{R}^N \to \mathbb{R}$ given by $V(z) = \langle z, Pz \rangle$ is a Lyapunov function for the linear system (5.12) and, moreover, $\langle (\nabla V)(z), Az \rangle = -\langle z, Qz \rangle < 0$ for all $z \neq 0$. This Lyapunov function is sometimes described as a *quadratic* Lyapunov function (because it is given by a quadratic form).

Proof of Theorem 5.26

Necessity. Assume that A is Hurwitz. Then, by Proposition 5.25, there exist constants $M \geq 1$ and $\alpha > 0$ such that

$$\| \exp At \| \leq M e^{-\alpha t} \quad \forall\, t \geq 0 \,.$$

Let $Q \in \mathbb{R}^{N \times N}$ be a symmetric positive-definite matrix. Then

$$P := \int_0^\infty \exp(A^* t) Q \exp(At) \, \mathrm{d}t$$

is a well-defined symmetric positive-definite matrix in $\mathbb{R}^{N \times N}$ satisfying

$$PA + A^* P = \int_0^\infty \frac{\mathrm{d}}{\mathrm{d}t} \big(\exp(A^* t) Q \exp(At) \big) \mathrm{d}t = -Q \,.$$

Sufficiency. Let $Q \in \mathbb{R}^{N \times N}$ be a symmetric positive-definite matrix and let $P \in \mathbb{R}^{N \times N}$ be a symmetric positive-definite solution of the Lyapunov equation (5.13). Defining

$$V \colon \mathbb{R}^N \to \mathbb{R}, \ z \mapsto \langle z, Pz \rangle \,,$$

we have $V(0) = 0$, $V(z) > 0$ for all $z \neq 0$ and (invoking Exercise 5.18)

$$\langle (\nabla V)(z), Az \rangle = 2\langle Pz, Az \rangle = \langle z, (PA + A^* P)z \rangle = -\langle z, Qz \rangle < 0 \quad \forall\, z \neq 0 \,.$$

By Corollary 5.17, 0 is an asymptotically stable equilibrium of (5.12) and so, by Proposition 5.25, A is Hurwitz. $\qquad\square$

Exercise 5.19

Consider the damped linear harmonic oscillator described by $\ddot{y} + \dot{y} + y = 0$. Written as a first-order system, this becomes

$$\dot{x}_1 = x_2, \quad \dot{x}_2 = -x_1 - x_2 \,.$$

(a) By solving (5.13) with $Q = I$, find a Lyapunov function $V \colon \mathbb{R}^2 \to \mathbb{R}$ for the above system with the property that the derivative of V along non-zero solutions is negative.

(b) Show that $V \colon \mathbb{R}^2 \to \mathbb{R}$, $z \mapsto \|z\|^2$ is a Lyapunov function for the system. Compare this Lyapunov function with that found in part (a).

5.5 Nonlinearly perturbed linear systems

Let $G \subset \mathbb{R}^N$ be a nonempty open subset with $0 \in G$. Consider the differential equation

$$\dot{x} = Ax + h(x),\tag{5.14}$$

where $A \in \mathbb{R}^{N \times N}$ and $h \colon G \to \mathbb{R}^N$ is a continuous function satisfying

$$\lim_{z \to 0} \frac{h(z)}{\|z\|} = 0.\tag{5.15}$$

In particular, $h(0) = 0$ and so 0 is an equilibrium for (5.14). The next result confirms that, if the equilibrium 0 of the linear system $\dot{x} = Ax$ is asymptotically stable, then the latter property persists under the perturbation h.

Theorem 5.27

Assume that $h \colon G \to \mathbb{R}^N$ is continuous and satisfies (5.15). If A is Hurwitz, then 0 is an asymptotically stable equilibrium of (5.14).

Proof

By Theorem 5.26, there exists a positive symmetric matrix $P \in \mathbb{R}^{N \times N}$ satisfying $PA + A^*P + I = 0$. Furthermore, by openness of G and (5.15) there exists $\varepsilon > 0$ such that $U := \mathbb{B}(0, \varepsilon) \subset G$ and

$$\|z\| < \varepsilon \quad \Longrightarrow \quad \|h(z)\| \le \|z\|/(4\|P\|)\,.$$

Defining $V \colon U \to \mathbb{R}$, $z \mapsto \langle z, Pz \rangle$ and $f \colon G \to \mathbb{R}^N$, $z \mapsto Az + h(z)$, we have, for all $z \in U$,

$$V_f(z) = 2\langle Pz, Az + h(z) \rangle \le -\|z\|^2 + 2\|h(z)\|\|P\|\|z\| \le -\|z\|^2/2.$$

Therefore, by Exercise 5.13, 0 is asymptotically stable. $\qquad\square$

Example 5.28

Consider the second-order equation (of van der Pol type)

$$\ddot{y} + (1 - y^2)\dot{y} + y = 0$$

which, on writing $x = (x_1, x_2) = (y, \dot{y})$, can be expressed as

$$\begin{pmatrix} \dot{x}_1 \\ \dot{x}_2 \end{pmatrix} = \begin{pmatrix} x_2 \\ -x_1 - x_2 + x_1^2 x_2 \end{pmatrix} = \begin{pmatrix} 0 & 1 \\ -1 & -1 \end{pmatrix} \begin{pmatrix} x_1 \\ x_2 \end{pmatrix} + \begin{pmatrix} 0 \\ x_1^2 x_2 \end{pmatrix} = Ax + h(x),$$

where

$$A = \begin{pmatrix} 0 & 1 \\ -1 & -1 \end{pmatrix} \quad \text{and} \quad h \colon \mathbb{R}^2 \to \mathbb{R}^2, \quad \begin{pmatrix} z_1 \\ z_2 \end{pmatrix} \mapsto \begin{pmatrix} 0 \\ z_1^2 z_2 \end{pmatrix}.$$

Clearly, 0 is an equilibrium of the system and $h(z)/\|z\| \to 0$ as $z \to 0$. Since $\sigma(A) = \{(-1 \pm i\sqrt{3})/2\} \subset \mathbb{C}_-$, it follows from Theorem 5.27 that 0 is an asymptotically stable equilibrium of the nonlinear system.   △

Exercise 5.20

Consider the system, with $G = \mathbb{R}^3$,

$$\dot{x}_1 = -2x_1 + x_1^2 |x_3| + x_2,$$
$$\dot{x}_2 = x_1 \sin x_3 - x_2 + 4x_3,$$
$$\dot{x}_3 = x_1 x_2 - x_2 x_3 - x_3.$$

Show that 0 is an asymptotically stable equilibrium.

5.6 Linearization of nonlinear systems

Throughout this section, we assume that $f \colon G \to \mathbb{R}^N$ in (5.1) is continuous, with $0 \in G$ and $f(0) = 0$. In addition, we assume that f is differentiable at 0 and define

$$A := (Df)(0) = ((\partial_j f_i)(0))_{1 \leq i, j \leq N}. \tag{5.16}$$

By definition of the (Fréchet) derivative $(Df)(0)$ of f at 0 (see Appendix A.3), we have $\lim_{z \to 0} (f(z) - Az)/\|z\| = 0$.

The aim of this section is to investigate situations in which stability/instability properties of the equilibrium 0 of (5.1) may be deduced from those of the *linearized system* $\dot{x} = Ax$.

The following is a simple consequence of Theorem 5.27.

Corollary 5.29

If A is Hurwitz, then 0 is an asymptotically stable equilibrium of the nonlinear system (5.1).

Proof

Defining $h\colon G \to \mathbb{R}^N$ by $h(z) := f(z) - Az$, we have $f(z) = Az + h(z)$ for all $z \in G$ and $h(z)/\|z\| \to 0$ as $z \to 0$. An application of Theorem 5.27 yields the claim. $\qquad\square$

Exercise 5.21

Consider the linear controlled and observed system

$$\dot{x} = Ax + bu, \quad y = c^* x,$$

where $A \in \mathbb{R}^{N \times N}$ and $b, c \in \mathbb{R}^N$. Application of the feedback $u = -ky$, where $k \in \mathbb{R}$ is a constant, leads to the feedback system $\dot{x} = (A - kbc^*)x$. Assume that this system is asymptotically stable for every $k \in (\alpha, \beta)$, where $\alpha < \beta$. Show that, for every continuously differentiable $\psi\colon \mathbb{R} \to \mathbb{R}$ with $\psi(0) = 0$ and $\psi'(0) \in (\alpha, \beta)$, the nonlinear feedback $u = -\psi(y)$ renders the equilibrium 0 of the nonlinear feedback system $\dot{x} = Ax - b\psi(c^*x)$ asymptotically stable.

Exercise 5.22

Consider the system in Exercise 5.20. Show that all (nine) partial derivatives of the right-hand side exist at the origin. Explain why Corollary 5.29 cannot be applied to this system in order to establish asymptotic stability of the equilibrium 0.

If 0 is a stable, but not asymptotically stable, equilibrium of the linearized system, then we cannot infer stability of the equilibrium 0 of the nonlinear system, as the following example shows.

Example 5.30

Let $a \in \mathbb{R}$ and consider $\dot{x} = ax^3$ on $G = \mathbb{R}$. The linearized equation is $\dot{x} = 0$ and so 0 is a stable equilibrium of the linearized system. However, separation of variables shows that the unique solution of the differential equation $\dot{x} = ax^3$ satisfying the initial condition $x(0) = \xi$ is given by

$$x(t) = \frac{\xi}{\sqrt{1 - 2a\xi^2 t}},$$

implying that the stability of the nonlinear system depends on the sign of a: the equilibrium 0 is stable if, and only if, $a \leq 0$. $\triangle$

On the other hand, if 0 is an *unstable* equilibrium of the linearized system (in the sense that A has at least one eigenvalue with positive real part), then we may infer that 0 is an unstable equilibrium of the nonlinear system. This is the content of the following theorem.

Theorem 5.31

If A has an eigenvalue with positive real part, then 0 is an unstable equilibrium of the nonlinear system (5.1).

We preface the proof with a technicality.

Lemma 5.32

If $A \in \mathbb{R}^{N \times N}$ has an eigenvalue with positive real part, then there exist constant $\mu > 0$, symmetric $P \in \mathbb{R}^{N \times N}$ and symmetric positive-definite $Q \in \mathbb{R}^{N \times N}$ such that

$$PA + A^*P = \mu P + Q \quad \text{and} \quad \langle z, Pz \rangle > 0 \text{ for some } z \in \mathbb{R}^N. \tag{5.17}$$

Proof

Define

$$\mu := \min\{\operatorname{Re}\lambda \colon \lambda \in \sigma(A) \cap \mathbb{C}_+\} > 0.$$

Then $M := A - (\mu/2)I$ has no eigenvalues on the imaginary axis and has at least one eigenvalue with positive real part. As a consequence of the Jordan form theorem (see Theorem A.9), there exists an invertible matrix T such that

$$TMT^{-1} = \begin{pmatrix} M_+ & 0 \\ 0 & M_- \end{pmatrix}$$

where the eigenvalues of M_+ (respectively, M_-) have positive (respectively, negative) real parts. Of course, if every eigenvalue of A has positive real part, then M_- is vacuous, $T = I$ and $M_+ = M$. Theorem 5.26 guarantees the existence of real, symmetric, positive-definite matrices (of appropriate dimensions) P_+ and P_- such that

$$-P_+M_+ - M_+^*P_+ + I = 0 \quad \text{and} \quad P_-M_- + M_-^*P_- + I = 0.$$

Defining symmetric $P \in \mathbb{R}^{N \times N}$ and symmetric positive-definite $Q \in \mathbb{R}^{N \times N}$ by

$$P := T^* \begin{pmatrix} P_+ & 0 \\ 0 & -P_- \end{pmatrix} T \quad \text{and} \quad Q := T^*T$$

we have

$$PA + A^*P = PM + M^*P + \mu P = T^*T + \mu P = Q + \mu P.$$

Finally, let $z_+ \in \mathbb{R}^{N_+} \setminus \{0\}$, where $N_+ \times N_+$ is the format of the matrix M_+. Since T is invertible, there exists $z \in \mathbb{R}^N \setminus \{0\}$ such that

$$Tz = \begin{pmatrix} z_+ \\ 0 \end{pmatrix}.$$

Thus,

$$\langle z, Pz \rangle = \left\langle Tz, \begin{pmatrix} P_+ & 0 \\ 0 & P_- \end{pmatrix} Tz \right\rangle = \langle z_+, P_+ z_+ \rangle > 0,$$

completing the proof. $\qquad \square$

We now provide the proof of Theorem 5.31.

Proof of Theorem 5.31

The proof is based on an application of Theorem 5.7 in conjunction with Lemma 5.32. First, we write f in the form

$$f(z) = Az + h(z) \ \forall z \in G,$$

where $A := (Df)(0)$. Then $h(z) = f(z) - Az$ for all $z \in G$ and

$$\lim_{z \to 0} \frac{h(z)}{\|z\|} = 0. \tag{5.18}$$

Moreover, by Lemma 5.32, there exist symmetric $P \in \mathbb{R}^{N \times N}$, a constant $\mu > 0$ and symmetric, positive-definite $Q \in \mathbb{R}^{N \times N}$ such that (5.17) holds. Since the function $z \mapsto \langle z, Qz \rangle$ is continuous and the sphere $\{z \in \mathbb{R}^N : \|z\| = 1\}$ is compact, it follows (see statement (2) of Proposition A.18) that

$$q := \inf_{\|z\|=1} \langle z, Qz \rangle > 0.$$

Thus,

$$q\|z\|^2 \le \langle z, Qz \rangle \ \forall z \in \mathbb{R}^N.$$

Let $\varepsilon > 0$ be such that $U := \mathbb{B}(0, \varepsilon) \subset G$ and

$$2\|h(z)\|\|P\| \le q\|z\| \ \forall z \in U.$$

The existence of such an $\varepsilon > 0$ follows from (5.18). Consequently,

$$2\|h(z)\|\|Pz\| \le 2\|h(z)\|\|P\|\|z\| \le q\|z\|^2 \le \langle z, Qz \rangle \quad \forall\, z \in U. \tag{5.19}$$

Defining $V \colon U \to \mathbb{R}$ by setting $V(z) := \langle z, Pz \rangle$ for all $z \in U$, it follows from (5.17) that

$$V_f(z) = 2\langle Pz, Az + h(z) \rangle = \langle z, (PA + A^*P)z \rangle + 2\langle Pz, h(z) \rangle$$
$$\ge \mu V(z) + \langle z, Qz \rangle - 2\|h(z)\|\|Pz\| \quad \forall\, z \in U.$$

Invoking (5.19), we conclude that if $z \in U$ and $V(z) > 0$, then $V_f(z) > 0$. Therefore, hypothesis (1) of Theorem 5.7 holds. Finally, by (5.17), there exists $z \ne 0$ such that $\langle z, Pz \rangle > 0$. Let $\delta > 0$ and choose $c > 0$ sufficiently small so that $\xi = cz \in U$ and $\|\xi\| < \delta$. Then $V(\xi) = V(cz) = c^2\langle z, Pz \rangle > 0$ and so hypothesis (2) of Theorem 5.7 also holds and so we may infer that 0 is an unstable equilibrium for (5.1).   □

Exercise 5.23

Consider the Lorenz system, with $G = \mathbb{R}^3$,

$$\dot{x}_1 = \sigma(x_2 - x_1), \quad \dot{x}_2 = rx_1 - x_2 - x_1x_3, \quad \dot{x}_3 = x_1x_2 - bx_3\,,$$

with three positive parameters b, r and σ (see also Exercise 5.16). Show that, if $r > 1$, then the equilibrium 0 is unstable.

Theorem 5.31 shows that, if $0 \in G$, $f(0) = 0$ and $A := (Df)(0)$ has at least one eigenvalue with positive real part, then the equilibrium 0 is unstable. We can say more if *all* eigenvalues of A have positive real part. In particular, we will show that the unstable equilibrium is *repelling* in the following sense. Assume that $0 \in G$, $f(0) = 0$ and f is locally Lipschitz with associated local flow φ. Then the equilibrium 0 of (5.1) is *repelling* if there exists a neighbourhood $U \subset G$ of 0 such that, for each $\xi \in U \backslash \{0\}$, there exists $\tau \in I_\xi$ such that $\varphi(t, \xi) \notin U$ for all $t \in [\tau, \infty) \cap I_\xi$. In words, 0 is a repelling equilibrium if it has a neighbourhood such that every non-zero solution starting in the neighbourhood must, in forwards time, ultimately exit the neighbourhood and not return.

The next result identifies conditions under which 0 is a repelling equilibrium of (5.1).

Theorem 5.33

Assume that f is locally Lipschitz and $\mathrm{Re}\,(\lambda) > 0$ for all $\lambda \in \sigma(A)$. Then 0 is a repelling equilibrium of the nonlinear system (5.1).

Proof

Define $h\colon G \to \mathbb{R}^N$, $z \mapsto h(z) := f(z) - Az$. By hypothesis, the matrix $-A$ is Hurwitz and so, by Theorem 5.26, there exists symmetric positive-definite $P \in \mathbb{R}^{N \times N}$ such that

$$PA + A^*P = 2I.$$

Define $V\colon \mathbb{R}^N \to \mathbb{R}_+$, $z \mapsto V(z) := \langle z, Pz \rangle$ and $\mu := \|P\| > 0$, and note that $V(z) \leq \mu \|z\|^2$ for all $z \in \mathbb{R}^N$.

Since $h(z)/\|z\| \to 0$ as $z \to 0$, we may choose $\varepsilon > 0$ sufficiently small so that

$$C := \{z \in \mathbb{R}^N : V(z) \leq \varepsilon\} \subset G \quad \text{and} \quad 2\mu \|h(z)\| \leq \|z\| \ \forall\, z \in C.$$

Observe that

$$
\begin{aligned}
\langle (\nabla V)(z), f(z) \rangle = 2\langle Pz, (Az + h(z)) \rangle &= \langle (A^*P + PA)z, z \rangle + 2\langle Pz, h(z) \rangle \\
&\geq 2\|z\|^2 - 2\|P\| \|z\| \|h(z)\| \geq \|z\|^2 \\
&\geq V(z)/\mu \ \forall\, z \in C.
\end{aligned}
\tag{5.20}
$$

In particular, we have $\langle (\nabla V)(z), f(z) \rangle \geq \varepsilon/\mu > 0$ for all z with $V(z) = \varepsilon$ and so the set $D := \{z \in G : V(z) \geq \varepsilon\}$ is positively invariant under the local flow φ. Therefore, to establish that 0 is a repelling equilibrium, it suffices to show that 0 has a neighbourhood $U \subset G$ such that $U \cap D = \emptyset$ and $O^+(\xi) \cap D \neq \emptyset$ for all $\xi \in U \backslash \{0\}$. Define $U := \{z \in \mathbb{R}^N : V(z) < \varepsilon\} \subset C \subset G$. Seeking a contradiction, suppose that there exists $\xi \in U \backslash \{0\}$ such that $O^+(\xi) \cap D = \emptyset$. Then $O^+(\xi)$ is contained $U \subset C$. By compactness of C, we have $\mathbb{R}_+ \subset I_\xi$ and, by (5.20),

$$(V \circ x)'(t) = \langle (\nabla V)(x(t)), f(x(t)) \rangle \geq (V \circ x)(t)/\mu \ \forall\, t \in \mathbb{R}_+.$$

Therefore,

$$\frac{d}{dt}\left(e^{-t/\mu}(V \circ x)(t)\right) = e^{-t/\mu}\left((V \circ x)'(t) - (V \circ x)(t)/\mu\right) \geq 0 \ \forall\, t \in \mathbb{R}_+,$$

which, on integration, yields $e^{-t/\mu}(V \circ x)(t) \geq (V \circ x)(0)$ for all $t \in \mathbb{R}_+$. We now have

$$\varepsilon > V(x(t)) \geq V(\xi)e^{t/\mu} \ \forall\, t \in \mathbb{R}_+$$

which is impossible since $V(\xi) > 0$ and $\mu > 0$. Therefore, our supposition is false and so 0 is a repelling equilibrium. $\qquad\square$

5.7 Nonlinear systems and exponential stability

Let $G = \mathbb{R}^N$, let $f : \mathbb{R}^N \to \mathbb{R}^N$ be locally Lipschitz, with $f(0) = 0$ and let φ be the local flow generated by

$$\dot{x}(t) = f(x(t)). \tag{5.21}$$

The following is a "nonlinear" counterpart of Definition 5.24.

Definition 5.34

The equilibrium 0 of system (5.21) is said to be *exponentially stable* if there exist constants $M \geq 1$ and $\alpha > 0$ such that

$$\|\varphi(t, \xi)\| \leq M e^{-\alpha t} \|\xi\| \quad \forall \xi \in \mathbb{R}^N, \ \forall t \in I_\xi \cap \mathbb{R}_+. \tag{5.22}$$

Note that, if the equilibrium 0 of system (5.21) is exponentially stable, then $\mathbb{R}_+ \subset I_\xi$ for all $\xi \in \mathbb{R}^N$. The next result encapsulates sufficient conditions for exponential stability.

Theorem 5.35

Let $f : \mathbb{R}^N \to \mathbb{R}^N$ be locally Lipschitz with $f(0) = 0$. If there exists a continuously differentiable function $V : \mathbb{R}^N \to \mathbb{R}$ such that, for some positive constants $a_1, a_2, a_3 > 0$,

$$a_1 \|z\|^2 \leq V(z) \leq a_2 \|z\|^2, \quad V_f(z) \leq -a_3 \|z\|^2; \quad \forall z \in \mathbb{R}^N,$$

then 0 is an exponentially stable equilibrium for (5.21).

Proof

Let φ be the local flow generated by (5.21). By Theorem 5.22, 0 is a globally asymptotically stable equilibrium and so, for all $\xi \in \mathbb{R}^N$, $\mathbb{R}_+ \subset I_\xi$. Let $\xi \in \mathbb{R}^N$ be arbitrary. If $\xi = 0$, then $\varphi(t, \xi) = 0$ for all $t \in \mathbb{R}_+$ and so (5.22) holds trivially (for every $M \geq 1$ and $\alpha > 0$). Now assume that $\xi \neq 0$ and write $x(t) := \varphi(t, \xi) \neq 0$ for all $t \in \mathbb{R}_+$. Observing that

$$(V \circ x)'(t) = V_f(x(t)) \leq -a_3 \|x(t)\|^2 \quad \forall t \in \mathbb{R}_+,$$

and setting $\alpha := a_3/(2a_2)$, we conclude

$$(V \circ x)'(t) \leq -2\alpha V(x(t)) \quad \forall t \in \mathbb{R}_+.$$

Dividing both sides of this inequality by $V(x(t)) > 0$ and integrating, we obtain

$$\ln V(x(t)) \leq \ln V(\xi) - 2\alpha t = \ln\left(e^{-2\alpha t}V(\xi)\right) \quad \forall t \in \mathbb{R}_+.$$

This leads to

$$a_1\|x(t)\|^2 \leq V(x(t)) \leq e^{-2\alpha t}V(\xi) \leq a_2 e^{-2\alpha t}\|\xi\|^2 \quad \forall t \in \mathbb{R}_+.$$

We may now conclude that

$$\|x(t)\| = \|\varphi(t,\xi)\| \leq Me^{-\alpha t}\|\xi\| \quad \forall t \in \mathbb{R}_+,$$

where $M := \sqrt{a_2/a_1}$. Therefore, (5.22) holds and so 0 is an exponentially stable equilibrium for (5.21). $\qquad\square$

Exercise 5.24

Consider again the Lorenz system

$$\dot{x}_1 = \sigma(x_2 - x_1), \quad \dot{x}_2 = rx_1 - x_2 - x_1x_3, \quad \dot{x}_3 = x_1x_2 - bx_3,$$

see also Exercises 5.16 and 5.23. The three real parameters b, r and σ are all positive. Assume that $0 < r < 1$. In view of Exercise 5.16, we know that 0 is a globally asymptotically stable equilibrium. By reconsidering the function V given in the hint for Exercise 5.16 in conjunction with Theorem 5.35, deduce that 0 is an exponentially stable equilibrium.

5.8 Input-to-state stability

This section comprises a study of stability-type questions pertaining to systems with input u (which, on the one hand, may be an extraneous disturbance or, on the other hand, may be a control open to choice):

$$\dot{x}(t) = f(x(t), u(t)), \quad x(0) = \xi. \tag{5.23}$$

It is assumed that $f: \mathbb{R}^N \times \mathbb{R}^M \to \mathbb{R}^N$ is locally Lipschitz and the input u is piecewise continuous. Input-to-state stability investigates properties of the map

$$(\xi, u(\cdot)) \mapsto x(\cdot)$$

via a concept that encompasses intuitively appealing modes of dynamic behaviour such as:

the *bounded-input bounded-state* (BIBS) property

$$u \text{ bounded} \implies x \text{ bounded},$$

and the *0-convergent-input 0-convergent-state* (0-CICS) property

$$u(t) \to 0 \text{ as } t \to \infty \implies x(t) \to 0 \text{ as } t \to \infty.$$

5.8.1 Linear prototype

By way of motivation, we first consider the linear initial-value problem

$$\dot{x}(t) = Ax(t) + Bu(t), \quad x(0) = \xi, \quad A \in \mathbb{R}^{N \times N}, \; B \in \mathbb{R}^{N \times M}. \tag{5.24}$$

We will be concerned only with behaviour in forwards time, that is, we restrict attention to input functions u defined on $\mathbb{R}_+$. By Section 2.2, for every $\xi \in \mathbb{R}^N$ and every piecewise continuous input $u \colon \mathbb{R}_+ \to \mathbb{R}^M$, (5.24) has a unique (piecewise continuously differentiable) solution $x \colon \mathbb{R}_+ \to \mathbb{R}^N$ given by

$$x(t) = \exp(At)\xi + \int_0^t \exp(A(t-s))Bu(s)\,\mathrm{d}s \quad \forall\, t \in{\geq}\, 0. \tag{5.25}$$

If we assume that, with zero input $u = 0$, the equilibrium 0 is (globally) asymptotically stable, or equivalently, that A is Hurwitz (see Proposition 5.25), then there exist $M \geq 1$ and $\alpha > 0$ such that

$$\|\exp(At)\| \leq Me^{-\alpha t} \quad \forall\, t \geq 0.$$

Therefore, a straightforward estimate of the right-hand side of (5.25) gives

$$\|x(t)\| \leq Me^{-\alpha t}\|\xi\| + M\|B\| \sup_{s \in [0,t]} \|u(s)\| \int_0^t e^{-\alpha(t-s)}\,\mathrm{d}s \quad \forall\, t \geq 0$$

and so, on writing $\gamma := M\|B\|/\alpha$, we have

$$\|x(t)\| \leq Me^{-\alpha t}\|\xi\| + \gamma \sup_{s \in [0,t]} \|u(s)\| \quad \forall\, t \geq 0. \tag{5.26}$$

It immediately follows that (5.24) has the BIBS property.

Now assume that $u(t) \to 0$ as $t \to \infty$ and define

$$U := \sup\{\|u(s)\| \colon s \in \mathbb{R}_+\} \quad \text{and} \quad U_t := \sup\{\|u(s)\| \colon s \in [t/2, \infty)\}.$$

Obviously, $U < \infty$ and $U_t \to 0$ as $t \to \infty$. Decomposing the integral on the right-hand side of (5.25) in the form

$$\int_0^t \exp(A(t-s))Bu(s)\,\mathrm{d}s = \left(\int_0^{t/2} + \int_{t/2}^t\right) \exp(A(t-s))Bu(s)\,\mathrm{d}s$$

and invoking estimates similar to those leading to (5.26) yields

$$\|x(t)\| \leq Me^{-\alpha t}\|\xi\| + \gamma e^{-\alpha t/2}U + \gamma U_t \quad \forall\, t \geq 0,$$

showing that $x(t) \to 0$ as $t \to \infty$ and so (5.24) has the 0-CICS property:

$$u(t) \to 0 \text{ as } t \to \infty \implies x(t) \to 0 \text{ as } t \to \infty.$$

Thus, for the linear system (5.24), if the matrix A is Hurwitz, then the BIBS property and the 0-CICS property hold.

Exercise 5.25

Assume that in (5.24) the matrix A is Hurwitz and let $u: \mathbb{R}_+ \to \mathbb{R}^M$ be piecewise continuous input such that the limit $\lim_{t \to \infty} u(t) =: u^\infty$ exists. Show that, for every $\xi \in \mathbb{R}^N$, the solution x of (5.24) satisfies

$$\lim_{t \to \infty} x(t) = -A^{-1} B u^\infty.$$

Exercise 5.25 shows that if A is Hurwitz, then the linear system (5.24) has not only the 0-CICS property, but the "stronger" convergent-input-convergent-state property

$$u(t) \text{ converges as } t \to \infty \implies x(t) \text{ coverges as } t \to \infty,$$

where the limit of x depends only on the limit of u, but not on the initial state.

5.8.2 Nonlinear systems

Consider now the nonlinear system

$$\dot{x}(t) = f(x(t), u(t)), \quad x(0) = \xi, \tag{5.27}$$

where $f: \mathbb{R}^N \times \mathbb{R}^M \to \mathbb{R}^N$ is assumed to be locally Lipschitz, that is, for every $z \in \mathbb{R}^N \times \mathbb{R}^M$, there exists an open neighbourhood Z of z and $L > 0$ such that $\|f(v) - f(w)\| \le L\|v - w\|$ for all $v, w \in Z$.

Exercise 5.26

Let J be an interval and let $u : J \to \mathbb{R}^M$ be piecewise continuous. Assume that $f: \mathbb{R}^N \times \mathbb{R}^M \to \mathbb{R}^N$ is locally Lipschitz. Show that the function $f_u : J \times \mathbb{R}^N \to \mathbb{R}^N$ defined by $f_u(t, z) := f(z, u(t))$ satisfies Assumption **A** of Section 4.4.

As in Section 5.8.1, we will be concerned only with behaviour in forwards time and henceforth will assume that the piecewise continuous u is defined on $\mathbb{R}_+$. Exercise 5.26 combined with Theorem 4.22 shows that for every piecewise continuous $u : \mathbb{R}_+ \to \mathbb{R}^M$, the initial-value problem (5.27) has a unique maximal solution $x : [0, \omega) \to \mathbb{R}^N$, where $0 < \omega \le \infty$; moreover, by Theorem 4.25, if $\omega < \infty$, then $\|x(t)\| \to \infty$ as $t \to \omega$.

The intuitive counterpart of the condition that the matrix A in (5.24) is Hurwitz is the property that, with zero input $u = 0$, the origin 0 in $\mathbb{R}^N$ is an equilibrium of the autonomous system

$$\dot{x}(t) = f(x(t), 0)$$

(that is, $f(0,0) = 0$) and this equilibrium is globally asymptotically stable. We refer to this property by the acronym 0-GAS, which, in the context of the linear system (5.24) is equivalent to the Hurwitz property (and so ensures the BIBS and 0-CICS properties).

In the context of the nonlinear system (5.27), the 0-GAS property does not, in general, imply either the BIBS or the 0-CICS property, as illustrated by the following examples.

Example 5.36

Consider the scalar system given by $\dot{x} = -x + x^2 u$ which clearly has the 0-GAS property: however, with initial data $x(0) = 1$ and bounded and 0-convergent input $u: t \mapsto 2e^{-t}$, the system has unbounded solution $x: t \mapsto e^t$ and so both the BIBS and 0-CICS properties fail to hold. $\triangle$

Example 5.37

For 0-GAS nonlinear systems, the BIBS and/or 0-CICS properties may fail to hold in a manner more dramatic than that of the previous example. In particular, with input $u : \mathbb{R}_+ \to \mathbb{R}$ given by

$$u(t) := \left\{ \begin{array}{ll} 2, & t \in [0,1] \\ 0, & t > 1, \end{array} \right.$$

and initial data $x(0) = 1$, the scalar system $\dot{x} = (u - 1)x|x|$ has solution $t \mapsto x(t) = 1/(1 - t)$ on $[0,1)$, with $x(t) \to \infty$ as $t \uparrow 1$: thus, this system, which has the 0-GAS property (see Exercise 5.27) can exhibit finite-time blow-up of the state for a bounded and 0-convergent input u. $\triangle$

Exercise 5.27

Show that the system $\dot{x} = (u - 1)x|x|$ has the 0-GAS property.

What separates the linear case from the nonlinear case? The distinguishing feature of the former is the fact that, in the context of (5.24), we have equivalence of the following:

(a) A is Hurwitz,

(b) there exist $M \geq 1$, $\alpha > 0$ and $\gamma > 0$ such that (5.26) holds for all $\xi \in \mathbb{R}^N$ and all piecewise continuous inputs u.

We have already established that (a) implies (b). To establish the reverse implication, simply observe that, on setting $u = 0$, (b) implies exponential stability of the homogeneous system $\dot{x} = Ax$, and so (a) holds (see Proposition 5.25).

In the nonlinear case, we seek a counterpart to (5.26) which implies (but is not necessarily implied by) the 0-GAS property and, moreover, ensures both the BIBS and 0-CICS properties. This requires further notation and terminology. In particular, in addition to the classes $\mathcal{K}$ and $\mathcal{K}_\infty$ of comparison functions introduced in Section 5.1, it is convenient to introduce so-called $\mathcal{KL}$ functions: a function $b\colon \mathbb{R}_+ \times \mathbb{R}_+ \to \mathbb{R}_+$ is said to be a $\mathcal{KL}$ function or a function of class $\mathcal{KL}$ if, for each $t \in \mathbb{R}_+$, $b(\cdot, t) \in \mathcal{K}$ and, for each $s \in \mathbb{R}_+$, $b(s, \cdot)$ is decreasing with $b(s, t) \to 0$ as $t \to \infty$. We now introduce a generalization of (5.26).

Definition 5.38

System (5.27) is *input-to-state stable* (ISS) if there exist $\beta \in \mathcal{KL}$ and $\gamma \in \mathcal{K}_\infty$ such that, for all $\xi \in \mathbb{R}^N$ and all piecewise continuous $u\colon \mathbb{R}_+ \to \mathbb{R}^M$, the unique maximal solution $x\colon [0, \omega) \to \mathbb{R}^N$ of (5.27) is such that

$$\|x(t)\| \le \beta(\|\xi\|, t) + \gamma\left(\sup_{s \in [0, t]} \|u(s)\|\right) \quad \forall t \in [0, \omega). \tag{5.28}$$

(Note that (5.28) implies that $\omega = \infty$.)

Exercise 5.28

Show system (5.27) is ISS if, and only if, there exist $\beta \in \mathcal{KL}$ and $\gamma \in \mathcal{K}_\infty$ such that, for all $\xi \in \mathbb{R}^N$ and all piecewise continuous $u\colon \mathbb{R}_+ \to \mathbb{R}^M$, the unique maximal solution $x\colon [0, \omega) \to \mathbb{R}^N$ of (5.27) satisfies

$$\|x(t)\| \le \max\left\{\beta(\|\xi\|, t),\, \gamma\left(\sup_{s \in [0, t]} \|u(s)\|\right)\right\} \quad \forall t \in [0, \omega).$$

Examples 5.36 and 5.37 serve to confirm that the 0-GAS property does not imply that (5.27) is ISS. However, the reverse implication is true, and furthermore, ISS implies the BIBS and 0-CICS properties, as we shall now show.

Theorem 5.39

If system (5.27) is ISS, then it has the 0-GAS, BIBS and 0-CICS properties.

Proof

Assume that system (5.27) is ISS. To show that it has the 0-GAS property, let φ be the local flow generated by the system $\dot{x} = f(x, 0)$. Then, by (5.28) (with $u = 0$), we may infer that

$$\|\varphi(t, \xi)\| \le \beta(\|\xi\|, t) \quad \forall (t, \xi) \in \mathbb{R}_+ \times \mathbb{R}^N.$$

Therefore, $\|\varphi(t,0)\| \le \beta(0,t) = 0 \ \forall t \in \mathbb{R}_+$ and so $f(0,0) = 0$. Moreover, since $\beta(\|\xi\|,t) \downarrow 0$ as $t \to \infty$ for all ξ, it follows that the equilibrium 0 is globally attractive. To show that this equilibrium is stable, let $\varepsilon > 0$ be arbitrary. Since $\beta(\cdot,0) \in \mathcal{K}$, there exists $\delta > 0$ such that $\beta(\delta,0) \le \varepsilon$ and thus, for all $\xi \in \mathbb{R}^N$ with $\|\xi\| \le \delta$, we have

$$\|\varphi(t,\xi)\| \le \beta(\|\xi\|,t) \le \beta(\|\xi\|,0) \le \beta(\delta,0) \le \varepsilon \ \forall t \in \mathbb{R}_+,$$

and so the equilibrium 0 of $\dot{x} = f(x,0)$ is stable. Therefore, (5.27) has the 0-GAS property.

The BIBS property of (5.27) is an immediate consequence of the estimate (5.28). In particular, if $\|u(t)\| \le U$ for all $t \in \mathbb{R}_+$, then $\|x(t)\| \le \beta(\|\xi\|,0)+\gamma(U)$ for all $t \in \mathbb{R}_+$.

Finally, to show that (5.27) has the 0-CICS property, let $\xi \in \mathbb{R}^N$ and assume that $u\colon \mathbb{R}_+ \to \mathbb{R}^M$ is piecewise continuous and such that $u(t) \to 0$ as $t \to \infty$. Let $x : [0,\omega) \to \mathbb{R}^N$ be the corresponding maximal solution. Since (5.27) is ISS, it follows that $\omega = \infty$. Let $\varepsilon > 0$ be arbitrary. It suffices to show that there exists $T_\varepsilon > 0$ such that $\|x(t)\| \le \varepsilon$ for all $t \ge T_\varepsilon$. Since γ is a $\mathcal{K}_\infty$ function, it has an inverse function $\gamma^{-1} \in \mathcal{K}_\infty$. Define $\delta := \gamma^{-1}(\varepsilon/2)$ and choose $T > 0$ sufficiently large so that $\|u(t)\| \le \delta$ for all $t \ge T$. Define $\tilde{\xi} := x(T)$ and $\tilde{u}\colon \mathbb{R}_+ \to \mathbb{R}^M$, $t \mapsto \tilde{u}(t) := u(t+T)$. Then $\|\tilde{u}(t)\| \le \delta$ for all $t \ge 0$ and the unique maximal solution $\tilde{x}\colon \mathbb{R}_+ \to \mathbb{R}^N$ of the initial value problem

$$\dot{y}(t) = f(y(t),\tilde{u}(t)), \quad y(0) = \tilde{\xi}$$

is given by $\tilde{x}(t) = x(t+T)$ for all $t \ge 0$. By ISS, (5.28) holds with $\tilde{x}, \tilde{\xi}$ and $\tilde{u}$ replacing x, ξ and u, respectively, and thus,

$$\|\tilde{x}(t)\| \le \beta(\|\tilde{\xi}\|,t) + \gamma(\delta) = \beta(\|\tilde{\xi}\|,t) + \frac{\varepsilon}{2} \ \forall t \ge 0.$$

Now choose $S > 0$ sufficiently large so that $\beta(\|\tilde{\xi}\|,t) \le \varepsilon/2$ for all $t \ge S$ and define $T_\varepsilon := S + T$. We may now infer that

$$\|x(t+T_\varepsilon)\| = \|\tilde{x}(t+S)\| \le \frac{\varepsilon}{2} + \frac{\varepsilon}{2} = \varepsilon \ \forall t \ge 0,$$

or, equivalently, $\|x(t)\| \le \varepsilon$ for all $t \ge T_\varepsilon$. This completes the proof. $\qquad\square$

Example 5.40

Let $f\colon \mathbb{R}^N \times \mathbb{R}^M \to \mathbb{R}^N$ and $g\colon \mathbb{R}^M \to \mathbb{R}^M$ be locally Lipschitz functions. Assume that the system $\dot{x} = f(x,u)$ is ISS and that 0 is a globally asymptotically stable equilibrium of the system $\dot{y} = g(y)$. Consider the cascade of these two systems given by the interconnection equation $u = y$, see Figure 5.2. The

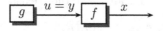

Figure 5.2 Cascade of two systems

cascade system is described by

$$\dot{y}(t) = g(y(t)), \quad \dot{x}(t) = f(x(t), y(t)). \tag{5.29}$$

Claim. The origin $(0,0)$ is a globally asymptotically stable equilibrium of the cascade system (5.29).

By ISS of $\dot{x} = f(x, u)$, it is clear that $f(0,0) = 0$. Moreover, $g(0) = 0$, and so $(0,0)$ is an equilibrium of (5.29). To show that this equilibrium is globally asymptotically stable, we first confirm that $(0,0)$ is globally attractive. Let $(\eta, \xi) \in \mathbb{R}^M \times \mathbb{R}^N$ be arbitrary. Let ψ denote the local flow generated by $\dot{y} = g(y)$. By global asymptotic stability of 0, we have that $\mathbb{R}_+ \times \mathbb{R}^M \subset \mathrm{dom}(\psi)$ and $\psi(t, \eta) \to 0$ as $t \to \infty$. By Theorem 5.39, the ISS property of the system $\dot{x} = f(x, u)$ guarantees that the 0-CICS property holds and so the unique maximal solution x of the initial-value problem

$$\dot{x}(t) = f(x(t), \psi(t, \eta)), \quad x(0) = \xi \tag{5.30}$$

is such that $x(t) \to 0$ as $t \to \infty$. Therefore, the equilibrium $(0,0)$ of the cascade (5.29) is globally attractive.

It remains to show that $(0,0)$ is stable. By the ISS property of $\dot{x} = f(x, u)$, there exist $\beta \in \mathcal{KL}$ and $\gamma \in \mathcal{K}_\infty$ such that, for each $(\eta, \xi) \in \mathbb{R}^M \times \mathbb{R}^N$, the unique maximal solution x of the initial-value problem (5.30) satisfies

$$\|x(t)\| \le \beta(\|\xi\|, t) + \gamma\big(\sup_{s \in [0,t]} \|\psi(t, \eta)\|\big) \quad \forall t \in \mathbb{R}_+.$$

Let $\varepsilon > 0$ be arbitrary. Since 0 is a stable equilibrium of $\dot{y} = g(y)$, there exists $\delta_1 > 0$ such

$$\|\eta\| \le \delta_1 \implies \|\psi(t, \eta)\| \le \min\{\varepsilon/3, \gamma^{-1}(\varepsilon/3)\} \quad \forall t \in \mathbb{R}_+.$$

Now choose $\delta \in (0, \delta_1]$ sufficiently small so that $\beta(\delta, 0) \le \varepsilon/3$. We may now infer that the unique maximal solution (x, y) of the initial-value problem for the cascade

$$\dot{y} = g(y), \quad \dot{x} = f(x, y), \quad (y(0), x(0)) = (\eta, \xi) \in \mathbb{R}^M \times \mathbb{R}^N$$

is such that, for all $t \in \mathbb{R}_+$ and all $(\eta, \xi) \in \mathbb{R}^M \times \mathbb{R}^N$ with $\|(\eta, \xi)\| \le \delta$,

$$\|(y(t), x(t))\| \le \|x(t)\| + \|\psi(t, \eta)\| \le \beta(\delta, 0) + \gamma\big(\gamma^{-1}(\varepsilon/3)\big) + \varepsilon/3 \le \varepsilon.$$

Thus, we have shown that, for all $\varepsilon > 0$, there exists $\delta > 0$ such that

$$\|(\eta, \xi)\| \leq \delta \implies \|(y(t), x(t))\| \leq \varepsilon \ \forall t \in \mathbb{R}_+,$$

and so $(0,0)$ is a stable equilibrium of the cascade (5.29). $\triangle$

Next, we provide a sufficient condition for input-to-state stability formulated in terms of a Lyapunov-like function.

Theorem 5.41

Let $f \colon \mathbb{R}^N \times \mathbb{R}^M \to \mathbb{R}^N$ be locally Lipschitz. If there exist a continuously differentiable function $V \colon \mathbb{R}^N \to \mathbb{R}$ and functions $a_1, a_2, a_3, a_4 \in \mathcal{K}_\infty$ such that

$$a_1(\|z\|) \leq V(z) \leq a_2(\|z\|) \ \forall z \in \mathbb{R}^N \tag{5.31}$$

$$V_f(z, v) := \langle (\nabla V)(z), f(z, v) \rangle \leq -a_3(\|z\|) + a_4(\|v\|) \ \forall (z, v) \in \mathbb{R}^N \times \mathbb{R}^M, \tag{5.32}$$

then system (5.27) is ISS.

We first present two lemmas which are used in the proof of Theorem 5.41.

Lemma 5.42

Let $a \in \mathcal{K}_\infty$. Then there exists a continuously differentiable $b \in \mathcal{K}_\infty$ such that $b(s) < a(s)$ for all $s > 0$.

Proof

Define $c \colon \mathbb{R}_+ \to \mathbb{R}_+$ by

$$c(s) := \frac{2}{s} \int_{s/2}^s a(t) \mathrm{d}t \ \forall s > 0, \quad c(0) := 0.$$

Since $a \in \mathcal{K}_\infty$, we have $a(s/2) < c(s) < a(s)$ for all $s > 0$ and it is clear that c is continuous and $\lim_{s\to\infty} c(s) = \infty$. Moreover, by the fundamental theorem of calculus, c is continuously differentiable on $(0, \infty)$ with derivative

$$c'(s) = \frac{1}{s}\big(2a(s) - a(s/2) - c(s)\big) > 0 \ \forall s > 0. \tag{5.33}$$

In particular, c is strictly increasing, and so $c \in \mathcal{K}_\infty$. Now define $b \colon \mathbb{R}_+ \to \mathbb{R}_+$ by $b(s) := c(s)d(s)$, where

$$d(s) = \begin{cases} s(2 - s), & s \in [0, 1] \\ 1, & s > 1. \end{cases}$$

Note that $0 \leq d(s) \leq 1$ for all $s \in \mathbb{R}_+$, d is non-decreasing and d is continuously differentiable. Clearly, $b \in \mathcal{K}_\infty$, $b(s) < a(s)$ for all $s > 0$ and b is continuously differentiable on $(0, \infty)$. Also, b is also differentiable at 0 with derivative

$$b'(0) = \lim_{s \to 0} \frac{1}{s} c(s) d(s) = \lim_{s \to 0} c(s)(2 - s) = 0.$$

Finally, invoking (5.33), we have for all $s \in (0, 1)$,

$$b'(s) = c'(s)d(s) + d'(s)c(s) = (2 - s)\big(2a(s) - a(s/2) - c(s)\big) + 2(1 - s)c(s).$$

Hence, $\lim_{s \to 0} b'(s) = 0 = b'(0)$, which shows that b is continuously differentiable on $[0, \infty)$, completing the proof. $\qquad\square$

Lemma 5.43

Consider the scalar differential equation $\dot{x} = g(x)$, where $g \colon \mathbb{R} \to \mathbb{R}$ is locally Lipschitz, and let ψ denote the corresponding local flow. Let $\xi \in \mathbb{R}$ and $\tau > 0$ be such that $\tau \in I_\xi$, where I_ξ denotes the maximal interval of existence of the initial-value problem $\dot{x} = g(x)$, $x(0) = \xi$. Assume that $y \colon [0, \tau] \to \mathbb{R}$ is piecewise continuously differentiable and such that

$$\dot{y}(t) \leq g(y(t)) \ \forall t \in [0, \tau] \backslash E, \quad y(0) = \xi,$$

where E is the set of points in $[0, \tau]$ at which y fails to be differentiable. Then $y(t) \leq \psi(t, \xi)$ for all $t \in [0, \tau]$.

Proof

For notational convenience, write $x(t) = \psi(t, \xi)$ for all $t \in [0, \tau]$. Seeking a contradiction, suppose that there exists $t_1 \in [0, \tau]$ such that $y(t_1) > x(t_1)$. Define the set $T := \{t \in [0, t_1] \colon y(t) \leq x(t)\}$ (non-empty, since $0 \in T$) and write $t_0 := \sup T$. Then $y(t_0) = x(t_0)$ and $y(t) > x(t)$ for all $t_0 < t \leq t_1$. Therefore, for $t \in [t_0, t_1]$,

$$y(t) - x(t) = |y(t) - x(t)| \leq \left| \int_{t_0}^t \big(g(y(s)) - g(x(s))\big) ds \right|$$

$$\leq \int_{t_0}^t |g(y(s)) - g(x(s))| ds.$$

Let K be a compact interval such that $y(t), x(t) \in K$ for all $t \in [t_0, t_1]$. By the local Lipschitz property of g and Proposition 4.15, there exists $L \geq 0$ such that $|g(z_1) - g(z_2)| \leq L|z_1 - z_2|$ for all $z_1, z_2 \in K$. Therefore,

$$|y(t) - x(t)| \leq L \int_{t_0}^t |y(s) - x(s)| ds \ \forall t \in [t_0, t_1].$$

By Gronwall's lemma (Lemma 2.4), it follows that $y(t) = x(t)$ for all $t \in [t_0, t_1]$ which contradicts the fact that $y(t_1) > x(t_1)$. $\square$

We are now in the position to prove Theorem 5.41.

Proof of Theorem 5.41

Let $\xi \in \mathbb{R}^N$ and piecewise continuous $u \colon \mathbb{R}_+ \to \mathbb{R}^M$ be arbitrary. Let $x \colon [0, \omega) \to \mathbb{R}^N$ be the unique maximal solution of (5.27). Since x is piecewise continuously differentiable, so is $V \circ x$ and, by (5.32),

$$\frac{d}{dt}(V \circ x)(t) \leq a_4(\|u(t)\|) \quad \forall t \in [0, \omega) \backslash E, \tag{5.34}$$

where E is the set of all points in $[0, \omega)$ at which x fails to be differentiable. We claim that $\omega = \infty$. Let $\tau \in (0, \infty)$ be such that $[0, \tau) \subset [0, \omega)$. Clearly, the piecewise continuous function $u \colon \mathbb{R}_+ \to \mathbb{R}^M$ is bounded on the bounded interval $[0, \tau)$ and so, by continuity of a_4, there exists $c > 0$ such that $a_4(\|u(t)\|) \leq c$ for all $t \in [0, \tau)$. By (5.31) and (5.34), $a_1(\|x(t)\|) \leq V(x(t)) \leq V(\xi) + c\tau$ for all $t \in [0, \tau)$, whence

$$\|x(t)\| \leq a_1^{-1}\big(V(\xi) + c\tau\big) \quad \forall t \in [0, \tau). \tag{5.35}$$

Note that, if $\omega < \infty$, then (5.35) holds for $\tau = \omega$, which is impossible by Theorem 4.25. Therefore, $\omega = \infty$.

Let $b \in \mathcal{K}_\infty$ be continuously differentiable and such that $b(s) \leq (a_3 \circ a_2^{-1})(s)$ for all $s \in \mathbb{R}_+$ (such a function exists by Lemma 5.42). Define the locally Lipschitz function $g \colon \mathbb{R} \to \mathbb{R}$ by

$$g(s) := \begin{cases} -b(s)/2, & s \geq 0 \\ 0, & s < 0. \end{cases}$$

and let ψ be the local flow generated by the scalar system $\dot{y} = g(y)$. Obviously, 0 is an equilibrium of this system and $g(s)s < 0$ for all $s > 0$. It follows in particular that $\mathbb{R}_+ \times \mathbb{R}_+ \subset \text{dom}(\psi)$ and the function $\mathbb{R}_+ \times \mathbb{R}_+ \to \mathbb{R}_+$, $(r, t) \mapsto \psi(t, r)$ is of class $\mathcal{KL}$ (see Exercise 5.29). We will show that the ISS estimate (5.28) holds with $\beta \in \mathcal{KL}$ and $\gamma \in \mathcal{K}_\infty$ defined by

$$\beta(r, t) := a_1^{-1}\big(\psi(t, a_2(r))\big), \quad \gamma(s) := \big(a_1^{-1} \circ a_2 \circ a_3^{-1}\big)(2a_4(s)).$$

Let $U \colon \mathbb{R}_+ \to \mathbb{R}_+$ be given by $U(t) := \sup_{\sigma \in [0,t]} \|u(\sigma)\|$ and define

$$T_1 := \{t \in \mathbb{R}_+ : V(x(t)) \leq \big(a_2 \circ a_3^{-1}\big)\big(2a_4(U(t))\big)\},$$
$$T_2 := \mathbb{R}_+ \backslash T_1 = \{t \in \mathbb{R}_+ : V(x(t)) > \big(a_2 \circ a_3^{-1}\big)\big(2a_4(U(t))\big)\}.$$

Invoking (5.31), we have

$$\|x(t)\| \le \gamma(U(t)) \quad \forall\, t \in T_1,\tag{5.36}$$

and

$$2a_4(U(t)) < (a_3 \circ a_2^{-1})(V(x(t))) \le a_3(\|x(t)\|) \quad \forall\, t \in T_2.\tag{5.37}$$

Invoking (5.32), we have

$$(V \circ x)'(t) \le -a_3(\|x(t)\|) + a_4(U(t)) \quad \forall\, t \in \mathbb{R}_+ \backslash E,$$

which, in conjunction with (5.37), gives

$$(V \circ x)'(t) \le -\frac{a_3(\|x(t)\|)}{2} \quad \forall\, t \in T_2 \backslash E.$$

Observe that, for all $t \in T_2$,

$$0 > 2g(V(x(t))) = -b(V(x(t))) \ge -(a_3 \circ a_2^{-1})(V(x(t))) \ge -a_3(\|x(t)\|)$$

and so

$$(V \circ x)'(t) \le g((V \circ x)(t)) < 0 \quad \forall\, t \in T_2 \backslash E.\tag{5.38}$$

Next, we claim that

$$t \in T_2 \implies [0,t] \subset T_2.\tag{5.39}$$

Fix $t \in T_2$ arbitrarily, write $c := (a_2 \circ a_3^{-1})(2a_4(U(t)))$ and note that $V(x(t)) > c$. . Since the function U is non-decreasing, to conclude that $[0,t] \in T_2$ it suffices to show that $V(x(s)) > c$ for all $s \in [0,t]$. Seeking a contradiction, suppose otherwise. Then, by continuity of $V \circ x$, there exists at least one point $s \in [0,t]$ such that $V(x(s)) = c$. Let σ by the supremum of the set of all such points s, that is, $\sigma := \sup\{s \in [0,t]\colon V(x(s)) = c\}$. Then $\sigma \in [0,t)$, $(\sigma,t] \subset T_2$ and $V(x(t)) - V(x(\sigma)) > 0$. The latter inequality contradicts the fact that, by (5.38),

$$(V \circ x)'(t) < 0 \quad \forall\, t \in (\sigma,t] \backslash E$$

Therefore, property (5.39) holds. Combining (5.38) and (5.39), we obtain

$$(V \circ x)'(s) \le g((V \circ x)(s)) \quad \forall\, s \in [0,t] \backslash E \quad \forall\, t \in T_2.$$

Therefore, by Lemma 5.43, $(V \circ x)(s) \le \psi(s, V(\xi))$ for all $s \in [0,t]$ and all $t \in T_2$. In particular, we have

$$V(x(t)) \le \psi(t, V(\xi)) \quad \forall\, t \in T_2.$$

Finally, invoking (5.31), we conclude that

$$\|x(t)\| \le \beta(\|\xi\|, t) \quad \forall\, t \in T_2,$$

which, in conjunction with (5.36), now yields

$$\|x(t)\| \le \beta(\|\xi\|, t) + \gamma(\sup_{s \in [0,t]} \|u(s)\|) \quad \forall\, t \in \mathbb{R}_+,$$

completing the proof. $\qquad\square$

Exercise 5.29

Consider the local flow ψ generated by the (scalar) differential equation $\dot{y} = g(y)$ in the proof of Theorem 5.41. Show that

(a) $\mathbb{R}_+ \times \mathbb{R}_+ \subset \mathrm{dom}(\psi)$;

(b) $\psi(t, r) \in \mathbb{R}_+$ for all $(t, r) \in \mathbb{R}_+ \times \mathbb{R}_+$;

(c) the function $\theta \colon \mathbb{R}_+ \times \mathbb{R}_+ \to \mathbb{R}_+$, $(r, t) \mapsto \theta(r, t) := \psi(t, r)$ is of class $\mathcal{KL}$.

Sometimes the following corollary is easier to use than Theorem 5.41.

Corollary 5.44

Let $f \colon \mathbb{R}^N \times \mathbb{R}^M \to \mathbb{R}^N$ be locally Lipschitz and assume that $f(0, 0) = 0$. If there exist a continuously differentiable function $V \colon \mathbb{R}^N \to \mathbb{R}$ and functions $a_1, a_2, b_1, b_2 \in \mathcal{K}_\infty$ such that (5.31) holds and

$$\|z\| \geq b_1(\|v\|) \implies V_f(z, v) = \langle (\nabla V)(z), f(z, v) \rangle \leq -b_2(\|z\|), \qquad (5.40)$$

then system (5.27) is ISS.

Proof

By Theorem 5.41, it is sufficient to show that there exist $a_3, a_4 \in \mathcal{K}_\infty$ such that (5.32) holds. To this end, define $b_3 \colon \mathbb{R}_+ \to \mathbb{R}_+$ by

$$b_3(s) := \sup\{|V_f(z, w)| \colon \|z\| \leq b_1(s), \|w\| \leq s\} + b_2(b_1(s)) \qquad (5.41)$$

Then $b_3 \in \mathcal{K}_\infty$ (see Exercise 5.30) and, invoking (5.40), shows that

$$V_f(z, v) \leq -b_2(\|z\|) + b_3(\|v\|) \ \forall \, (z, v) \in \mathbb{R}^N \times \mathbb{R}^M.$$

Hence (5.32) holds with $a_3 = b_2$ and $a_4 = b_3$. $\square$

Exercise 5.30

Show that the function b_3 defined by (5.41) is in $\mathcal{K}_\infty$.

Example 5.45

Consider system (5.27) with $f \colon \mathbb{R} \times \mathbb{R} \to \mathbb{R}$, $(z, v) \mapsto -z^3 + z^2 v$. Let $V \colon \mathbb{R} \to \mathbb{R}$ be defined by $V(z) = z^2$. Then

$$V_f(z, v) = V'(z) f(z) = -2z^4 + 2z^3 v \ \forall \, (z, v) \in \mathbb{R}^2. \qquad (5.42)$$

Note that, if $|z| \geq 2|v|$, then

$$V_f(z,v) \leq -2|z|^4 + 2|v||z|^3 \leq -|z|^4 - 2|v||z|^3 + 2|v||z|^3 = -|z|^4.$$

By Corollary 5.44, with $a_1(s) = a_2(s) = s^2$, $b_1(s) = 2s$ and $b_2(s) = s^4$, it follows that the system is ISS.

Alternatively, Theorem 5.41 may be invoked. Indeed, by making the additional observation that, if $|z| < 2|v|$, then

$$V_f(z,v) \leq -|z|^4 + 16|v|^4,$$

we may conclude that

$$V_f(z,v) \leq -|z|^4 + 16|v|^4 \quad \forall\, (z,v) \in \mathbb{R}^2.$$

Hence, an application of Theorem 5.41 (with $a_1(s) = a_2(s) = s^2$, $a_3(s) = s^4$ and $a_4(s) = 16s^4$) shows that the system is ISS. $\triangle$

In applications of Theorem 5.41 and Corollary 5.44 to higher-dimensional examples, the following lemma is useful.

Lemma 5.46

Let $W : \mathbb{R}^N \to \mathbb{R}_+$ be continuous and such that $W(0) = 0$, $W(z) > 0$ for every $z \neq 0$ and W is radially unbounded, that is, $W(z) \to \infty$ as $\|z\| \to \infty$. Then there exist functions $w_1, w_2 \in \mathcal{K}_\infty$ such that $w_1(\|z\|) \leq W(z) \leq w_2(\|z\|)$ for all $z \in \mathbb{R}^N$.

Proof

Define $a_1 : \mathbb{R}_+ \to \mathbb{R}_+$ by

$$a_1(s) := \inf\{W(y) : \|y\| \geq s\}.$$

Then a_1 is a non-decreasing function with the properties $a_1(0) = 0$, $a_1(s) > 0$ for $s > 0$, $\lim_{s \to \infty} a_1(s) = \infty$ and $W(z) \geq a_1(\|z\|)$ for $z \in \mathbb{R}^N$. Moreover, a_1 is continuous (see Exercise 5.31). Since a_1, in general, is not strictly increasing we cannot conclude that a_1 is in $\mathcal{K}_\infty$. However, the function $w_1 : \mathbb{R}_+ \to \mathbb{R}_+$ given by $w_1(s) := (1 - e^{-s})a_1(s)$ is strictly increasing and such that $w_1(s) \leq a_1(s)$. Consequently, $w_1 \in \mathcal{K}_\infty$ and $w_1(\|z\|) \leq W(z)$ for all $z \in \mathbb{R}^N$.

Similarly, defining $a_2 : \mathbb{R}_+ \to \mathbb{R}_+$

$$a_2(s) := \sup\{W(y) : \|y\| \leq s\},$$

we have that a_2 is non-decreasing, with the properties $a_2(0) = 0$, $a_2(s) > 0$ for all $s > 0$, $\lim_{s \to \infty} a_2(s) = \infty$ and $W(z) \leq a_2(\|z\|)$ for all $z \in \mathbb{R}^N$. Moreover,

a_2 is continuous (see Exercise 5.31). Now define $w_2 \colon \mathbb{R}_+ \to \mathbb{R}_+$ by $w_2(s) := e^s a_2(s)$, for which the requisite properties hold: $w_2 \in \mathcal{K}_\infty$ and $W(z) \le w_2(\|z\|)$ for all $z \in \mathbb{R}^N$.   □

Exercise 5.31

Prove continuity of the functions a_1 and a_2 defined in the proof of Lemma 5.46.

Example 5.47

To show that the system

$$\dot{x}_1 = -x_1 + x_2, \quad \dot{x}_2 = -x_1^3 - x_2 + u$$

is ISS, we consider the function $V \colon \mathbb{R}^2 \to \mathbb{R}$ given by $V(z) = V(z_1, z_2) = z_1^4 + 2z_2^2$. Obviously, by Lemma 5.46, there exist $a_1, a_2 \in \mathcal{K}_\infty$ such that

$$a_1(\|z\|) \le V(z) \le a_2(\|z\|) \ \forall\, z \in \mathbb{R}^2.$$

Setting
$$f(z,v) = (-z_1 + z_2, -z_1^3 - z_2 + v) \ \forall\, (z,v) \in \mathbb{R}^2 \times \mathbb{R},$$

we obtain,

$$V_f(z,v) = -4z_1^4 - 4z_2^2 + 4z_2 v \le -2V(z) + 4z_2 v \ \forall\, (z,v) \in \mathbb{R}^2 \times \mathbb{R}.$$

Noting that $4z_2 v \le 2z_2^2 + 2v^2 \le V(z) + 2v^2$ for all $(z,v) \in \mathbb{R}^2 \times \mathbb{R}$, we have $V_f(z,v) \le -V(z) + 2v^2$ for all $(z,v) \in \mathbb{R}^2 \times \mathbb{R}$. Therefore,

$$\|z\| \ge a_1^{-1}(4v^2) \implies V(z) \ge 4v^2 \implies V_f(z,v) \le -V(z)/2.$$

An application of Corollary 5.44 (with $b_1, b_2 \in \mathcal{K}_\infty$ given by $b_1(s) := a_1^{-1}(4s^2)$ and $b_2(s) = a_1(s)/2$) shows that the system is ISS.   △

Exercise 5.32

Show that the following systems are ISS.

(a) $\dot{x} = -x(1 + 2x^2) + (1 + x^2)u^2$.

(b) $\dot{x}_1 = -x_1 + x_2^2, \quad \dot{x}_2 = -x_2 + u$.

(c) $\dot{x}_1 = -x_1 - x_2 + u_1, \quad \dot{x}_2 = x_1 - x_2^3 + u_2$.

Stability of feedback systems and stabilization

By way of motivation, consider the linear one-dimensional controlled system

$$\dot{x} = ax + u, \quad x(0) = \xi \in \mathbb{R}, \tag{6.1}$$

with real parameter $a > 0$. With zero input $u = 0$, this system has solution x given by $x(t) = e^{at}\xi$ and so, for $\xi \neq 0$, $|x(t)|$ diverges to infinity exponentially fast. Deeming this behaviour as undesirable, the aim is to choose the input u in such a way that the state $x(t)$ converges to 0 as $t \to \infty$. We describe and discuss two approaches to this problem: precomputed control (also called "open-loop control") and feedback (or "closed-loop") control. The discussion is predicated on the realistic assumption that, in practical situations, neither the system parameter a nor the initial state ξ will be precisely known: instead, only estimates $\hat{a}$ and $\hat{\xi}$ of the true values may be available to the controller.

Pre-computed control. Consider the choice of input function u given by

$$u(t) = -(\hat{a} + 1)e^{-t}\hat{\xi} \quad \forall t \in \mathbb{R}_+, \tag{6.2}$$

which incorporates only the estimated parameter values $\hat{a}$ and $\hat{\xi}$. Assume initially that the estimates are exact, that is, $\hat{a} = a$ and $\hat{\xi} = \xi$. Then a routine calculation, involving the variation of parameters formula, shows that the solution of (6.1) is given by $x(t) = e^{-t}\xi$, and so indeed $x(t) \to 0$ as $t \to \infty$. Now consider the more realistic scenario wherein the estimates are inexact, that is, at least one of the following hold: $\hat{a} \neq a$ or $\hat{\xi} \neq \xi$.

H. Logemann and E. P. Ryan, *Ordinary Differential Equations*,
Springer Undergraduate Mathematics Series,
DOI: 10.1007/978-1-4471-6398-5_6, © Springer-Verlag London 2014

Applying the control (6.2) to system (6.1) gives

$$\dot{x}(t) = ax(t) - (\hat{a} + 1)e^{-t}\hat{\xi}, \quad x(0) = \xi.$$

A straightforward calculation, again invoking the variation of parameters formula, gives

$$(a + 1)x(t) = \big((a + 1)\xi - (\hat{a} + 1)\hat{\xi}\big)e^{at} + (\hat{a} + 1)\hat{\xi}e^{-t} \quad \forall t \in \mathbb{R}_+. \qquad (6.3)$$

Since $a > 0$, it follows that, for all estimates $\hat{a}$ and $\hat{\xi}$ such that

$$(\hat{a} + 1)\hat{\xi} \neq (a + 1)\xi,$$

not only does the state $x(t)$, given by (6.3), fail to converge to 0 but, in fact, $|x(t)| \to \infty$ exponentially fast as $t \to \infty$. Equivalently, each estimate pair $(\hat{a}, \hat{\xi})$ in the set $U = \{(\hat{a}, \hat{\xi}) \in \mathbb{R}^2 : (\hat{a} + 1)\hat{\xi} \neq (a + 1)\xi\}$ leads to an exponential growth of $|x|$. The set $U \subset \mathbb{R}^2$ is "large" in the sense that it is open and dense in $\mathbb{R}^2$. Therefore, "almost all" estimate pairs lead to exponentially fast diverging solutions. In this sense, the pre-computed control (6.2) is highly non-robust with respect to inaccuracies in the estimates of the parameter a and initial state ξ.

Feedback control. Now assume that the input $u(t)$ to system (6.1) at time t is a multiple (referred to as the control gain) of the current state $x(t)$ of the real system. Here, we are assuming that the instantaneous state $x(t)$ is available for feedback without error: in due course, we will also consider the case wherein the instantaneous state

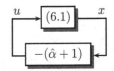

Figure 6.1 Feedback or closed-loop control

is available only to within a bounded error term. In particular, we consider the feedback control (see Figure 6.1)

$$u(t) = -(\hat{a} + 1)x(t) \qquad (6.4)$$

which incorporates only the estimate $\hat{a}$ of the true parameter a. If $\hat{a} = a$, that is, if the estimate is exact, then the conjunction of (6.1) and (6.4) gives $\dot{x} = -x$ with initial condition $x(0) = \xi$. The solution of this initial-problem is given by $x(t) = e^{-t}\xi$, which is identical to the solution corresponding to the open-loop control (6.2) with the exact estimates $\hat{a} = a$ and $\hat{\xi} = \xi$. If $\hat{a} \neq a$, then the feedback controlled system is

$$\dot{x} = (a - \hat{a} - 1)x, \quad x(0) = \xi,$$

with solution x given by $x(t) = e^{(a-\hat{a}-1)t}\xi$, which converges exponentially fast to 0 as $t \to \infty$ provided that $\hat{a} > a - 1$. [1]

Thus, we see that, in contrast to the pre-computed control control (6.2), for all estimates $\hat{a}$ satisfying $\hat{a} > a - 1$, application of the feedback control (6.4) to system (6.1) guarantees exponentially decaying solutions for all $\xi \in \mathbb{R}$.

Note that the pre-computed control function (6.2) depends only on the system parameter a and the initial value $x(0) = \xi$ and, in contrast with the feedback control (6.4), does not depend on $x(t)$ for $t > 0$ and hence does not receive any up-to-date information from the system it is supposed to control. The example illustrates a general phenomenon: a feedback law tends to attenuate perturbations that are not accounted for in the mathematical model, whilst a pre-computed control does not.

There is, of course, a logical inconsistency in the above analysis of feedback control: it is assumed that knowledge of the initial state ξ is imprecise, whereas the state $x(t)$ for $t > 0$ is assumed to be precisely known. In reality, only an estimate $\hat{x}(t)$ of the true state $x(t)$ will be available to the controller, in which case the feedback control is given by

$$u(t) = -(\hat{a} + 1)\hat{x}(t) \quad \forall t \in \mathbb{R}_+. \tag{6.5}$$

We assume that the true state can be estimated to within a bounded error and that the estimate $\hat{x}$ is piecewise continuous. In particular, defining $d := \hat{x} - x$, then d is piecewise continuous and, for some $\delta > 0$, $|d(t)| \leq \delta$ for all $t \in \mathbb{R}_+$. Applying (6.5) to system (6.1) yields

$$\dot{x}(t) = (a - \hat{a} - 1)x(t) - (\hat{a} + 1)d(t), \quad x(0) = \xi. \tag{6.6}$$

Invoking the variation of parameters formula, the solution x of (6.6) is given by

$$x(t) = e^{(a-\hat{a}-1)t}\xi - (\hat{a} + 1)e^{(a-\hat{a}-1)t} \int_0^t e^{(1+\hat{a}-a)s}d(s)\mathrm{d}s.$$

Since the only available information on d is that $|d(t)| \leq \delta$ for all $t \in \mathbb{R}_+$, we cannot determine x exactly. However, assuming that $c := a - \hat{a} - 1 < 0$, an upper bound on $|x(t)|$ is easily obtained:

$$|x(t)| \leq e^{ct}|\xi| + |\hat{a} + 1|e^{ct} \int_0^t e^{-cs}|d(s)|\mathrm{d}s$$

$$\leq e^{ct}|\xi| + \frac{\delta|\hat{a} + 1|}{|c|} \quad \forall t \in \mathbb{R}_+.$$

---

[1] This is not a strong restriction. If we assume that an upper bound for the system parameter $a > 0$ is known, that is, $a \leq b$ for some known $b > 0$, then adopting the estimate $\hat{a} = b$, we have $a - \hat{a} - 1 \leq -1$ and so exponential decay of solutions is ensured.

Since $c < 0$, the above inequality shows that there exists $T > 0$ such that

$$|x(t)| \leq \frac{2\delta|\hat{a} + 1|}{|c|} \quad \forall t \in [T, \infty).$$

In summary, if $a - \hat{a} - 1 < 0$ and the error in the measurement of the state is bounded, then the resulting solution of the feedback system is bounded. Moreover, for all sufficiently large $t \geq 0$, $|x(t)|$ is bounded by a constant that is proportional to the state measurement error bound $\delta > 0$.

The benign behaviour of the system under feedback (in particular, its ability to tolerate errors in the system model and in the measurement of the system state) is in stark contrast with the highly non-robust behaviour of the system under open-loop control.

We adopt the feedback viewpoint throughout this chapter: in particular, we investigate the stability properties of feedback systems and address the synthesis of stabilizing feedback laws for a variety of linear and nonlinear systems.

6.1 Linear systems and state feedback

Consider the controlled linear system

$$\dot{x} = Ax + Bu, \quad x(0) = \xi \in \mathbb{R}^N, \tag{6.7}$$

where $A \in \mathbb{R}^{N \times N}$ and $B \in \mathbb{R}^{N \times M}$. Application of feedback of the form

$$u(t) = Fx(t) + v(t), \quad \text{where } F \in \mathbb{R}^{M \times N}, \, v \in PC(\mathbb{R}_+, \mathbb{R}^M), \tag{6.8}$$

leads to

$$\dot{x} = (A + BF)x + Bv, \quad x(0) = \xi \tag{6.9}$$

which describes the *closed-loop* or *feedback system* shown in Figure 6.2. Feedback of the form (6.8) is also referred to as *state feedback*, because the whole state vector is used for feedback. The function v represents a signal impinging on the feedback system, for example a disturbance or reference signal. In the following, v is usually assumed to be identically equal to 0.

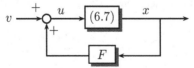

Figure 6.2 Feedback system

The aim is to choose F in such a way that the dynamics of the feedback system behave in a prescribed manner. For example, if the equilibrium 0 of the un-controlled system $\dot{x} = Ax$ is not exponentially stable, or, equivalently, A is not Hurwitz (see Proposition 5.25), we might wish to "stabilize" the system, that is, to choose F such that the equilibrium 0 of the system $\dot{x} = (A + BF)x$ is ex-ponentially stable, or, equivalently, $A + BF$ is Hurwitz. In particular, if $A + BF$ is Hurwitz, then, for every initial-condition $\xi \in \mathbb{R}^N$ and every $v \in PC(\mathbb{R}_+, \mathbb{R}^M)$ such that $v(t)$ converges to 0 as $t \to \infty$, the solution of the feedback system (6.9) converges to 0. Underlying this is the following question: to what extent can the spectrum $\sigma(A + BF)$ be influenced through choice of the feedback ma-trix F? We proceed to investigate this question of *eigenvalue assignment* (also referred to as the *pole placement* or *pole shifting* problem).

6.1.1 Eigenvalue assignment by state feedback

In the sequel, the following lemma for single-input systems ($M = 1$) will prove useful.

Lemma 6.1

Let $(A, b) \in \mathbb{R}^{N \times N} \times \mathbb{R}^N$, write $\det(sI - A) = s^N + \alpha_{N-1}s^{N-1} + \cdots + \alpha_0$ and set

$$A_c := \begin{pmatrix} 0 & 1 & 0 & \cdots & 0 \\ 0 & 0 & 1 & \cdots & 0 \\ \vdots & \vdots & \vdots & \ddots & \vdots \\ 0 & 0 & 0 & \cdots & 1 \\ -\alpha_0 & -\alpha_1 & -\alpha_2 & \cdots & -\alpha_{N-1} \end{pmatrix}, \quad b_c := \begin{pmatrix} 0 \\ 0 \\ \vdots \\ 0 \\ 1 \end{pmatrix}. \tag{6.10}$$

The pair (A, b) is controllable if, and only if, there exists $S \in GL(N, \mathbb{R})$ such that

$$A = S^{-1}A_cS, \quad b = S^{-1}b_c. \tag{6.11}$$

Proof

A routine application (see Exercise 6.1) of the rank condition (Theorem 3.6) establishes controllability of the pair (A_c, b_c).

Assume the existence of $S \in GL(N, \mathbb{R})$ such that (6.11) holds. Observe that

$$\text{rk}\,\mathcal{C}(A, b) = \text{rk}\big(S^{-1}\mathcal{C}(A_c, b_c)\big) = \text{rk}\,\mathcal{C}(A_c, b_c) = N$$

and so, by Theorem 3.6, (A, b) is controllable (this can also be established from first principles – recall Exercise 3.6).

Conversely, assume that (A, b) is controllable. Then $\mathcal{C}(A, b) \in \mathbb{R}^{N \times N}$ is invertible. Let $q \in \mathbb{R}^N$ be such that q^* forms the N-th row of the inverse $\mathcal{C}(A, b)^{-1}$ and define

$$S := \begin{pmatrix} q^* \\ q^* A \\ \vdots \\ q^* A^{N-1} \end{pmatrix} \in \mathbb{R}^{N \times N}.$$

First we show that $S \in GL(N, \mathbb{R})$. To this end, note that

$$q^* \mathcal{C}(A, b) = (0, 0, \ldots, 0, 1),$$

and thus

$$q^* b = q^* A b = \cdots = q^* A^{N-2} b = 0, \quad q^* A^{N-1} b = 1.$$

Consequently,

$$SC(A, b) = \begin{pmatrix} 0 & 0 & 0 & \cdots & 0 & 1 \\ 0 & 0 & 0 & \cdots & 1 & \star \\ \vdots & & & & \vdots & \vdots \\ 0 & 1 & \star & \cdots & \star & \star \\ 1 & \star & \star & \cdots & \star & \star \end{pmatrix}, \tag{6.12}$$

(the subdiagonal entries are immaterial here) showing that $SC(A, b)$ is invertible which in turn implies that $S \in GL(N, \mathbb{R})$. The identity (6.12) also shows that $Sb = b_c$, whence $b = S^{-1} b_c$.

Furthermore, since $\sum_{i=0}^{N-1} (-\alpha_k) A^k = A^N$ by the Cayley-Hamilton theorem (see Theorem A.6),

$$A_c S = \begin{pmatrix} q^* A \\ q^* A^2 \\ \vdots \\ q^* A^{N-1} \\ q^* \sum_{k=0}^{N-1} (-\alpha_k) A^k \end{pmatrix} = \begin{pmatrix} q^* A \\ q^* A^2 \\ \vdots \\ q^* A^{N-1} \\ q^* A^N \end{pmatrix} = SA.$$

Thus, $A = S^{-1} A_c S$. $\square$

Exercise 6.1

Prove that (A_c, b_c) is controllable.

For a controllable pair $(A, b) \in \mathbb{R}^{N \times N} \times \mathbb{R}^N$, the corresponding pair (A_c, b_c) is called the *controller form* of (A, b). The map $(A, b) \mapsto (A_c, b_c)$ that takes a controllable pair (A, b) to its controller form, is a a so-called *canonical form*. We make precise the meaning of this in the following exercise.

Exercise 6.2

Let S be a non-empty set with an equivalence relation $\sim$ defined thereon. Assume that $\Gamma\colon S \to S$ is a map with the properties: (i) $\Gamma(s) \sim s$ for all $s \in S$; (ii) if $s_1, s_2 \in S$ and $s_1 \sim s_2$, then $\Gamma(s_1) = \Gamma(s_2)$. Such a map is called a *canonical form* (relative to the relation $\sim$).

(a) Show that, for each $s \in S$, the equivalence class $[s] := \{\sigma \in S\colon \sigma \sim s\}$ has precisely one representative in $\Gamma(S)$: this unique representative is termed the canonical form for $s \in S$ (relative to the relation $\sim$). Thus, loosely speaking, a canonical form chooses a unique "natural" representative from each equivalence class.

(b) Set $S := \{(A, b) \in \mathbb{R}^{N \times N} \times \mathbb{R}^N \colon (A, b)$ is controllable$\}$ and let $\Gamma\colon S \to S$ be the map which takes a point $(A, b) \in S$ into its controller form (A_c, b_c). Define a relation $\sim$ on S by $(A_1, b_1) \sim (A_2, b_2)$ if, and only if, there exists $S \in GL(N, \mathbb{R})$ such that $A_2 = S^{-1} A_1 S$ and $b_2 = S^{-1} b_1$. Show that $\sim$ is an equivalence relation on S and deduce that Γ is a canonical form.

Consider a polynomial

$$P(s) = \alpha_N s^n + \alpha_{N-1} s^{N-1} + \cdots + \alpha_0, \quad a_N \neq 0$$

of degree N. We say that P is *monic* if $a_N = 1$. The polynomial P is said to be *real* if all coefficients α_j are real. For $A \in \mathbb{R}^{N \times N}$, let $P_A(s) := \det(sI - A)$ denote the characteristic polynomial of A. Then P_A is a monic real polynomial of degree N.

Definition 6.2

Let $(A, B) \in \mathbb{R}^{N \times N} \times \mathbb{R}^{N \times M}$. We say that a polynomial P can be *assigned* (or is *assignable*) to the pair (A, B) if there exists $F \in \mathbb{R}^{M \times N}$ such that $P_{A+BF} = P$.

Of course, if a polynomial P can be assigned to a pair $(A, B) \in \mathbb{R}^{N \times N} \times \mathbb{R}^{N \times M}$, then P is monic, real and of degree N. The next result, which implies that the dynamics of a controllable system can be changed drastically by state feedback, is one of the "highlights" of linear control theory.

Theorem 6.3 (Eigenvalue assignment)

Let $(A, B) \in \mathbb{R}^{N \times N} \times \mathbb{R}^{N \times M}$. Every monic real polynomial of degree N can be assigned to (A, B) if, and only if, (A, B) is controllable.

Proof

We use contraposition to show that controllability is necessary for every monic real polynomial of degree N to be assignable. Assume that (A, B) is not controllable. Then, by Theorem 3.11, there exists $\lambda \in \mathbb{C}$ such that $\operatorname{rk}(\lambda I - A, B) < N$. Consequently, there exists $z \in \mathbb{C}^N$, $z \neq 0$, such that $z^*(\lambda I - A, B) = 0$. Hence $z^*(\lambda I - A) = 0$ and $z^* B = 0$. Therefore,

$$z^*(\lambda I - A - BF) = 0 \quad \forall F \in \mathbb{R}^{M \times N},$$

showing that λ is an eigenvalue of $A + BF$ for every $F \in \mathbb{R}^{M \times N}$. We conclude that every polynomial P such that $P(\lambda) \neq 0$ cannot be assigned to (A, B).

To show that controllability is sufficient for every monic real polynomial of degree N to be assignable, assume that (A, B) is controllable. We first prove the claim if $M = 1$, in which case, we write $b := B$ to emphasize the temporary single-input context. By Lemma 6.1, there exists $S \in GL(N, \mathbb{R})$ such that $A_c := SAS^{-1}$ and $b_c := Sb$ take the form (6.10), wherein the α_j, $j = 0, \ldots, N - 1$, are the coefficients of P_A. Given any monic real polynomial of the form

$$P(s) = s^N + \beta_{N-1} s^{N-1} + \cdots + \beta_0,$$

define $f_c \in \mathbb{R}^N$ by

$$f_c^* := (\alpha_0 - \beta_0, \alpha_1 - \beta_1, \ldots, \alpha_{N-1} - \beta_{N-1}).$$

Then

$$A_c + b_c f_c^* = \begin{pmatrix} 0 & 1 & 0 & \cdots & 0 \\ 0 & 0 & 1 & \cdots & 0 \\ \vdots & & & & \vdots \\ 0 & 0 & 0 & \cdots & 1 \\ -\beta_0 & -\beta_1 & -\beta_2 & \cdots & -\beta_{N-1} \end{pmatrix}.$$

Therefore (see Exercise 6.3),

$$P_{A_c + b_c f_c^*}(s) = s^n + \beta_{N-1} s^{N-1} + \cdots + \beta_0 = P(s).$$

Defining $f := S^* f_c$, we have

$$A + bf^* = S^{-1}(A_c + b_c f_c^*)S,$$

and so

$$P_{A+bf^*}(s) = \det(sI - A - bf^*) = \det(sI - A_c - b_c f_c^*) = P_{A_c + b_c f_c^*}(s) = P(s),$$

showing that P can be assigned to (A, b).

Now let $M \in \mathbb{N}$ be arbitrary. Pick any vector $v \in \mathbb{R}^M$ such that $b := Bv \neq 0$ (such a vector exists by the controllability of (A, B)). We claim that it is sufficient to show the existence of a matrix $E \in \mathbb{R}^{M \times N}$ such that $(A + BE, b)$ is controllable. To see this, assume that E is such that $(A + BE, b)$ is controllable. By the result in the case when $M = 1$, for every monic real polynomial P of degree N, there exists $f \in \mathbb{R}^N$ such that

$$P(s) = \det(sI - A - BE - bf^*) = \det(sI - A - B(E + vf^*)).$$

Setting $F := E + vf^*$, we see $\det(sI - A - BF) = P(s)$, showing that P can be assigned to (A, B).

It remains to prove the existence of E such that $(A + BE, b)$ is controllable. Let $\{x_1, \ldots, x_m\}$, $1 \leq m \leq N$, be the largest set of linearly independent vectors in $\mathbb{R}^N$ with the properties

$$x_1 = b = Bv, \quad x_i - Ax_{i-1} \in \operatorname{im} B \quad i = 2, \ldots, m. \tag{6.13}$$

We show first that $m = N$. Consider

$$V := \operatorname{span}\{x_1, \ldots, x_m\} \subset \mathbb{R}^N.$$

By maximality of m,

$$Ax_m + Bu \in V \quad \forall u \in \mathbb{R}^M, \tag{6.14}$$

because otherwise $Ax_m + Bw \notin V$ for some $w \in \mathbb{R}^M$ and $\{x_1, \ldots, x_{m+1}\}$, where $x_{m+1} := Ax_m + Bw$, would be a set of linearly independent vectors satisfying (6.13) with m replaced by $m+1$ (contradicting maximality of m). In particular, setting $u = 0$ in (6.14) gives,

$$Ax_m \in V. \tag{6.15}$$

By (6.14) and (6.15),

$$\operatorname{im} B \subset V - Ax_m \subset V, \tag{6.16}$$

and so (6.13) implies

$$Ax_i \in V, \quad i = 1, \ldots, m - 1. \tag{6.17}$$

Combining (6.15)–(6.17), shows that V is an A-invariant subspace containing $\operatorname{im} B$. It follows that

$$\mathbb{R}^N = \operatorname{im} \mathcal{C}(A, B) \subset V = \operatorname{span}\{x_1, \ldots, x_m\},$$

where the first equality is due to the controllability of (A, B) and the inclusion is a consequence of Proposition 3.8. We conclude that $m = N$.

Finally, choose $u_1, \ldots, u_{N-1} \in \mathbb{R}^M$ such that

$$x_i - Ax_{i-1} = Bu_{i-1}, \quad i = 2, \ldots, N,$$

which is possible by (6.13). Define $E \in \mathbb{R}^{M \times N}$ by

$$Ex_i = u_i, \quad i = 1, \ldots, N-1, \quad\quad Ex_N \quad \text{arbitrary}.$$

Then

$$(A + BE)x_{i-1} = x_i, \quad i = 2, \ldots, N.$$

Since $\mathcal{C}(A + BE, b) = \mathcal{C}(A + BE, Bv) = \mathcal{C}(A + BE, x_1)$, it follows that

$$
\begin{aligned}
\mathcal{C}(A + BE, b) &= \left(x_1, (A + BE)x_1, (A + BE)^2 x_1, \ldots, (A + BE)^{N-1} x_1\right) \\
&= \left(x_1, x_2, (A + BE)x_2, \ldots, (A + BE)^{N-2} x_2\right) \\
&= \left(x_1, x_2, x_3, \ldots, (A + BE)^{N-3} x_3\right) \\
&\;\;\vdots \\
&= \left(x_1, x_2, x_3, \ldots, x_N\right),
\end{aligned}
$$

showing that $\mathcal{C}(A + BE, b)$ is invertible. Hence, $(A + BE, b)$ is controllable. $\quad\square$

Exercise 6.3

A matrix $M \in \mathbb{C}^{N \times N}$ is said to be a *companion matrix* (or to be in *companion form*) if

$$
M = \begin{pmatrix}
0 & 1 & 0 & \ldots & 0 \\
0 & 0 & 1 & \ldots & 0 \\
\vdots & & & & \vdots \\
0 & 0 & 0 & \ldots & 1 \\
-m_0 & -m_1 & -m_2 & \ldots & -m_{N-1}
\end{pmatrix},
$$

where $m_j \in \mathbb{C}$ for $j = 0, \ldots, N - 1$. Show (by induction on N) that

$$P_M(s) = s^N + m_{N-1} s^{N-1} + \cdots + m_1 s + m_0.$$

Next, we discuss some algorithmic aspects of the computation of F. The following result shows that, in the single-input case (i.e., $M = 1$), the feedback achieving the desired eigenvalue assignment is unique and given by Ackermann's[2] formula.

Proposition 6.4 (Ackermann's formula)

Let $(A, b) \in \mathbb{R}^{N \times N} \times \mathbb{R}^N$ be controllable and let P be a monic real polynomial of degree N. There exists a unique $f \in \mathbb{R}^N$ such that $P_{A+bf^*} = P$ and f is given by

$$f^* = -(0, \ldots, 0, 1)\mathcal{C}(A, b)^{-1} P(A). \tag{6.18}$$

---

[2] Jürgen Ackermann (born 1936), German.

Exercise 6.4

Prove Proposition 6.4.

(*Hint.* By Theorem 6.3, there exists $f \in \mathbb{R}^N$ such that $P_{A+bf^*} = P$. Set $\hat{A} := A + bf^*$ and show, by induction, that

$$\hat{A}^k = A^k + \sum_{i=0}^{k-1} A^{k-1-i} bf^* \hat{A}^i, \quad k \in \mathbb{N}.$$

Use the above formula for $\hat{A}^k$, together with the fact that $P(\hat{A}) = 0$ (by the Cayley-Hamilton theorem), to infer that

$$P(A) = - \left(bg_0^* + Abg_1^* + \ldots + A^{n-2} bg_{n-2}^* + A^{n-1} bf^* \right)$$

for some $g_i \in \mathbb{R}^N$ (the specific form of the vectors g_i is not important here). Now deduce (6.18).)

Exercise 6.5

Consider the linearized satellite example (1.10), with $\omega = \sigma = 1$ in which case we have

$$A = \begin{pmatrix} 0 & 1 & 0 & 0 \\ 3 & 0 & 0 & 2 \\ 0 & 0 & 0 & 1 \\ 0 & -2 & 0 & 0 \end{pmatrix}, \qquad B = \begin{pmatrix} 0 & 0 \\ 1 & 0 \\ 0 & 0 \\ 0 & 1 \end{pmatrix}.$$

(a) First consider the problem with tangential thrust only, that is, replace B by its second column

$$b := \begin{pmatrix} 0 \\ 0 \\ 0 \\ 1 \end{pmatrix}.$$

Find the unique feedback law $u = f^* x$, where $f \in \mathbb{R}^4$, which places the eigenvalues of $A + bf^*$ at -1 and -2 such that each of these eigenvalues has multiplicity 2.

(b) Returning to the satellite problem with two inputs (radial and tangential thrust), find $F \in \mathbb{R}^{2 \times 4}$ such that the feedback law $u = Fx$ places the eigenvalues of $A + BF$ at -1 and -2 such that each of these eigenvalues has multiplicity 2 (same locations and multiplicities as in part (a)).

For controllable multi-input systems, the feedback matrix achieving the desired eigenvalue assignment is usually not unique. The following conceptual algorithm is based on Proposition 6.4 and the proof of Theorem 6.3.

Data. $A \in \mathbb{R}^{N \times N}$ and $B \in \mathbb{R}^{N \times M}$ with (A, B) controllable and $P(s)$, a monic real polynomial of degree N.

Required. $F \in \mathbb{R}^{M \times N}$ such that $P_{A+BF} = P$.

Algorithm. *Step 1.* Find a vector $v \in \mathbb{R}^M$ such that $b := Bv \neq 0$.

Step 2. Find $E \in \mathbb{R}^{M \times N}$ such that $(A + BE, b)$ is controllable. (It was shown in the proof of Theorem 6.3 that such a matrix E exists).

Step 3. Compute $f \in \mathbb{R}^N$ from

$$f^* = -(0, \ldots, 0, 1)\mathcal{C}(A + BE, b)^{-1}P(A + BE).$$

(Note that, by Proposition 6.4, $P_{A+BE+bf^*} = P$.)

Step 4. Compute $F = E + vf^*$. (Note that $A + BF = A + BE + bf^*$.)

Result. F is the desired feedback matrix.

The procedure for finding the matrix E in Step 2 is considerably easier than the construction carried out in the proof of Theorem 6.3 suggests: it turns out that, if the matrix E is chosen using a random number generator, then we can be "almost sure" that $(A + BE, b)$ is a controllable pair. We now explain this.

A set $S \subset \mathbb{R}^L$ is an *algebraic set* if there exists a real polynomial Γ in L variables such that

$$S = \{(z_1, z_2, \ldots, z_L)^* \in \mathbb{R}^L : \Gamma(z_1, z_2, \ldots, z_L) = 0\}.$$

Recall that Γ is a real polynomial in L variables if Γ is of the form

$$\Gamma(s_1, s_2, \ldots, s_L) = \sum_{(j_1, j_2, \ldots, j_L) \in \mathbb{N}_0^L} \gamma_{(j_1, j_2, \ldots, j_L)} s_1^{j_1} s_2^{j_2} \cdots s_L^{j_L},$$

where at most finitely many of the real coefficients $\gamma_{(j_1, j_2, \ldots, j_L)}$ are not equal to 0 (for example, $\Gamma(s_1, s_2, s_3) = s_1^2 + 4s_1 s_2^3 s_3^5 + (3/2)s_2 s_3^7 + \sqrt{2}$ is a real polynomial of three variables). If an algebraic set S is not equal to all of $\mathbb{R}^L$ (equivalently, if the coefficients of Γ are not all equal to zero), then we say that S is a *proper* algebraic set. It can be shown (see Exercise 6.6) that a proper algebraic set is a "very small" set. Specifically, if S is a proper algebraic set, then:

– the complement $S^c := \mathbb{R}^L \backslash S$ is open and dense in $\mathbb{R}^L$:

– S has *zero Lebesgue measure*, that is, for every $\varepsilon > 0$, there exist hyper-rectangles $R_j \subset \mathbb{R}^L$, $j \in \mathbb{N}$, such that

$$S \subset \bigcup_{j=1}^{\infty} R_j \quad \text{and} \quad \sum_{j=1}^{\infty} V(R_j) \leq \varepsilon,$$

where a *hyper-rectangle* $R \subset \mathbb{R}^L$ is a set of the form $R = (a_1, b_1) \times \cdots \times (a_L, b_L)$ and the *volume* $V(R)$ of R is defined by

$$V(R) := \prod_{i=1}^{L}(b_i - a_i).$$

Intuitively, these two properties mean that, if we choose a point $z \in \mathbb{R}^L$ "at random", then the likelihood that z belongs to S is vanishingly small. It is useful to think that consequently "typical" elements of $\mathbb{R}^L$ belong to the complement S^c. As an illustration, consider the unit circle U in $\mathbb{R}^2$ given by

$$U = \{(z_1, z_2) \in \mathbb{R}^2 : z_1^2 + z_2^2 - 1 = 0\}.$$

Obviously, U is a proper algebraic set and, if a point in the plane $\mathbb{R}^2$ were to be chosen randomly, then "almost surely" that point would not lie in the unit circle U.

Exercise 6.6

Let $S \subset \mathbb{R}^L$ be a proper algebraic set.

(a) Show that the complement $\mathbb{R}^L \backslash S$ is open and dense in $\mathbb{R}^L$.

(b) Prove that S has zero Lebesgue measure. (*Hint.* This part of the exercise requires some familiarity with basic results in the theory of Lebesgue measure. Use induction over L by writing a polynomial Γ in $L + 1$ variables in the form $\Gamma(s_1, \ldots, s_{L+1}) = \sum_{i=0}^{k} \Delta_i(s_1, \ldots, s_L) s_{L+1}^i$, where the Δ_i, $0 \le i \le k$, are polynomials in L variables.)

It remains to show that the set of matrices $E \in \mathbb{R}^{M \times N}$ such that $(A + BE, b)$ is controllable (see Step 2 of the above algorithm) is the complement of a proper algebraic set. This is an immediate consequence of the following lemma.

Lemma 6.5

Let $(A, B) \in \mathbb{R}^{N \times N} \times \mathbb{R}^{N \times M}$ be controllable. Choose $v \in \mathbb{R}^M$ such that $b = Bv \ne 0$. Then the set

$$S := \{K \in \mathbb{R}^{M \times N} : (A + BK, b) \text{ is not controllable}\},$$

viewed as a subset of $\mathbb{R}^{MN}$, is a proper algebraic set.

Proof

Noting that $S = \{K \in \mathbb{R}^{M \times N} : \det \mathcal{C}(A + BK, b) = 0\}$, it follows that S is an algebraic set, since the equation $\det \mathcal{C}(A + BK, b) = 0$, with A, B and b fixed, defines a polynomial in the MN entries of the matrix K. Moreover, in the proof of Theorem 6.3, it was shown that there exists a matrix E such that the pair $(A + BE, b)$ is controllable, or, equivalently, $\det \mathcal{C}(A + BE, b) \ne 0$, implying that S is a proper algebraic set. $\qquad\square$

Exercise 6.7

Consider the controlled system given by

$$A = \begin{pmatrix} 0 & 0 & 2 \\ 0 & 2 & 0 \\ 1 & 0 & 1 \end{pmatrix}, \qquad B = \begin{pmatrix} 1 & 0 \\ 0 & 1 \\ 0 & 1 \end{pmatrix}.$$

Apply the above algorithm to compute a matrix $F \in \mathbb{R}^{2 \times 3}$ such that the feedback law $u = Fx$ assigns the eigenvalues -1, -2 and -5 to $A + BF$.

6.1.2 Stabilizability of linear systems

The property of eigenvalue assignability is strong. It is easy to envisage situations wherein a weaker property may suffice, namely, that every eigenvalue of A with non-negative real part can be moved, by feedback, to a value with negative real part (in other words, a matrix A which is not Hurwitz can be transformed, by feedback, into a Hurwitz matrix $A + BF$ (and so the equilibrium 0 of the feedback system $\dot{x} = (A + BF)x$ is exponentially stable). This is the concept of stabilizability.

The system (6.7) (or the pair (A, B)) is said to be *stabilizable* if there exists $F \in \mathbb{R}^{M \times N}$ such that $A + BF$ is Hurwitz.

The following corollary is an immediate consequence of Theorem 6.3.

Corollary 6.6

If the pair $(A, B) \in \mathbb{R}^{N \times N} \times \mathbb{R}^{N \times M}$ is controllable, then it is also stabilizable.

Exercise 6.8

Show that the converse of Corollary 6.6 fails to hold by constructing an example of a stabilizable system that fails to be controllable.

Define

$$\mathbb{C}_+ := \{s \in \mathbb{C} \colon \operatorname{Re} s > 0\},$$

that is, the *open* right half complex plane[3] (with closure $\overline{\mathbb{C}}_+$). As before, $\mathbb{C}_-$ denotes the open left half complex plane $\{s \in \mathbb{C} \colon \operatorname{Re} s < 0\}$.

There now follows a necessary and sufficient condition for stabilizability.

---

[3] This notation is somewhat inconsistent with our adoption of $\mathbb{R}_+$ for the *closed* half real line. However, our definitions of $\mathbb{C}_+$ and $\mathbb{R}_+$ conform with standard usage in the literature and, for this reason, we tolerate the notational inconsistency (and hope that the reader will also do so).

Theorem 6.7 (Hautus criterion for stabilizability)

A pair $(A, B) \in \mathbb{R}^{N \times N} \times \mathbb{R}^{N \times M}$ is stabilizable if, and only if, rk $(sI - A, B) = N$ for all $s \in \overline{\mathbb{C}}_+$.

Proof

Assume first that (A, B) is stabilizable. If $\lambda \in \mathbb{C}$ is such that rk $(\lambda I - A, B) < N$, then, as in the proof of the eigenvalue-assignment theorem (Theorem 6.3), it follows that λ is an eigenvalue of $A + BF$ for every $F \in \mathbb{R}^{M \times N}$. This, together with the hypothesis of stabilizability, implies that $\lambda \in \mathbb{C}_-$. Therefore, rk $(sI - A, B) = N$ for all $s \in \overline{\mathbb{C}}_+$.

Now assume that the rank condition holds. If (A, B) is controllable, then, by Corollary 6.6, (A, B) is also stabilizable. So let us assume that (A, B) is not controllable. If $B = 0$, then, by the rank condition,

$$\text{rk}\,(sI - A) = \text{rk}\,(sI - A, 0) = N \quad \forall\, s \in \overline{\mathbb{C}}_+.$$

Consequently, $\sigma(A) \subset \mathbb{C}_-$, and so A is Hurwitz, which trivially implies that $(A, B) = (A, 0)$ is stabilizable. If $B \neq 0$, then $0 < Q := \text{rk}\,\mathcal{C}(A, B) < N$, and, by Lemma 3.10, there exists $S \in GL(N, \mathbb{R})$ such that

$$\tilde{A} := S^{-1} A S = \begin{pmatrix} A_1 & A_2 \\ 0 & A_3 \end{pmatrix}, \quad \tilde{B} := S^{-1} B = \begin{pmatrix} B_1 \\ 0 \end{pmatrix},$$

where the pair $(A_1, B_1) \in \mathbb{R}^{Q \times Q} \times \mathbb{R}^{Q \times M}$ is controllable. Thus, by Corollary 6.6, there exists $F_1 \in \mathbb{R}^{M \times Q}$ such that

$$\sigma(A_1 + B_1 F_1) \subset \mathbb{C}_-. \tag{6.19}$$

Now

$$(sI - A, B) = S(sI - \tilde{A}, \tilde{B}) \begin{pmatrix} S^{-1} & 0 \\ 0 & I \end{pmatrix}$$

and hence, by hypothesis,

$$\text{rk}\,(sI - \tilde{A}, \tilde{B}) = N \quad \forall\, s \in \overline{\mathbb{C}}_+. \tag{6.20}$$

We claim that

$$\sigma(A_3) \subset \mathbb{C}_-. \tag{6.21}$$

To this end, let $\lambda \in \sigma(A_3)$. Then there exists $v \in \mathbb{C}^{N-Q}$, $v \neq 0$, such that $v^*(\lambda I - A_3) = 0$. Setting

$$w := \begin{pmatrix} 0 \\ v \end{pmatrix} \in \mathbb{C}^N,$$

we obtain $w^*(\lambda I - \tilde{A}, \tilde{B}) = 0$ and thus $\operatorname{rk}(\lambda I - \tilde{A}, \tilde{B}) < N$. Consequently, by (6.20), $\operatorname{Re}\lambda < 0$ and (6.21) follows.

Defining

$$\tilde{F} := (F_1, 0) \in \mathbb{R}^{M \times N},$$

we have

$$\tilde{A} + \tilde{B}\tilde{F} = \begin{pmatrix} A_1 + B_1 F_1 & A_2 \\ 0 & A_3 \end{pmatrix}.$$

A standard result on the spectrum of a triangular block matrix (see Theorem A.7) gives

$$\sigma(\tilde{A} + \tilde{B}\tilde{F}) = \sigma(A_1 + B_1 F_1) \cup \sigma(A_3),$$

which, combined with (6.19) and (6.21), leads to

$$\sigma(\tilde{A} + \tilde{B}\tilde{F}) \subset \mathbb{C}_-. \tag{6.22}$$

Finally, setting $F := \tilde{F}S^{-1}$, we see that

$$A + BF = S\tilde{A}S^{-1} + S\tilde{B}\tilde{F}S^{-1} = S(\tilde{A} + \tilde{B}\tilde{F})S^{-1}.$$

Therefore, $\sigma(A + BF) = \sigma(\tilde{A} + \tilde{B}\tilde{F})$ and so, by (6.22), $\sigma(A + BF) \subset \mathbb{C}_-$, showing that (A, B) is stabilizable. $\qquad\square$

Exercise 6.9

Let

$$A = \begin{pmatrix} -1 & \alpha & 0 \\ 1 & 1 & 0 \\ 1 & 1 & 1 \end{pmatrix}, \qquad b = \begin{pmatrix} 1 \\ 0 \\ 1 \end{pmatrix}.$$

For which values of the real parameter α is the system (A, b) stabilizable?

6.2 Nonlinear systems and feedback

Here, we turn attention to controlled nonlinear systems of the form

$$\dot{x}(t) = g(x(t), u(t)), \quad g(0, 0) = 0, \tag{6.23}$$

with $g \in C^1(\mathbb{R}^N \times \mathbb{R}^M, \mathbb{R}^N)$ (the space of continuously differentiable functions $\mathbb{R}^N \times \mathbb{R}^M \to \mathbb{R}^N$).

It is clear that 0 is an equilibrium of the uncontrolled system, that is, the system with zero input $u = 0$ given by $\dot{x} = g(x, 0)$. The question to be addressed is stabilizability of the zero equilibrium by feedback: does there exist a feedback law $u = k(x)$

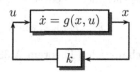

Figure 6.3 System with feedback

which renders 0 an asymptotically stable equilibrium of the feedback system $\dot{x} = g(x, k(x))$ (illustrated in Figure 6.3).

Definition 6.8

System (6.23) is C^1-*stabilizable* if there exists $k \in C^1(\mathbb{R}^N, \mathbb{R}^M)$, with $k(0) = 0$, such that 0 is an asymptotically stable equilibrium of the feedback system $\dot{x} = g(x, k(x))$.

6.2.1 Stabilizability and linearization

Not surprisingly, the issue of C^1-stabilizability of (6.23) is related to stabilizability of its *linearization* about $(0, 0)$, that is, the linear system $\dot{x} = Ax + Bu$ with the pair (A, B) defined by

$$A := (D_1 g)(0, 0) \in \mathbb{R}^{N \times N}, \quad B := (D_2 g)(0, 0) \in \mathbb{R}^{N \times M}, \qquad (6.24)$$

where $D_1 g$ (respectively, $D_2 g$) denotes the matrix of partial derivatives of components of g with respect to components of its first argument (respectively, second argument). See Appendix A.3.

Theorem 6.9

System (6.23) is C^1-stabilizable if its linearization (A, B), given by (6.24), is stabilizable.

Proof

Assume that the pair (A, B), given by (6.24), is stabilizable. Then there exists $F \in \mathbb{R}^{N \times M}$ such that $A + BF$ is Hurwitz. By Theorem 5.26, there exists a symmetric positive-definite $P \in \mathbb{R}^{N \times N}$ such that

$$P(A + BF) + (A + BF)^* P + I = 0.$$

Applying the linear feedback $u = Fx$ to the nonlinear system (6.23) leads to

$$\dot{x} = f(x), \quad \text{where } f(z) := g(z, Fz) \text{ for all } z \in \mathbb{R}^N.$$

By the chain rule (Theorem A.33),

$$(Df)(0) = \big((D_1 g)(0,0)\,,\,(D_2 g)(0,0)\big) \begin{pmatrix} I \\ F \end{pmatrix} = (D_1 g)(0,0) + \big((D_2 g)(0,0)\big) F$$

$$= A + BF$$

and therefore, by Corollary 5.29, it follows that 0 is an asymptotically stable equilibrium of the feedback system.   □

Is the converse of the above result true? If the linearization (about $(0,0)$) fails to be stabilizable, does it follow that (6.23) fails to be C^1-stabilizable? The answer to this question is no – as the following scalar counterexample shows.

Example 6.10

Consider the scalar system

$$\dot{x}(t) = u(t)x^2(t), \quad x(t), u(t) \in \mathbb{R}.$$

Linearizing about $(0,0)$ yields the system $\dot{x}(t) = 0$, which is evidently not stabilizable. However, the linear feedback $u(t) = -x(t)$ applied to the nonlinear system yields the feedback system $\dot{x}(t) = -x^3(t)$ for which 0 is an asymptotically stable equilibrium. Note that the linearization only "narrowly" fails to be stabilizable: the eigenvalue is located at the origin and so the equilibrium is stable but not asymptotically stable.   △

It is convenient to introduce the concept of an *uncontrollable eigenvalue*. To this end, let $(A, B) \in \mathbb{R}^{N \times N} \times \mathbb{R}^{N \times M}$ and $\lambda \in \sigma(A)$. We say that λ is *uncontrollable* if $\mathrm{rk}\,(\lambda I - A, B) < N$. Exercise 6.10 below shows that the uncontrollable eigenvalues of A are precisely the eigenvalues which cannot be moved by feedback (whence the terminology "uncontrollable eigenvalue").

Exercise 6.10

Let $(A, B) \in \mathbb{R}^{N \times N} \times \mathbb{R}^{N \times M}$ and $\lambda \in \sigma(A)$. Show that λ is uncontrollable if, and only if, $\lambda \in \sigma(A + BF)$ for all $F \in \mathbb{R}^{M \times N}$

Note that in Example 6.10 the eigenvalue zero is uncontrollable. We will now consider the case wherein the linearization has at least one uncontrollable eigenvalue with positive real part: we will show that in this case the converse of Theorem 6.9 holds.

Theorem 6.11

Assume that the linearization (A, B), given by (6.24), of the nonlinear system (6.23) has at least one uncontrollable eigenvalue with positive real part. Then system (6.23) is not C^1-stabilizable.

Proof

Let A and B be as in (6.24). By hypothesis, there exists an uncontrollable eigenvalue λ of A with $\operatorname{Re}\lambda > 0$. Let k be any continuously differentiable function $\mathbb{R}^N \to \mathbb{R}^M$ with $k(0) = 0$. Setting $F := (Dk)(0) \in \mathbb{R}^{M \times N}$ and defining $f\colon \mathbb{R}^N \to \mathbb{R}^N$ by $f(z) = g(z, k(z))$, it follows that $(Df)(0) = A + BF$. By Exercise 6.10, λ is an eigenvalue of $A + BF = (Df)(0)$. Invoking Theorem 5.31, it follows that 0 is an unstable equilibrium of $\dot{x} = f(x)$. Therefore, (6.23) is not C^1-stabilizable. $\qquad\square$

6.2.2 Feedback stabilization of smooth input-affine systems

We consider the class of nonlinear systems given by

$$\dot{x} = a(x) + u b(x), \tag{6.25}$$

where u is a scalar-valued input function and $a, b\colon \mathbb{R}^N \to \mathbb{R}^N$ are smooth functions with $a(0) = 0 = b(0)$. Throughout this section, the term *smooth* is synonymous with C^∞ (functions with continuous partial derivatives of all orders). The right-hand side of (6.25) exhibits affine linear dependence on the input and therefore the system is said to be *input affine*. Defining $g\colon \mathbb{R}^N \times \mathbb{R} \to \mathbb{R}^N$ by $g(z, v) := a(z) + v b(z)$ for all $(z, v) \in \mathbb{R}^N \times \mathbb{R}$, it is clear that (6.25) is a special case of system (6.23).

We introduce some convenient notation. Let $V\colon \mathbb{R}^N \to \mathbb{R}$ be a smooth function. In the notation of Chapter 5, $V_b\colon \mathbb{R}^N \to \mathbb{R}$ denotes the directional derivative of V along b, given by

$$V_b(z) = \langle (\nabla V)(z), b(z) \rangle \quad \forall z \in \mathbb{R}^N.$$

Extending this notation, $V_{ba}\colon \mathbb{R}^N \to \mathbb{R}$ denotes the directional derivative of the smooth function V_b along a:

$$V_{ba}(z) = \langle (\nabla V_b)(z), a(z) \rangle \quad \forall z \in \mathbb{R}^N.$$

If $b = a$, then we write V_{a^2} in place of V_{aa}. Proceeding inductively, for $n \in \mathbb{N}$ and $n \geq 2$, V_{a^n} denotes the directional derivative of the smooth function $V_{a^{n-1}}$ along a.

Theorem 6.12

Let $a, b \colon \mathbb{R}^N \to \mathbb{R}^N$ be smooth with $a(0) = 0 = b(0)$. Assume that there exists a smooth function $V \colon \mathbb{R}^N \to \mathbb{R}$ such that $V(0) = 0$, $V(z) > 0$ for all $z \neq 0$, $V(z) \to \infty$ as $\|z\| \to \infty$, and

$$V_a(z) \leq 0 \quad \forall z \in \mathbb{R}^N. \tag{6.26}$$

Assume further that

$$\Sigma := \left\{ z \in \mathbb{R}^N : V_a(z) = 0 = V_b(z), \ V_{ba^n}(z) = 0 \ \forall n \in \mathbb{N} \right\} = \{0\}. \tag{6.27}$$

Let $h \colon \mathbb{R} \to \mathbb{R}$ be any smooth function with the property that $yh(y) > 0$ for all $y \neq 0$. Then the smooth feedback law $u = -h(V_b(x))$, applied to (6.25), renders 0 a globally asymptotically stable equilibrium of the resulting feedback system

$$\dot{x} = a(x) - h(V_b(x))b(x). \tag{6.28}$$

Proof

Defining $f \colon \mathbb{R}^N \to \mathbb{R}^N$ by $f(z) := a(z) - h(V_b(z))b(z)$ for all $z \in \mathbb{R}^N$, the feedback system (6.28) can be written in the form $\dot{x} = f(x)$. It is clear that $f(0) = 0$ and thus, 0 is an equilibrium of (6.28). Let φ denote the local flow generated by f. Observe that

$$V_f(z) = V_a(z) - h(V_b(z))V_b(z) \leq 0 \quad \forall z \in \mathbb{R}^N,$$

and let Σ_0 denote the largest invariant subset (with respect to φ) of $V_f^{-1}(0) = \{z \in \mathbb{R}^N : V_f(z) = 0\}$. In view of Theorem 5.22, it suffices to show that $\Sigma_0 = \{0\}$. Let $\xi \in \Sigma_0$, write $x(\cdot) := \varphi(\cdot, \xi)$ and, as usual, let $I := I_\xi$ denote the corresponding (maximal) interval of existence. Then $x(t) \in \Sigma_0$ for all $t \in I$ and so,

$$0 = V_f(x(t)) = V_a(x(t)) - h(V_b(x(t)))V_b(x(t)) \quad \forall t \in I.$$

Since $yh(y) > 0$ for all $y \in \mathbb{R} \backslash \{0\}$ and invoking (6.26), we may infer that

$$V_a(x(t)) = 0 = V_b(x(t)) \quad \forall t \in I.$$

Therefore, $f(x(t)) = a(x(t))$, implying that $\dot{x}(t) = a(x(t))$ for all $t \in I$, and hence,

$$V_{ba}(x(t)) = \langle (\nabla V_b)(x(t)), a(x(t)) \rangle = \frac{\mathrm{d}}{\mathrm{d}t} V_b(x(t)) = 0 \quad \forall t \in I.$$

Proceeding inductively, we have

$$V_{ba^n}(x(t)) = \frac{\mathrm{d}^n}{\mathrm{d}t^n} V_b(x(t)) = 0 \quad \forall t \in I, \ \forall n \in \mathbb{N}.$$

Invoking (6.27), we may conclude that $x(t) = 0$ for all $t \in I$. Therefore, $\xi = 0$ and so $\Sigma_0 = \{0\}$, completing the proof. $\qquad \square$

Example 6.13

Consider the bilinear system

$$\dot{x}(t) = Ax(t) + u(t)Bx(t), \quad A = \begin{pmatrix} 0 & 1 \\ -1 & 0 \end{pmatrix}, \quad B = \begin{pmatrix} 0 & 1 \\ 1 & 0 \end{pmatrix}.$$

In this case, $a(z) = Az$ and $b(z) = Bz$ for all $z \in \mathbb{R}^2$. With V given by $V : \mathbb{R}^2 \to \mathbb{R}, z \mapsto \|z\|^2/2$, we have

$$V_a(z) = \langle z, Az \rangle = 0 \ \ \forall z \in \mathbb{R}^2,$$

and so hypothesis (6.26) of Theorem 6.12 holds. Moreover,

$$V_b(z) = \langle z, Bz \rangle = 2z_1 z_2, \quad V_{ba}(z) = 2(z_2^2 - z_1^1) \ \ \forall z = (z_1, z_2) \in \mathbb{R}^2.$$

Therefore,

$$\Sigma \subset \{z = (z_1, z_2) \in \mathbb{R}^2 : z_1 z_2 = 0 = z_2^2 - z_1^2\} = \{0\},$$

implying that $\Sigma = \{0\}$, and so hypothesis (6.27) of Theorem 6.12 also holds. We may now conclude that the feedback

$$u(t) = -V_b(x(t)) = -2x_1(t)x_2(t)$$

renders 0 a globally asymptotically stable equilibrium of the feedback-controlled bilinear system. For illustration, Figure 6.4 depicts the positive semi-orbit of

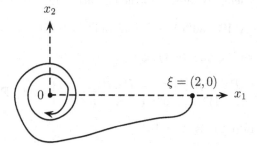

Figure 6.4 Typical behaviour of the feedback system

the feedback system with initial data $\xi = (x_1(0), x_2(0)) = (2, 0)$,. $\triangle$

The above example provides a prototype for the following treatment of a particular class of bilinear systems.

6.2.3 Feedback stabilization of bilinear systems

Consider the class of bilinear systems (with scalar-valued control u)

$$\dot{x}(t) = Ax(t) + u(t)Bx(t), \quad A, B \in \mathbb{R}^{N \times N}, \tag{6.29}$$

characterized by the property that there exists a symmetric positive definite matrix $P \in \mathbb{R}^{N \times N}$ such that

$$PA + A^*P = 0. \tag{6.30}$$

If, as in Example 6.13, A is skew-symmetric (that is, $A^* = -A$), then (6.30) holds with $P = I$. In terms of dynamics, the existence of a symmetric positive definite matrix $P \in \mathbb{R}^{N \times N}$ satisfying (6.30), means that the function

$$V \colon \mathbb{R}^N \to \mathbb{R}, \; z \mapsto \langle z, Pz \rangle / 2 \tag{6.31}$$

is constant along solutions of (6.29) with zero input $u = 0$ (see Exercise 6.11). In particular, the equilibrium 0 of the uncontrolled system $\dot{x} = Ax$ is stable.

Exercise 6.11

Let $P \in \mathbb{R}^{N \times N}$ be symmetric and positive definite. Show that P solves (6.30) if, and only if, $\langle (\nabla V)(z), Az \rangle = 0$ for all $z \in \mathbb{R}^N$, where V is the function given by (6.31).

We seek conditions on the pair (A, B) under which the (stable) equilibrium 0 can be rendered globally asymptotically stable by feedback.

With a view to applying Theorem 6.12, write $a(z) = Az$ and $b(z) = Bz$ for all $z \in \mathbb{R}^N$. Let $[X, Y] := XY - YX$ denote the commutator of matrices $X, Y \in \mathbb{R}^{N \times N}$, and define (recursively) the operation ad^n by

$$\text{ad}^1(X, Y) := [X, Y], \quad \text{ad}^n(X, Y) := \big[X, \text{ad}^{n-1}(X, Y)\big], \quad n = 2, 3, \dots$$

A routine calculation invoking (6.30) gives

$$\left. \begin{array}{l} V_a(z) = \langle Pz, Az \rangle, \quad V_b(z) = \langle Pz, Bz \rangle, \\ V_{ba^n}(z) = (-1)^n \langle Pz, \text{ad}^n(A, B)z \rangle \; \forall n \in \mathbb{N} \end{array} \right\} \; \forall z \in \mathbb{R}^N,$$

whence Σ defined in (6.27) is given by

$$\Sigma := \big\{ z \in \mathbb{R}^N \colon \langle Pz, Az \rangle = 0 = \langle Pz, Bz \rangle, \; \langle Pz, \text{ad}^n(A, B)z \rangle = 0 \; \forall n \in \mathbb{N} \big\}.$$

Observe that, if

$$\text{span}\{Az, Bz, \text{ad}^1(A, B)z, \text{ad}^2(A, B)z, \dots\} = \mathbb{R}^N \; \forall z \in \mathbb{R}^N \backslash \{0\},$$

then $\Sigma = \{0\}$. Thus, invoking Theorem 6.12, we arrive at the following result.

Corollary 6.14

Let $h\colon \mathbb{R} \to \mathbb{R}$ be any smooth function with the property that $yh(y) > 0$ for all $y \neq 0$. Assume that there exists a symmetric positive definite matrix $P \in \mathbb{R}^{N \times N}$ such that (6.30) holds. If

$$\mathrm{span}\{Az, Bz, \mathrm{ad}^1(A, B)z, \mathrm{ad}^2(A, B)z, \ldots\} = \mathbb{R}^N \quad \forall\, z \in \mathbb{R}^N \backslash \{0\},$$

then the smooth feedback law $u = -h(\langle x, PBx \rangle)$, applied to (6.29), renders 0 a globally asymptotically stable equilibrium of the resulting feedback system $\dot{x} = Ax - h(\langle x, PBx \rangle)Bx$.

Exercise 6.12

Consider system (6.29) with $N = 2$ and

$$A = \begin{pmatrix} 0 & 1 \\ -1 & 0 \end{pmatrix}, \quad B = \begin{pmatrix} 0 & 0 \\ 1 & 0 \end{pmatrix}.$$

Show that the feedback law $u(t) = -x_1(t)x_2(t)$ is globally asymptotically stabilizing.

In the following, a set $S \subset \mathbb{R}^N$ is said to be *positively* $\exp(At)$-*invariant*, if, for each $\xi \in S$, $\exp(At)\xi \in S$ for all $t \in \mathbb{R}_+$.

Exercise 6.13

Let $A \in \mathbb{R}^{N \times N}$ and let $S \subset \mathbb{R}^N$. The set S is said to be A-*invariant* if $Az \in S$ for all $z \in S$ and it is said to be $\exp(At)$-*invariant* if $\exp(At)z \in S$ for all $z \in S$ and all $t \in \mathbb{R}$.

(a) By finding counterexamples show that, in general, A-invariance does not imply positive $\exp(At)$-invariance and $\exp(At)$-invariance does not imply A-invariance.

(b) Assume that $S \subset \mathbb{R}^N$ is a subspace. Show that S is A-invariant if, and only if, S is positively $\exp(At)$-invariant.

(c) By finding a counterexample show that, in general, positive $\exp(At)$-invariance does not imply $\exp(At)$-invariance.

(d) Show that, if $S \subset \mathbb{R}^N$ is a subspace, then positive $\exp(At)$-invariance implies $\exp(At)$-invariance.

In Corollary 6.14, the spanning condition is required to hold on the set $\mathbb{R}^N \backslash \{0\}$. Next, we seek to refine this condition by identifying a smaller set Ω on which the spanning condition is sufficient for global asymptotic stabilizability. As we shall see, this is indeed the case, provided that $\{0\}$ is the only positively $\exp(At)$-invariant subset of the intersection of the complement $\mathbb{R}^N \backslash \Omega$ of Ω and the set $\Gamma := \{z \in \mathbb{R}^N \colon \langle z, PBz \rangle = 0\}$.

Theorem 6.15

Assume that there exists a symmetric positive definite matrix $P \in \mathbb{R}^{N \times N}$ such that (6.30) holds and define $\Gamma := \{z \in \mathbb{R}^N : \langle z, PBz \rangle = 0\}$. Let $\Omega \subset \mathbb{R}^N$ be such that

$$\text{span}\{Az, Bz, \text{ad}^1(A, B)z, \text{ad}^2(A, B)z, \ldots\} = \mathbb{R}^N \ \ \forall z \in \Omega \qquad (6.32)$$

and $\{0\}$ is the only positively $\exp(At)$-invariant subset of $(\mathbb{R}^N \backslash \Omega) \cap \Gamma$. Let $h \colon \mathbb{R} \to \mathbb{R}$ be any smooth function with the property that $yh(y) > 0$ for all $y \neq 0$. Then the feedback law $u = -h(\langle x, PBx \rangle)$, applied to (6.29), renders 0 a globally asymptotically stable equilibrium of the resulting feedback system $\dot{x} = Ax - h(\langle x, PBx \rangle) Bx$.

Observe that, on setting $\Omega = \mathbb{R}^N \backslash \{0\}$, we recover Corollary 6.14.

Proof of Theorem 6.15

As before, define V by (6.31) and $f : \mathbb{R}^N \to \mathbb{R}^N$ by $f(z) := Az - h(\langle z, PBz \rangle) Bz$ for all $z \in \mathbb{R}^N$. Moreover, set $k(z) := -h(\langle z, PBz \rangle)$, so that $f(z) = Az + k(z)Bz$ for all $z \in \mathbb{R}^N$. Then, invoking (6.30),

$$V_f(z) = \langle (\nabla V)(z), Az + k(z)Bz \rangle = \langle Pz, Az + k(z)Bz \rangle = k(z)\langle Pz, Bz \rangle$$
$$= -h(\langle z, PBz \rangle)\langle z, PBz \rangle \leq 0 \ \ \forall z \in \mathbb{R}^N. \qquad (6.33)$$

Furthermore, $V_f^{-1}(0) = \Gamma$, as follows from (6.33) in conjunction with the fact that $yh(y) > 0$ for all $y \in \mathbb{R} \backslash \{0\}$. Since V is radially unbounded, global asymptotic stability will follow from Theorem 5.22, provided that we can show that $\{0\}$ is the only invariant subset of Γ.

Let φ denote the local flow generated by f and let $\Gamma_0 \subset \Gamma$ be the largest invariant set in Γ. Let $\xi \in \Gamma_0$ be arbitrary. In view of (6.33), we see that $\mathbb{R}_+ \subset I_\xi$ and, by invariance of Γ_0, $\varphi(t, \xi) \in \Gamma_0$ for all $t \in \mathbb{R}_+$. Therefore, $\langle \varphi(t, \xi), PB\varphi(t, \xi) \rangle = 0$ for all $t \in \mathbb{R}_+$, and it follows that $\varphi(t, \xi) = \exp(At)\xi$ for all $t \in \mathbb{R}_+$. Consequently, Γ_0 is positively $\exp(At)$-invariant and, in particular,

$$\langle P \exp(At)\xi, B \exp(At)\xi \rangle = 0 \ \ \forall t \in \mathbb{R}_+, \qquad (6.34)$$

Moreover, it follows from (6.30) that

$$PA^n = (-A^*)^n P \ \ \forall n \in \mathbb{N}_0,$$

and so, $P \exp(At) = \exp(-A^*t)P$ for all $t \in \mathbb{R}$. Combining this with (6.34) leads to

$$\langle P\xi, \exp(-At)B \exp(At)\xi \rangle = 0 \ \ \forall t \in \mathbb{R}_+.$$

Repeated differentiation yields

$$\langle P\xi, \exp(-At)(-1)^n \mathrm{ad}^n(A, B) \exp(At)\xi \rangle = 0 \quad \forall t \in \mathbb{R}_+, \ \forall n \in \mathbb{N}.$$

Evaluation at $t = 0$ gives

$$\langle P\xi, (-1)^n \mathrm{ad}^n(A, B)\xi \rangle = 0 \quad \forall n \in \mathbb{N},$$

which, in conjunction with the fact that $\langle P\xi, A\xi \rangle = 0 = \langle P\xi, B\xi \rangle$ and invoking hypothesis (6.32), implies that ξ cannot be in the set Ω and so we may infer that $\xi \in \mathbb{R}^N \backslash \Omega$. Therefore, $\Gamma_0 \subset (\mathbb{R}^N \backslash \Omega) \cap \Gamma$. By hypothesis, $\{0\}$ is the only positively $\exp(At)$-invariant subset of $(\mathbb{R}^N \backslash \Omega) \cap \Gamma$, showing that $\Gamma_0 = \{0\}$. This completes the proof. $\qquad\square$

Exercise 6.14

Consider two scalar bilinear systems with common control u:

$$\ddot{y}(t) + u(t)\dot{y}(t) + y(t) = 0, \quad \dot{z}(t) - u(t)z(t) = 0.$$

Observe that the stabilization goals for these systems individually may conflict insofar as a constant asymptotically stabilizing control for one system (e.g. $u(t) = 1$ for all $t \geq 0$ in the first equation) will render the 0 equilibrium of the other system unstable. A natural question arises: does there exist a feedback law that simultaneously stabilizes both systems?

(a) Show that these two systems collectively can be written in the form (6.29) with $N = 3$ and

$$A = \begin{pmatrix} 0 & 1 & 0 \\ -1 & 0 & 0 \\ 0 & 0 & 0 \end{pmatrix}, \quad B = \begin{pmatrix} 0 & 0 & 0 \\ 0 & -1 & 0 \\ 0 & 0 & 1 \end{pmatrix}.$$

(b) Confirm that Corollary 6.14 fails to resolve the question of global asymptotic stabilizability by feedback.

(c) Apply Theorem 6.15 to show that the feedback control

$$u(t) = \dot{y}^2(t) - z^2(t)$$

simultaneously renders the zero equilibrium of each of the original two systems globally asymptotically stable.

6.3 Lur'e systems and absolute stability

Feedback interconnections of the form shown in Figure 6.5, consisting of a linear system L in the forward path and a static sector-bounded nonlinearity k in the negative feedback path, see Figure 6.6, are ubiquitous in control theory and practice. Such interconnections are frequently referred to as *systems of Lur'e type* (named after a Russian control theorist, A.I. Lur'e[4] who initiated the study of such systems in the 1940s) and the study of their stability properties constitutes *absolute stability theory*. Loosely speaking, the methodology of ab-

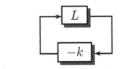

Figure 6.5 Lur'e system.

solute stability theory seeks to conclude stability of the feedback interconnection through the interplay or reciprocation of inherent properties of the transfer function of the linear component L and sector data for the nonlinearity k. In essence, if L and the sector data for k are matched in a sufficiently "nice" manner, then the interconnection is stable.

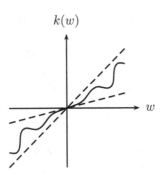

Figure 6.6 Sector-bounded nonlinearity

In particular, we consider nonlinear systems of the form

$$\dot{x} = Ax - bk(c^*x), \quad x(0) = \xi \in \mathbb{R}^N, \tag{6.35}$$

where $A \in \mathbb{R}^{N \times N}$, $b, c \in \mathbb{R}^N$ and $k \colon \mathbb{R} \to \mathbb{R}$ is a locally Lipschitz sector-bounded nonlinearity. We can think of (6.35) as the closed-loop system obtained from the following feedback interconnection

$$\dot{x} = Ax + bu, \quad x(0) = \xi; \quad y = c^*x, \tag{6.36}$$

$$u = -k(y). \tag{6.37}$$

---

[4] AnatolyIsaakovich Lur'e (1901-1980), Russian.

This interconnection of a linear (single-input single-output) system L with non-linear output feedback is illustrated in Figure 6.7.

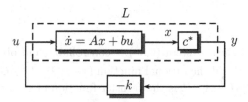

Figure 6.7 Nonlinear feedback applied to a linear system

In the following, we will present a result (a typical representative of a large number of results of similar character) which guarantees stability of the feedback system (6.35), provided that the transfer function of the linear system L and the nonlinearity k are suitably matched. A number of preparations are required. With reference to Figure 6.6, we say that k satisfies a *sector condition* if

$$\alpha w^2 \leq k(w)w \leq \beta w^2 \quad \forall\, w \in \mathbb{R}, \tag{6.38}$$

where $\alpha < \beta$, and we write $k \in S[\alpha, \beta]$. Geometrically speaking, (6.38) means that the graph of k is "sandwiched" between two lines, the slopes of which are α and β, see Figure 6.6. If the inequalities in (6.38) are strict for every $w \neq 0$, then we write $k \in S(\alpha, \beta)$.

A proper rational function $R \in \mathbb{R}(s)$ is said to be *positive real* if $\operatorname{Re} R(s) \geq 0$ for all $s \in \mathbb{C}_+$ which are not poles of R.

The following result contains a characterization of the positive real property in terms of the values of R on the imaginary axis.

Lemma 6.16

A proper rational function $R \in \mathbb{R}(s)$ is positive real if, and only if, the following conditions hold.

(1) R does not have any poles in $\mathbb{C}_+$.

(2) If, for $\omega \in \mathbb{R}$, $i\omega$ is a pole of R, then this pole is simple and the residue of R at $s = i\omega$ is positive, that is, $\lim_{s \to i\omega}(s - i\omega)R(s) > 0$.

(3) $\operatorname{Re} R(i\omega) \geq 0$ for all $\omega \in \mathbb{R}$ such that $i\omega$ is not a pole of R.

A proof of Lemma 6.16 can be found at the end of this chapter, in Section 6.4.

In the following, using notation introduced in Chapter 4, let I_ξ denote the interval of existence of the unique maximal solution $\varphi(\cdot, \xi)$ of the initial-value problem (6.35), where φ denotes the local flow generated by the locally Lipschitz function $f \colon \mathbb{R}^N \to \mathbb{R}^N$ given $f(z) := Az - bk(c^*z)$ for all $z \in \mathbb{R}^N$.

Theorem 6.17

Assume that the linear system (6.36) is controllable and observable. Assume further that $\alpha < \beta$ and the rational function $(1 + \beta\hat{G})/(1 + \alpha\hat{G})$ is positive real, where $\hat{G}(s) = c^*(sI - A)^{-1}b$ is the transfer function of system (6.36).

(1) If $k \in S[\alpha, \beta]$, then $\mathbb{R}_+ \subset I_\xi$ for all $\xi \in \mathbb{R}^N$ and there exists $\gamma > 0$ such that

$$\|\varphi(t, \xi)\| \le \gamma\|\xi\| \quad \forall t \in \mathbb{R}_+, \ \forall \xi \in \mathbb{R}^N. \tag{6.39}$$

In particular, the equilibrium 0 of (6.35) is stable.

(2) If $k \in S(\alpha, \beta)$, then the conclusions of statement (1) hold, and, moreover, the equilibrium 0 of (6.35) is globally asymptotically stable.

For the proof of Theorem 6.17, the following result is crucial. Known as the *positive real lemma* (or as the *Kalman-Yakubovich[5]-Popov[6] lemma*), it provides a characterization of positive realness of a proper rational function R via algebraic conditions on a minimal realization of the strictly proper rational function $R - d$, where $d := \lim_{|s| \to \infty} R(s)$.

Lemma 6.18 (Positive real lemma)

Let $R \in \mathbb{R}(s)$ be a proper rational function and set $d := \lim_{|s| \to \infty} R(s)$. Let $A \in \mathbb{R}^{N \times N}$ and $b, c \in \mathbb{R}^N$ be such that (A, b, c^*) is a minimal realization of the strictly proper rational function $s \mapsto R(s) - d$. Then R is positive real if, and only if, there exist a symmetric positive definite matrix $P \in \mathbb{R}^{N \times N}$ and a vector $l \in \mathbb{R}^N$ such that

$$PA + A^*P = -ll^*, \quad Pb = c - \sqrt{2d}\, l. \tag{6.40}$$

A proof of Lemma 6.18 can be found at the end of this chapter, in Section 6.4.

Proof of Theorem 6.17

Setting $A_\alpha := A - \alpha bc^*$, a simple calculation shows that

$$\hat{G}_\alpha(s) := \hat{G}(s)\big(1 + \alpha\hat{G}(s)\big)^{-1} = c^*(sI - A_\alpha)^{-1}b,$$

---

[5] Vladimir Andreevich Yakubovich (1926-2012), Russian.
[6] Vasile Mihai Popov (born 1928), Romanian.

that is, $\hat{G}_\alpha$ is the transfer function of the system given by (A_α, b, c^*). Moreover, defining a new nonlinearity $k_\alpha \colon \mathbb{R} \to \mathbb{R}$ by $k_\alpha(w) = k(w) - \alpha w$, we have

$$0 \le k_\alpha(w)w \le (\beta - \alpha)w^2 \quad \forall\, w \in \mathbb{R}.$$

Setting $d := 1/(\beta - \alpha)$, it follows that

$$dk_\alpha^2(w) \le k_\alpha(w)w \quad \forall\, w \in \mathbb{R}. \tag{6.41}$$

Let $\xi \in \mathbb{R}^N$. By Proposition 4.12, we may infer that $I_\xi = \mathbb{R}$. Set $x(t) := \varphi(t, \xi)$ for all $t \ge 0$. Since $A_\alpha z - bk_\alpha(c^*z) = Az - bk(c^*z) =: f(z)$ for all $z \in \mathbb{R}^N$, we see that x is also the unique (maximal) solution of the initial-value problem

$$\dot{x} = A_\alpha x - bk_\alpha(c^*x) = f(x), \quad x(0) = \xi.$$

Since $\beta - \alpha > 0$ and

$$(1 + \beta\hat{G})(1 + \alpha\hat{G})^{-1} = 1 + (\beta - \alpha)\hat{G}_\alpha,$$

it follows from the positive realness hypothesis that $d + \hat{G}_\alpha$ is positive real. Moreover, the minimality of (A, b, c^*) implies that (A_α, b, c^*) is a minimal realization of $\hat{G}_\alpha$ (see Exercise 6.16). Consequently, by Lemma 6.18, there exist a symmetric positive definite matrix $P \in \mathbb{R}^{N \times N}$ and a vector $l \in \mathbb{R}^N$ such that

$$PA_\alpha + A_\alpha^*P = -ll^*, \quad Pb = c - \sqrt{2d}\,l, \tag{6.42}$$

Defining $V \colon \mathbb{R}^N \to \mathbb{R}_+$ by $V(z) = \langle z, Pz \rangle$, we obtain that

$$V_f(z) = \langle (\nabla V)(z), A_\alpha z - bk_\alpha(c^*z) \rangle = \langle (PA_\alpha + A_\alpha^*P)z, z \rangle - 2\langle Pbk_\alpha(c^*z), z \rangle$$

Invoking (6.42) yields

$$V_f(z) = -(l^*z)^2 - 2\langle ck_\alpha(c^*z), z \rangle + 2\sqrt{2d}\langle lk_\alpha(c^*z), z \rangle$$

and thus

$$V_f(z) = -(l^*z)^2 - 2(c^*z)k_\alpha(c^*z) + 2(l^*z)\sqrt{2d}k_\alpha(c^*z).$$

Completing the square then gives

$$V_f(z) = -\left(l^*z - \sqrt{2d}k_\alpha(c^*z)\right)^2 - 2\left(k_\alpha(c^*z)(c^*z) - dk_\alpha^2(c^*z)\right), \tag{6.43}$$

which together with (6.41) shows that

$$V_f(z) \le 0 \quad \forall\, z \in \mathbb{R}^N. \tag{6.44}$$

Furthermore, since P is positive definite, $\mu := \min_{\|z\|=1} V(z) > 0$ and invoking the Cauchy-Schwarz inequality, we have that

$$\mu\|z\|^2 \le V(z) \le \|P\|\|z\|^2 \quad \forall\, z \in \mathbb{R}^N, \tag{6.45}$$

It now follows from Proposition 5.5, in conjunction with (6.44) and (6.45), that (6.39) holds with $\gamma = \sqrt{\|P\|/\mu}$, completing the proof of statement (1).

To prove statement (2), it is sufficient to show that the equilibrium 0 is globally attractive. To this end, note that the assumption $k \in S(\alpha, \beta)$ implies that strict inequality holds in (6.41):

$$dk_\alpha^2(w) < k_\alpha(w)w \quad \forall w \in \mathbb{R}, \ w \neq 0. \tag{6.46}$$

Therefore, by (6.43), the inequality in (6.44) is strict for every z such that $c^* z \neq 0$, showing that $V_f^{-1}(0) \subset \ker c^*$. It follows from Theorem 5.12 (LaSalle's invariance principle) that $x(t)$ approaches the largest invariant subset E in $\ker c^*$. It remains to show that $E = \{0\}$. If $\eta \in E$, then $c^* \varphi(t, \eta) = 0$ for all $t \in \mathbb{R}_+$, implying that $\varphi(t, \eta) = \exp(A_\alpha t)\eta$ for all $t \in \mathbb{R}_+$. Therefore $c^* \exp(A_\alpha t)\eta = 0$ for all $t \in \mathbb{R}_+$, whence $\eta = 0$ by observability of (c^*, A_α), completing the proof of statement (2). $\qquad\square$

Exercise 6.15

In the proof of Theorem 6.17, the positive constant μ was defined as

$$\mu = \min_{\|z\|=1} V(z) = \min_{\|z\|=1} \langle Pz, z \rangle.$$

Show that $\mu = 1/\|P^{-1}\|$.

Exercise 6.16

Let (A, b, c^*) be controllable and observable. Prove that $(A - \alpha bc^*, b, c^*)$ is controllable and observable for all $\alpha \in \mathbb{R}$.

Exercise 6.17

Consider system (6.35) with

$$A = \begin{pmatrix} 0 & 1 \\ 2 & -1 \end{pmatrix}, \quad b = \begin{pmatrix} 0 \\ 1 \end{pmatrix}, \quad c = \begin{pmatrix} 1 \\ 0 \end{pmatrix}.$$

Show that 0 is (a) a stable equilibrium if $k \in S[2, 3]$ and (b) a globally asymptotically stable equilibrium if $k \in S(2, 3)$.

Next we show that Theorem 6.17 extends to the case of "infinite" β. To make this precise, consider the condition

$$\alpha w^2 \le k(w)w \quad \forall w \in \mathbb{R}, \tag{6.47}$$

which is (6.38) with $\beta = \infty$. If k satisfies (6.47), then we write $k \in S[\alpha, \infty)$. Condition (6.47) allows k to have superlinear growth. For example, the cubic function given by $k(w) = w^3$ satisfies (6.47) with $\alpha = 0$. If the inequality in (6.47) is strict for every $w \neq 0$, then we write $k \in S(\alpha, \infty)$.

Theorem 6.19

Let $\alpha \in \mathbb{R}$, assume that the linear system (6.36) is controllable and observable and the rational function $\hat{G}/(1 + \alpha \hat{G})$ is positive real, where $\hat{G}(s) = c^*(sI - A)^{-1}b$ is the transfer function of system (6.36).

(1) If $k \in S[\alpha, \infty)$, then there exists $\gamma > 0$ such that (6.39) holds. In particular, the equilibrium 0 of (6.35) is stable.

(2) If $k \in S(\alpha, \infty)$, then the conclusion of statement (1) holds, and, moreover, the equilibrium 0 of (6.35) is globally asymptotically stable.

Exercise 6.18

Prove Theorem 6.19 by modifying the proof of Theorem 6.17.

Exercise 6.19

Consider again the example from circuit theory in Section 1.1.1:

$$\dot{x}_1(t) = x_2(t), \quad \dot{x}_2(t) = -\mu x_1(t) - g(x_2(t)),$$

wherein, for notational simplicity, we have taken $\mu = \mu_1 > 0$ and have absorbed the parameter μ_2 into the nonlinearity g. Apply Theorem 6.19 to deduce that 0 is (a) a stable equilibrium if $g \in S[0, \infty)$ and (b) a globally asymptotically stable equilibrium if $g \in S(0, \infty)$.

We conclude this section with an application of Theorem 6.17 to the problem of set point tracking. We consider a linear system with input nonlinearity:

$$\dot{x} = Ax + bk(u), \quad x(0) = \xi; \quad y = c^*x, \tag{6.48}$$

where $A \in \mathbb{R}^{N \times N}$, $b, c \in \mathbb{R}^N$ and $k \colon \mathbb{R} \to \mathbb{R}$ is a continuous nonlinearity. As usual, let $\hat{G}$ denote the transfer function of the linear system given by (A, b, c^*), that is, $\hat{G}(s) = c^*(sI - A)^{-1}b$. Let $\rho \in \mathbb{R}$ be the *set point* or *reference value*. The aim is to find a feedback control law, such that the output y of the resulting feedback system asymptotically *tracks* ρ, that is, $\lim_{t \to \infty} y(t) = \rho$. A common control strategy in this context is *integral control*:

$$u(t) = \zeta + \gamma \int_0^t (\rho - y(s)) ds,$$

or, equivalently,

$$\dot{u}(t) = \gamma(\rho - y(t)), \quad u(0) = \zeta,$$

where γ is a real parameter, the so-called *integrator gain*. Applying this control law to (6.48) leads to the following feedback system

$$\left. \begin{array}{l} \dot{x} = Ax + bk(u), \quad x(0) = \xi, \\ \dot{u} = \gamma(\rho - c^*x), \quad u(0) = \zeta. \end{array} \right\} \tag{6.49}$$

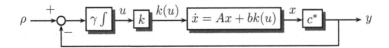

Figure 6.8 System (6.48) under integral control

The feedback system is shown in Figure 6.8.

Lemma 6.20

Assume that A is Hurwitz and $\hat{G}(0) > 0$. Then

$$\gamma^\sharp := \sup\{\gamma \geq 0 : 1 + \gamma\hat{G}(s)/s \text{ is positive real}\} > 0. \tag{6.50}$$

Proof

Defining the rational function R by $R(s) = (\hat{G}(s) - \hat{G}(0))/s$, it is clear that, by the Hurwitz property of A, R is bounded on $\mathbb{C}_+$, that is, $\sup_{s \in \mathbb{C}_+} |R(s)| < \infty$. Moreover,

$$1 + \gamma\frac{\hat{G}(s)}{s} = 1 + \gamma R(s) + \gamma\frac{\hat{G}(0)}{s}$$

and, since $\hat{G}(0) > 0$,

$$\operatorname{Re}\frac{\hat{G}(0)}{s} = \hat{G}(0)\frac{\operatorname{Re} s}{|s|^2} > 0 \quad \forall s \in \mathbb{C}_+.$$

Invoking the boundedness of R on $\mathbb{C}_+$, we may now conclude that $1 + \gamma\hat{G}(s)/s$ is positive real for all sufficiently small $\gamma > 0$, completing the proof. $\quad\square$

Theorem 6.21

Assume that the linear system given by (A, b, c^*) is controllable and observable, A is Hurwitz and $\hat{G}(0) > 0$. Furthermore, assume that k is non-decreasing, there exists $\lambda > 0$ such that

$$|k(w_1) - k(w_2)| \leq \lambda|w_1 - w_2| \quad \forall\, w_1, w_2 \in \mathbb{R}, \tag{6.51}$$

and the pre-image of $\rho/\hat{G}(0)$ under k is a singleton: $k^{-1}(\rho/\hat{G}(0)) = \{u^\rho\}$. Let $\gamma^\sharp$ be given by (6.50). Then, for every pair of initial conditions $(\xi, \zeta) \in \mathbb{R}^N \times \mathbb{R}$ and every $\gamma \in (0, \gamma^\#/\lambda)$, the interval of existence of the unique maximal solution (x, u) of (6.49) contains $\mathbb{R}_+$ and

$$\lim_{t \to \infty} x(t) = -A^{-1}b \frac{\rho}{\hat{G}(0)} \qquad \text{and} \qquad \lim_{t \to \infty} u(t) = u^\rho.$$

In particular, the output $y = c^*x$ satisfies $\lim_{t \to \infty} y(t) = \rho$, that is, the tracking objective is achieved.

The above result says that the control objective is achieved, provided that the positive gain parameter γ is sufficiently small. Consequently, the control strategy is sometimes described as "low-gain integral control". Note that condition (6.51) is assumed to hold for all $z_1, z_2 \in \mathbb{R}$: it is therefore also termed a global Lipschitz condition and k is said to be globally Lipschitz. If k is strictly increasing, then the assumption that $k^{-1}(\rho/\hat{G}(0))$ is a singleton is satisfied, provided that $\rho/\hat{G}(0) \in \text{im } k$. By contrast, Figure 6.9 shows a nonlinearity that is increasing (but not strictly so) and a point $\rho/\hat{G}(0) \in \text{im } k$ with the property that its preimage $k^{-1}(\rho/\hat{G}(0))$ is not a singleton.

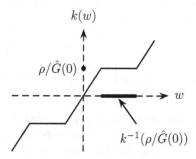

Figure 6.9 Example wherein $k^{-1}(\rho/\hat{G}(0))$ is not a singleton.

Proof of Theorem 6.21

Let I be the interval of existence of the unique maximal solution (x, u) of (6.49). Invoking (6.51), an application of Proposition 4.12 shows that $I = \mathbb{R}$ (see part (a) of Exercise 6.20) and, *a fortiori*, $\mathbb{R}_+ \subset I$.

Set $z(t) := x(t) + A^{-1}bk(u^\rho)$ and $v(t) := u(t) - u^\rho$ for all $t \in \mathbb{R}_+$, and $\tilde{k}(w) := k(w + u^\rho) - k(u^\rho)$ for all $w \in \mathbb{R}$. Since $k(u^\rho) = \rho/\hat{G}(0)$, a straightforward calculation (see part (b) of Exercise 6.20) shows that

$$\dot{z} = Az + b\tilde{k}(v), \quad \dot{v} = -\gamma c^* z. \tag{6.52}$$

Setting

$$\tilde{x} = \begin{pmatrix} z \\ v \end{pmatrix}, \quad \tilde{A} = \begin{pmatrix} A & 0 \\ -\gamma c^* & 0 \end{pmatrix}, \quad \tilde{b} = \begin{pmatrix} -b \\ 0 \end{pmatrix}, \quad \tilde{c} = \begin{pmatrix} 0 \\ 1 \end{pmatrix}, \quad (6.53)$$

the above system can be written in the form

$$\frac{d\tilde{x}}{dt} = \tilde{A}\tilde{x} - \tilde{b}\tilde{k}(\tilde{c}^*\tilde{x}). \tag{6.54}$$

The idea now is to apply statement (2) of Theorem 6.17 to (6.54). To do this, we need to verify that the relevant assumptions are satisfied. To this end, we note that the transfer function $\tilde{G}$ of the linear system given by $(\tilde{A}, \tilde{b}, \tilde{c}^*)$ satisfies

$$\tilde{G}(s) = \tilde{c}^*(sI - \tilde{A})^{-1}\tilde{b} = \gamma\frac{\hat{G}(s)}{s}.$$

Since $\gamma \in (0, \gamma^\sharp/\lambda)$, it is possible to choose $\tilde{\lambda} > \lambda$ such that

$$\tilde{\lambda}\gamma < \gamma^\sharp. \tag{6.55}$$

Consequently,

$$\mathrm{Re}\left(1 + \tilde{\lambda}\tilde{G}(s)\right) = 1 + \tilde{\lambda}\gamma\mathrm{Re}\,\frac{\hat{G}(s)}{s} \geq 0 \ \forall\, s \in \mathbb{C}_+,$$

showing that $1 + \tilde{\lambda}\tilde{G}$ is positive real. Moreover, since $k^{-1}(\rho/\hat{G}(0))$ is a singleton, $\tilde{k}(w) = 0$ if, and only if, $w = 0$. This, in conjunction with the assumption that k is non-decreasing and the global Lipschitz assumption (6.51), shows that

$$0 < \tilde{k}(w)w \leq \lambda w^2 < \tilde{\lambda}w^2 \ \forall\, w \in \mathbb{R}\backslash\{0\}. \tag{6.56}$$

Invoking the controllability and observability hypotheses relating to (A, b, c^*), applications of the Hautus criteria (Theorems 3.11 and 3.21) show that the linear system given by $(\tilde{A}, \tilde{b}, \tilde{c}^*)$ is controllable and observable (see parts (c) and (d) of Exercise 6.20). Consequently, in the context of system (6.54), all the assumptions of statement (2) of Theorem 6.17 are in place (with $\alpha = 0$ and $\beta = \tilde{\lambda}$), and thus, $z(t) \to 0$ and $v(t) \to 0$ as $t \to \infty$. Hence,

$$\lim_{t\to\infty} x(t) = -A^{-1}bk(u^\rho) = -A^{-1}b\frac{\rho}{\hat{G}(0)} \quad \text{and} \quad \lim_{t\to\infty} u(t) = u^\rho.$$

In particular, $y(t) = c^*x(t) \to -c^*A^{-1}b\rho/\hat{G}(0) = \rho$ as $t \to \infty$, completing the proof. $\qquad\qquad\square$

Exercise 6.20

The purpose of this exercise is to provide the details which are missing in the proof of Theorem 6.21. In the following, we use the notation of the statement and proof of Theorem 6.21 (in particular, (6.53))).

(a) Invoking the global Lipschitz condition (6.51) and Proposition 4.12, show that the unique maximal solution of (6.49) has interval of existence equal to $\mathbb{R}$.

(b) Show that the functions z and v satisfy (6.52).

(c) Invoking the controllability of (A, b) and the Hautus criterion for controllability (Theorem 3.11), show that $(\tilde{A}, \tilde{b})$ is controllable.

(d) Invoking the observability of (c^*, A) and the Hautus criterion for observability (Theorem 3.21), show that $(\tilde{c}^*, \tilde{A})$ is observable.

Example 6.22

Consider the system

$$\dot{x} = Ax + bk(u), \quad x(0) = \xi \in \mathbb{R}^2; \quad y = c^*x, \tag{6.57}$$

where

$$A = \begin{pmatrix} a_1 & a_2 \\ a_3 & a_4 \end{pmatrix}, \quad b = \begin{pmatrix} 0 \\ 1 \end{pmatrix}, \quad c = \begin{pmatrix} 1 \\ 0 \end{pmatrix},$$

and, with reference to Figure 6.10, k is a nonlinearity of saturation type,

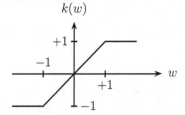

Figure 6.10 Saturation nonlinearity.

specifically, k is given by

$$k(w) = \begin{cases} w & \text{if } -1 \le w \le 1, \\ 1 & \text{if } w > 1, \\ -1 & \text{if } w < -1. \end{cases} \tag{6.58}$$

We assume that $a_2 > 0$, $\det A > 0$ and $\operatorname{tr} A < 0$, where $\operatorname{tr} A = a_1 + a_4$ is the trace of A. Since $a_2 \ne 0$, it is easy to show that (A, b) is controllable and (c^*, A) is observable. Moreover, together, the conditions $\det A > 0$ and $\operatorname{tr} A < 0$ are equivalent to the Hurwitz property of A. A straightforward calculation shows that

$$\hat{G}(s) = c^*(sI - A)^{-1}b = \frac{a_2}{\det(sI - A)} = \frac{a_2}{s^2 - (\operatorname{tr} A)s + \det A}.$$

In particular,

$$\hat{G}(0) = \frac{a_2}{\det A} > 0.$$

Also note that the nonlinearity k given by (6.58) satisfies the global Lipschitz condition (6.51) with $\lambda = 1$. Define $\gamma^\sharp$ by (6.50) and let $\gamma \in (0, \gamma^\sharp)$. It follows from Theorem 6.21 that, for every reference value ρ satisfying $|\rho| < a_2/\det A$, application of the control law $\dot{u} = \gamma(\rho - y)$, $u(0) = \zeta$, to (6.57) will result in a feedback system the output of which asymptotically tracks ρ, that is $y(t) \to \rho$ as $t \to \infty$. To find a suitable gain γ, we need to compute $\gamma^\sharp$ (computation of a positive lower bound for $\gamma^\sharp$ would be sufficient).

Set $D := \det A$ and $T := \operatorname{tr} A$ and note that

$$F(\omega) := \operatorname{Re} \frac{\hat{G}(i\omega)}{i\omega} = \frac{a_2 T}{(D - \omega^2)^2 + T^2 \omega^2} < 0 \quad \forall \omega \in \mathbb{R}.$$

Combining Lemmas 6.16 and 6.20, it follows that

$$\gamma^\sharp = \frac{1}{|\inf_{\omega \in \mathbb{R}} F(\omega)|} > 0.$$

We distinguish two cases.

Case 1: $2D > T^2$. Elementary calculus shows that $F(\omega)$ has a global minimum at $\omega = \pm\sqrt{(2D - T^2)/2}$, and so

$$\gamma^\sharp = \frac{1}{a_2}\left(|T|(D - T^2/4)\right).$$

Case 2: $2D \leq T^2$. In this case, $F(\omega)$ has a global minimum at $\omega = 0$, and so

$$\gamma^\sharp = \frac{D^2}{a_2|T|}.$$

For illustrative purposes, assume that $a_1 = 0$, $a_2 = 1$, $a_3 = -0.9$ and $a_4 = -1$, in which case we have $D = 0.9$, $T = -1$, $\hat{G}(0) = 1/D > 1$ and $\gamma^\sharp = 0.65$. For the reference signal, take $\rho = 1 < \hat{G}(0)$. Then, for each γ with $0 < \gamma < 0.65$, application of the feedback law $\dot{u} = \gamma(1 - y)$ will achieve the tracking objective: $y(t) \to 1$ as $t \to \infty$. Note that, for the chosen parameter values, we have

$$0.9 = D = \rho/\hat{G}(0) = k(u^\rho) = u^\rho.$$

With initial conditions $x_1(0) = 0 = x_2(0)$ and $u(0) = 0$, Figure 6.11 depicts the (MATLAB generated) output function y wherein the convergence to $\rho = 1$ is evident. Figure 6.12 depicts the control function $k(u)$: notice that the upper saturation level of the nonlinearity is attained by the control. Furthermore, convergence, as $t \to \infty$, of $k(u(t))$ to $k(u^\rho) = 0.9$ is evident.

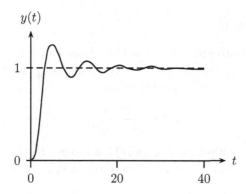

Figure 6.11 Output function y

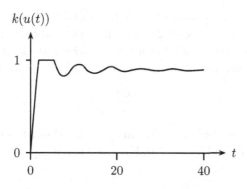

Figure 6.12 Control function $k(u)$ △

6.4 Proof of Lemmas 6.16 and 6.18

For the proof of Lemma 6.16, we need a standard result from complex analysis, the so-called *maximum modulus principle*.

Theorem 6.23 (Maximum modulus principle)

Let $U \subset \mathbb{C}$ be open, connected and bounded. Assume that $f : \overline{U} \to \mathbb{C}$ is continuous. If f is analytic in U, then $|f(s)|$ achieves its maximum on the boundary ∂U of U, that is

$$\max_{s \in \overline{U}} |f(s)| = \max_{s \in \partial U} |f(s)|.$$

A proof of this result can be found in any book on complex analysis (for example, see [5] or [16]).

Proof of Lemma 6.16

Assume that R is positive real. It is clear that condition (3) holds. Next we show that condition (1) is satisfied. Seeking a contradiction, suppose that $s_0 \in \mathbb{C}_+$ is a pole of R. Then R can be written in the form

$$R(s) = \frac{Q(s)}{(s - s_0)^m}$$

where $m \geq 1$ is the multiplicity of s_0 and Q is a rational function which does not have a pole at s_0 and $Q(s_0) \neq 0$. For sufficiently small $\varepsilon > 0$, $s = s_0 + \varepsilon e^{i\theta} \in \mathbb{C}_+$ for all $\theta \in [0, 2\pi)$ and

$$R(s_0 + \varepsilon e^{i\theta}) = \varepsilon^{-m} e^{-im\theta} Q(s_0 + \varepsilon e^{i\theta}). \tag{6.59}$$

Hence, since $Q(s_0 + \varepsilon e^{i\theta}) \to Q(s_0) \neq 0$ as $\varepsilon \to 0$ (uniformly in θ), it follows that there exist $\varepsilon > 0$ and $\theta \in [0, 2\pi)$ for which the real part of the right-hand side of (6.59) is negative, contradicting the positive realness of R.

To show that condition (2) holds, assume that $i\omega$ is a pole of R. Then

$$R(s) = \frac{Q(s)}{(s - i\omega)^m} \tag{6.60}$$

where $m \geq 1$ is the multiplicity of $i\omega$ and Q is a rational function which does not have a pole at $i\omega$ and $Q(i\omega) \neq 0$. For $\varepsilon > 0$ and $\theta \in [-\pi/2, \pi/2]$, it is clear that $i\omega + \varepsilon e^{i\theta} \in \overline{\mathbb{C}}_+$ and so,

$$0 \leq \operatorname{Re} R(i\omega + \varepsilon e^{i\theta}) = \operatorname{Re}\left(\varepsilon^{-m} e^{-im\theta} Q(i\omega + \varepsilon e^{i\theta}) \right). \tag{6.61}$$

Since $Q(i\omega + \varepsilon e^{i\theta}) \to Q(i\omega) \neq 0$ as $\varepsilon \to 0$ (uniformly in θ), we see that $m = 1$, because otherwise there would exist $\varepsilon > 0$ and $\theta \in [-\pi/2, \pi/2]$ such that the right-hand side of (6.61) is negative. Thus, by (6.61),

$$0 \leq (\operatorname{Re} Q(i\omega + \varepsilon e^{i\theta})) \cos\theta + (\operatorname{Im} Q(i\omega + \varepsilon e^{i\theta})) \sin\theta.$$

Considering the above inequality for $\theta = 0, \pm\pi/2$ and letting $\varepsilon \to 0$, shows that $\operatorname{Re} Q(i\omega) \geq 0$ and $\operatorname{Im} Q(i\omega) = 0$. Combining this with the fact that $Q(i\omega) \neq 0$, we conclude that $\operatorname{Re} Q(i\omega) > 0$. Since $m = 1$, it therefore follows from (6.60) that

$$\lim_{s \to i\omega} (s - i\omega) R(s) = Q(i\omega) = \operatorname{Re} Q(i\omega) > 0,$$

showing that condition (2) holds.

Conversely, assume that conditions (1)-(3) are satisfied. It follows from conditions (1) and (2) that R can be written in the form

$$R(s) = R_0(s) + \sum_{j=1}^{q} \frac{a_j s}{s^2 + \omega_j^2} \tag{6.62}$$

where $0 \le \omega_1 < \omega_2 < \ldots < \omega_q$, $a_j > 0$ and R_0 is in $\mathbb{R}(s)$ and does not have any poles in $\overline{\mathbb{C}}_+$. Since $a_j > 0$, it follows from a straightforward calculation that $(a_j s)/(s^2 + \omega_j^2)$ is positive real for all $j = 1, \ldots, q$. Thus, it suffices to show that R_0 is positive real. Note that, by (6.62), Re $R_0(i\omega) = $ Re $R(i\omega)$ for all $\omega \in \mathbb{R}$ such that $i\omega$ is not a pole of R. Consequently, by condition (3),

$$\text{Re } R_0(i\omega) \ge 0 \quad \forall \omega \in \mathbb{R}. \tag{6.63}$$

Since $R_0 \in \mathbb{R}(s)$ is proper, the limit $R_0(\infty) := \lim_{|s| \to \infty} R_0(s)$ exists and is real. It follows from (6.63) that

$$R_0(\infty) := \lim_{|s| \to \infty} \text{Re } R_0(s) = \lim_{|\omega| \to \infty} \text{Re } R_0(i\omega) \ge 0. \tag{6.64}$$

For $r > 0$, define the closed semidisc $\Delta_r := \{s \in \overline{\mathbb{C}}_+ : |s| \le r\}$ and denote its boundary by $\partial \Delta_r$. Appealing to (6.63) and (6.64), we obtain

$$\limsup_{r \to \infty} \max_{s \in \partial \Delta_r} e^{-\text{Re } R_0(s)} \le 1. \tag{6.65}$$

By Theorem 6.23,

$$\max_{s \in \Delta_r} e^{-\text{Re } R_0(s)} = \max_{s \in \Delta_r} \left| e^{-R_0(s)} \right| = \max_{s \in \partial \Delta_r} \left| e^{-R_0(s)} \right| = \max_{s \in \partial \Delta_r} e^{-\text{Re } R_0(s)}.$$

Letting $r \to \infty$ and invoking (6.65) leads to

$$\sup_{s \in \mathbb{C}_+} e^{-\text{Re } R_0(s)} \le 1,$$

which in turn is equivalent to the positive realness of R_0. $\qquad \square$

We now give a proof of the positive real lemma (Lemma 6.18) which played a crucial role in the proof of Theorem 6.17.

Proof of Lemma 6.18

To prove sufficiency, assume that there exist a symmetric positive definite matrix $P \in \mathbb{R}^{N \times N}$ and a vector $l \in \mathbb{R}^N$ such that (6.40) holds. Since

$$2\text{Re } R(s) = R(s) + \bar{R}(s) = 2d + c^*(sI - A)^{-1}b + b^*(\bar{s}I - A^*)^{-1}c$$

it follows from (6.40) that

$$2\text{Re } R(s) = 2d + b^*\big((\bar{s}I - A^*)^{-1}P + P(sI - A)^{-1}\big)b$$
$$+ \sqrt{2d}\big(b^*(\bar{s}I - A^*)^{-1}l + l^*(sI - A)^{-1}b\big),$$

and hence,

$$2\operatorname{Re} R(s) = 2d + b^*(\bar{s}I - A^*)^{-1}(2(\operatorname{Re} s)P - PA - A^*P)(sI - A)^{-1}b$$
$$+ \sqrt{2d}\big(b^*(\bar{s}I - A^*)^{-1}l + l^*(sI - A)^{-1}b\big).$$

Again using (6.40), we obtain

$$2\operatorname{Re} R(s) = 2d + b^*(\bar{s}I - A^*)^{-1}(2(\operatorname{Re} s)P + ll^*)(sI - A)^{-1}b$$
$$+ \sqrt{2d}\big(b^*(\bar{s}I - A^*)^{-1}l + l^*(sI - A)^{-1}b\big),$$

and thus

$$2\operatorname{Re} R(s) = \big(\sqrt{2d} + b^*(\bar{s}I - A^*)^{-1}l\big)\big(\sqrt{2d} + l^*(sI - A)^{-1}b\big)$$
$$+ 2(\operatorname{Re} s)b^*(\bar{s}I - A^*)^{-1}P(sI - A)^{-1}b.$$

Since the right-hand side of the last identity is nonnegative for all $s \in \mathbb{C}_+$, it follows that R is positive real.

To prove necessity, assume that R is positive real. We proceed in two steps.

Step 1. In this step, we assume that R does not have any poles on the imaginary axis. Therefore, by Lemma 6.16, all poles of R have negative real parts. Setting $\delta(s) := \det(sI - A)$ and $\nu_0(s) := \delta(s)c^*(sI - A)^{-1}b$, it is clear that ν_0 is a polynomial and $c^*(sI - A)^{-1}b = \nu_0(s)/\delta(s)$. Since (A, b, c^*) is minimal, it follows from Proposition 3.29 that the polynomials δ and ν_0 are coprime. We note that, as a consequence,

$$\sigma(A) \subset \mathbb{C}_-. \tag{6.66}$$

Defining a new polynomial $\nu := \nu_0 + d\delta$, we have that δ and ν are coprime and, furthermore

$$R(s) = c^*(sI - A)^{-1}b + d = \frac{\nu(s)}{\delta(s)}$$

Setting $\mu(s) := \nu(s)\delta(-s) + \delta(s)\nu(-s)$ and noting that

$$\operatorname{Re} R(i\omega) = \frac{\mu(i\omega)}{2|\delta(i\omega)|^2}, \quad \forall\omega \in \mathbb{R},$$

it follows from the positive-real property that

$$\mu(i\omega) \geq 0, \quad \forall\omega \in \mathbb{R}. \tag{6.67}$$

We claim that

$$\text{purely imaginary zeros of } \mu \text{ have even multiplicity.} \tag{6.68}$$

Assume that, for some real ω_0, $i\omega_0$ is a zero of μ with multiplicity m. Then

$$\mu(i\omega) = i^m(\omega - \omega_0)^m\lambda(i\omega), \quad \forall\omega \in \mathbb{R},$$

where λ is a polynomial (with complex coefficients) and $\lambda(i\omega_0) \neq 0$. Since $\mu(i\omega)$ is real for all $\omega \in \mathbb{R}$, we see that the polynomial $i^m \lambda(i\omega)$ is also real for all $\omega \in \mathbb{R}$. If m were odd, then, using the properties of λ, we see that $\mu(i\omega)$ changes sign at $\omega = \omega_0$, which is impossible by (6.67).

Combining (6.68) with the symmetry properties

$$\bar{\mu}(s) = \mu(\bar{s}), \quad \mu(s) = \mu(-s); \quad \forall s \in \mathbb{C},$$

shows that there exists a polynomial ρ of degree less or equal to N and with real coefficients such that

$$\mu(s) = \rho(s)\rho(-s), \quad \forall s \in \mathbb{C}; \quad \rho(s) \neq 0 \text{ if } \operatorname{Re} s < 0. \tag{6.69}$$

Consequently,

$$R(s) + R(-s) = \frac{\rho(s)}{\delta(s)} \frac{\rho(-s)}{\delta(-s)} \tag{6.70}$$

Setting $\rho_N := \lim_{|s|\to\infty} \rho(s)/s^N$, it follows that $\rho_N^2 = 2d$, so that $\rho_N = \pm\sqrt{2d}$. Without loss of generality we may assume that $\rho_N = \sqrt{2d}$. Consequently, there exists a real polynomial $\kappa(s) = k_{N-1}s^{N-1} + \ldots + k_1 s + k_0$ such that

$$\frac{\rho(s)}{\delta(s)} = \sqrt{2d} + \frac{\kappa(s)}{\delta(s)}.$$

Writing $\delta(s) = \det(sI - A) = s^N + a_{n-1}s^{N-1} + \ldots + a_1 s + a_0$ and setting

$$A_c := \begin{pmatrix} 0 & 1 & 0 & \cdots & 0 \\ 0 & 0 & 1 & \cdots & 0 \\ \vdots & & & & \vdots \\ 0 & 0 & 0 & \cdots & 1 \\ -a_0 & -a_1 & -a_2 & \cdots & -a_{N-1} \end{pmatrix}, \quad b_c := \begin{pmatrix} 0 \\ 0 \\ \vdots \\ 0 \\ 1 \end{pmatrix}, \quad k := \begin{pmatrix} k_0 \\ k_1 \\ \vdots \\ 0 \\ k_{N-1} \end{pmatrix},$$

we obtain from Proposition 3.29 that (A_c, b_c, k^*) is a minimal realization of κ/δ. On the other hand, since (A, b) is controllable, Lemma 6.1 guarantees the existence of a matrix $S \in GL(N, \mathbb{R})$ such that $A = S^{-1}A_cS$ and $b = S^{-1}b_c$. Setting $l := S^*k$, we conclude that (A, b, l^*) is a minimal realization of κ/δ. We are now in the position to express (6.70) in state-space terms:

$$2d + c^*(sI - A)^{-1}b + b^*(-sI - A^*)^{-1}c$$
$$= \left(\sqrt{2d} + l^*(sI - A)^{-1}b\right)\left(\sqrt{2d} + b^*(-sI - A^*)^{-1}l\right),$$

and thus,

$$c^*(sI - A)^{-1}b + b^*(-sI - A^*)^{-1}c$$
$$= \sqrt{2d}\, l^*(sI - A)^{-1}b + \sqrt{2d}\, b^*(-sI - A^*)^{-1}l$$
$$+ b^*(-sI - A^*)^{-1}ll^*(sI - A)^{-1}b. \tag{6.71}$$

Using that A is Hurwitz (see (6.66)), it follows that

$$P := \int_0^\infty \exp(A^*t)ll^* \exp(At)dt$$

is a well-defined, symmetric and positive semi-definite matrix in $\mathbb{R}^{N \times N}$. Since (A, b, l^*) is a minimal realization (of κ/δ), it follows from Theorem 3.30 that (l^*, A) is observable. Consequently, the kernel of P is trivial and we conclude that P is positive definite. Furthermore,

$$PA + A^*P = \int_0^\infty \frac{\mathrm{d}}{\mathrm{d}t}\big(\exp(A^*t)ll^* \exp(At)\big)\mathrm{d}t,$$

and thus

$$PA + A^*P = -ll^*. \tag{6.72}$$

This identity allows us to write the last term on the right-hand side of (6.71) as

$$b^*(-sI - A^*)^{-1}ll^*(sI - A)^{-1}b$$
$$= b^*(-sI - A^*)^{-1}\big(P(sI - A) + (-sI - A^*)P\big)(sI - A)^{-1}b$$
$$= b^*(-sI - A^*)^{-1}Pb + b^*P(sI - A)^{-1}b$$

Combining this with (6.71) yields

$$\big(c^* - b^*P - \sqrt{2d}\,l^*\big)(sI - A)^{-1}b + b^*(-sI - A^*)^{-1}\big(c - Pb - \sqrt{2d}\,l\big) = 0.$$

This identity holds for all $s \in \mathbb{C}$ and, invoking (6.66), we conclude that each of the two terms vanishes for all $s \in \mathbb{C}$. Therefore,

$$b^*(sI - A^*)^{-1}\big(c - Pb - \sqrt{2d}\,l\big) = 0 \quad \forall\, s \in \mathbb{C}$$

Now $(sI - A^*)^{-1}$ is the Laplace transform of e^{A^*t} and thus

$$b^*e^{A^*t}\big(c - Pb - \sqrt{2d}\,l\big) = 0 \quad \forall\, t \in \mathbb{R}_+.$$

Since (A, b, c^*) is minimal, we have that (A, b) is controllable (by Theorem 3.30), implying that (b^*, A^*) is observable. Hence it follows that $c - Pb - \sqrt{2d}\,l = 0$, and so

$$Pb = c - \sqrt{2d}\,l.$$

Together with (6.72) this yields the claim.

Step 2. We now remove the restriction on the position of the poles of R. Note that, by Lemma 6.16, R can be written in the form

$$R(s) = R_0(s) + \sum_{j=1}^q \frac{r_j s}{s^2 + \omega_j^2} \tag{6.73}$$

where $0 \leq \omega_1 < \omega_2 < \ldots < \omega_q$, $r_j > 0$ and R_0 is in $\mathbb{R}(s)$ and does not have any poles in $\overline{\mathbb{C}}_+$. Noting that $\mathrm{Re}\, R_0(i\omega) = \mathrm{Re}\, R(i\omega)$ for all $\omega \in \mathbb{R}$ such that $\omega \neq \pm\omega_j$, $1 \leq j \leq q$, it follows from the positive-realness of R that $\mathrm{Re}\, R_0(i\omega) \geq 0$ for all $\omega \in \mathbb{R}$. Consequently, by Lemma 6.16, R_0 is positive real. Let $A_0 \in \mathbb{R}^{N_0 \times N_0}$ and $b_0, c_0 \in \mathbb{R}^{N_0}$ be such that (A_0, b_0, c_0^*) is a minimal realization of $R_0(s) - d$. Note that $N_0 = N - 2q$ if $\omega_1 > 0$ and $N_0 = N - (2q - 1)$ if $\omega_1 = 0$. By Step 1, there exist a symmetric positive definite matrix $P_0 \in \mathbb{R}^{N_0 \times N_0}$ and $l_0 \in \mathbb{R}^{N_0}$ such that

$$P_0 A_0 + A_0^* P_0 = -l_0 l_0^*, \quad P_0 b_0 = c_0 - \sqrt{2d}\, l_0. \tag{6.74}$$

If $\omega_1 > 0$, then we set, for all $j = 1, \ldots, q$,

$$A_j := \begin{pmatrix} 0 & -\omega_j \\ \omega_j & 0 \end{pmatrix}, \quad b_j = c_j := \begin{pmatrix} \sqrt{r_j} \\ 0 \end{pmatrix}.$$

If $\omega_1 = 0$, then $A_1 := 0$, $b_1 = c_1 := \sqrt{r_1}$ and, for $j = 2, \ldots, q$, we define A_j, b_j and c_j as before. In any case, for every $j = 1, \ldots, q$, (A_j, b_j, c_j^*) is a minimal realization of $(r_j s)/(s^2 + \omega_j^2)$. Furthermore, if $\omega_1 > 0$, then we set, for all $j = 1, \ldots, q$,

$$P_j := \begin{pmatrix} 1 & 0 \\ 0 & 1 \end{pmatrix}, \quad l_j := \begin{pmatrix} 0 \\ 0 \end{pmatrix}.$$

If $\omega_1 = 0$, then we set $P_1 := 1$, $l_1 := 0$ and, for $j = 2, \ldots, q$, we define P_j and l_j as before. In any case, we have that

$$P_j A_j + A_j^* P_j = -l_j l_j^* = 0, \quad P_j b_j = c_j - \sqrt{2d}\, l_j \quad j = 1, \ldots, q. \tag{6.75}$$

Defining $\tilde{A} \in \mathbb{R}^{N \times N}$ and $\tilde{b}, \tilde{c} \in \mathbb{R}^N$ by

$$\tilde{A} := \begin{pmatrix} A_0 & 0 & 0 & \cdots & 0 \\ 0 & A_1 & 0 & \cdots & 0 \\ \vdots & \vdots & \vdots & \ddots & \vdots \\ 0 & 0 & 0 & \cdots & A_q \end{pmatrix}, \quad \tilde{b} := \begin{pmatrix} b_0 \\ b_1 \\ \vdots \\ b_q \end{pmatrix}, \quad \tilde{c} := \begin{pmatrix} c_0 \\ c_1 \\ \vdots \\ c_q \end{pmatrix},$$

and invoking (6.73), we see that $(\tilde{A}, \tilde{b}, \tilde{c}^*)$ is a realization of $R(s) - d$. Furthermore, setting

$$\tilde{P} := \begin{pmatrix} P_0 & 0 & 0 & \cdots & 0 \\ 0 & P_1 & 0 & \cdots & 0 \\ \vdots & \vdots & \vdots & \ddots & \vdots \\ 0 & 0 & 0 & \cdots & P_q \end{pmatrix} \in \mathbb{R}^{N \times N}, \quad \tilde{l} := \begin{pmatrix} l_0 \\ 0 \\ \vdots \\ 0 \end{pmatrix} \in \mathbb{R}^N,$$

we note that P is symmetric and positive definite. It follows from (6.74) and (6.75) that

$$\tilde{A}\tilde{P} + (\tilde{A})^* \tilde{P} = -\tilde{l}(\tilde{l})^*, \quad \tilde{P}\tilde{b} = \tilde{c} - \sqrt{2d}\, \tilde{l}. \tag{6.76}$$

The dimension of the realization $(\tilde{A}, \tilde{b}, \tilde{c}^*)$ is N and is therefore minimal. Consequently, by Theorem 3.31, there exists $S \in GL(n, \mathbb{R})$ such that $\tilde{A} = SAS^{-1}$, $\tilde{b} = Sb$ and $c = S^*\tilde{c}$. Finally, setting $P := S^*\tilde{P}S$ and $l := S^*\tilde{l}$, it is clear that P is symmetric and positive definite and it follows from (6.76) that

$$PA + A^*P = -ll^*, \quad Pb = c - \sqrt{2d}l,$$

completing the proof. $\qquad\qquad\square$

A

Appendix

In this appendix, we assemble a compendium of background concepts and results pertaining to linear algebra and matrix theory, metric and normed spaces, analysis, Laplace transforms, and Zorn's lemma which underpin the presentation and development of the material in the main body of the book. Most of the material is standard and can be found in many texts (see, for example, [5]–[7], [9]–[13], [16] and [17]) and so is presented without proof; results which we deem mildly non-standard are provided with proof.

A.1 Linear algebra and matrix theory

In the following let $\mathbb{F} = \mathbb{R}$ or $\mathbb{F} = \mathbb{C}$. We denote, by $\mathbb{F}^N$, the set of all ordered N-tuples z with components $z_1, \ldots, z_N$ in $\mathbb{F}$. An element $z \in \mathbb{F}^N$ can be viewed as a row $(1 \times N$ matrix), namely, $z = (z_1, \ldots, z_N)$ or, equivalently, as a column $(N \times 1$ matrix), namely,

$$z = \begin{pmatrix} z_1 \\ \vdots \\ z_N \end{pmatrix}. \tag{A.1}$$

In this section, we will be dealing with matrix manipulation in which context the column form is appropriate. We can give $\mathbb{F}^N$ the structure of a *vector space* (or *linear space*) over $\mathbb{F}$ by defining the operations of addition and multiplication by scalars (that is, elements in $\mathbb{F}$) in the standard componentwise sense.

H. Logemann and E. P. Ryan, *Ordinary Differential Equations*,
Springer Undergraduate Mathematics Series,
DOI: 10.1007/978-1-4471-6398-5, © Springer-Verlag London 2014

Also, in this section, we will not consider *abstract* vector spaces. We restrict our attention to the space $\mathbb{F}^N$ (because that is all we need). We say $x_1, \ldots, x_P$ in $\mathbb{F}^N$ are *linearly dependent*, if there exist scalars $\alpha_1, \ldots, \alpha_P$ in $\mathbb{F}$ – not all zero – such that $\alpha_1 x_1 + \cdots + \alpha_P x_P = 0$. Otherwise, $x_1, \ldots, x_P$ are said to be *linearly independent*.

A subset $S \subset \mathbb{F}^N$ is a *subspace* of the vector space $\mathbb{F}^N$ if it is itself a vector space (under the operations of addition and scalar multiplication inherited from $\mathbb{F}^N$), that is, for all $x, y \in S$ and for all $\alpha, \beta \in \mathbb{F}$, $\alpha x + \beta y \in S$. Let $S \subset \mathbb{F}^N$ be a subspace of $\mathbb{F}^N$. If $x_1, \ldots, x_P$ in S are linear independent and if these vectors *span* S, that is, for every $x \in S$, there exist $\alpha_1, \ldots, \alpha_P$ in $\mathbb{F}$ such that $x = \alpha_1 x_1 + \cdots + \alpha_P x_P$, then we say that $x_1, \ldots, x_P$ is a *basis* of S and P is the *dimension* of S (different bases have the same number of elements). We write $\dim S = P$. If we select a basis $b_1, b_2, \ldots, b_P$ of S, then each $x \in S$ has unique representation relative to this basis, that is, there exists unique $\beta_j \in \mathbb{F}$, $j = 1, 2, \ldots, P$, (called the *coordinates* of x with respect to the chosen basis) such that

$$x = \beta_1 b_1 + \cdots + \beta_P b_P .$$

The dimension of the vector space $\mathbb{F}^N$ is equal to N. If the chosen basis $b_1, \ldots, b_N$ of $\mathbb{F}^N$ is the *canonical basis*, that is, if

$$b_1 = e_1 := (1, 0, \ldots, 0, 0), \ldots, b_N = e_N := (0, 0, \ldots, 0, 1)$$

then the coordinates β_j of the vector x given by (A.1) with respect to the canonical basis are $\beta_j = x_j$.

The notion of *orthogonality* of vectors in $\mathbb{F}^N$ is captured by means of the *inner product*:

$$\langle x, y \rangle := x_1 \overline{y}_1 + x_2 \overline{y}_2 + \cdots + x_N \overline{y}_N, \quad x, y \in \mathbb{F}^N, \tag{A.2}$$

where x_j and y_j denote the components of x and y respectively and $\overline{\lambda}$ denotes the complex conjugate of λ. We observe that in the case $\mathbb{F} = \mathbb{R}$, the complex conjugation may be ignored. Two vectors $x, y \in \mathbb{F}^N$ are said to be *orthogonal* if $\langle x, y \rangle = 0$.

Let $Z \subset \mathbb{F}^N$ be a non-empty set. The *span* of Z is the set of all finite linear combinations of elements of Z, that is

$$\operatorname{span} Z := \left\{ \sum_{j=1}^{n} \alpha_j z_j : z_j \in Z, \ \alpha_j \in \mathbb{F}, \ n \in \mathbb{N} \right\}.$$

If $S_1, S_2 \subset \mathbb{F}^N$ are subspaces, then $S_1 + S_2$ is the subspace defined by

$$S_1 + S_2 = \{ x + y : x \in S_1, \ y \in S_2 \}.$$

If $S_1 \cap S_2 = \{0\}$, then the sum $S_1 + S_2$ is called the *direct sum* of S_1 and S_2, written $S_1 \oplus S_2$. If $z \in S_1 \oplus S_2$, then there exist unique $x \in S_1$ and $y \in S_2$ such that $z = x + y$: x is the projection of z on S_1 along S_2; y is the projection of z on S_2 along S_1.

The *orthogonal complement* $S^\perp \subset \mathbb{F}^N$ of a subspace $S \subset \mathbb{F}^N$ is the subspace of all vectors orthogonal to S, that is,

$$S^\perp = \{x \in \mathbb{F}^N : \langle x, y \rangle = 0 \ \forall \, y \in S\}.$$

Two important facts:

$$\text{(i)} \ \ S \oplus S^\perp = \mathbb{F}^N, \ \ \text{(ii)} \ (S^\perp)^\perp = S.$$

Let $M \in \mathbb{F}^{N \times P}$. The *image* of M is the set

$$\operatorname{im} M := \{Mx : x \in \mathbb{F}^P\},$$

which is a subspace of $\mathbb{F}^N$. The *kernel* of M is the set

$$\ker M := \{x \in \mathbb{F}^P : Mx = 0\},$$

which is a subspace of $\mathbb{F}^P$. Writing $M = (m_{ij})$ and denoting the columns of M by $c_j(M)$, that is

$$c_j(M) = \begin{pmatrix} m_{1j} \\ m_{2j} \\ \vdots \\ m_{Nj} \end{pmatrix} \in \mathbb{F}^N, \quad j = 1, 2, \dots, P,$$

the product Mx of M and the vector $x \in \mathbb{F}^P$ (with components x_j) can be written in the form

$$Mx = \begin{pmatrix} \sum_{j=1}^{P} m_{1j} x_j \\ \sum_{j=1}^{P} m_{2j} x_j \\ \vdots \\ \sum_{j=1}^{P} m_{Nj} x_j \end{pmatrix} = \sum_{j=1}^{P} x_j c_j(M). \tag{A.3}$$

It follows that

$$\operatorname{im} M = \operatorname{span}\{c_1(M), c_2(M), \dots, c_P(M)\} = \left\{ \sum_{j=1}^{P} \alpha_j c_j(M) : \alpha_j \in \mathbb{F} \right\}. \tag{A.4}$$

We record the *dimension formula* for $M \in \mathbb{F}^{N \times P}$:

$$\dim \operatorname{im} M + \dim \ker M = P. \tag{A.5}$$

Let $M \in \mathbb{F}^{N \times N}$, a square matrix. If there exists a matrix $T \in \mathbb{F}^{N \times N}$ such that $TM = MT = I$, then $T =: M^{-1}$ is unique and is called the *inverse* of the matrix M (which is said to be *invertible*). Furthermore, M is invertible if, and only if, $\dim \operatorname{im} M = N$. Consequently, by the above dimension formula, M is invertible if, and only if, $\ker M = \{0\}$.

Let $M \in \mathbb{C}^{N \times N}$, a square complex matrix. An *eigenvalue* λ of M is a complex number such that $Mv = \lambda v$ has a nonzero solution $v \in \mathbb{C}^N$. Any such vector v is called an *eigenvector* associated with λ. Therefore, $\lambda \in \mathbb{C}$ is an eigenvalue of M if, and only if, $\ker(\lambda I - M) \neq \{0\}$; an eigenvector associated with λ is any non-zero vector in the subspace $\ker(\lambda I - M)$. This subspace is said to be the *eigenspace* associated with λ and its dimension is said to be the *geometric multiplicity* of λ. The set of all eigenvalues of M is called the *spectrum* of M and will be denoted by $\sigma(M)$.

Let $M \in \mathbb{F}^{N \times N}$. A subspace $S \subset \mathbb{F}^N$ is called *M-invariant* if

$$MS \subset S, \quad \text{i.e. } Mx \in S \quad \forall x \in S.$$

If $\lambda \in \sigma(M)$, then it is easy to check that the eigenspace $\ker(\lambda I - M)$ associated with λ is an M-invariant subspace.

Let $M = (m_{ij}) \in \mathbb{C}^{N \times P}$. We define the *Hermitian transpose* (or *conjugate transpose*) $M^* \in \mathbb{C}^{P \times N}$ of the matrix M by

$$M^* := \begin{pmatrix} \overline{m}_{11} & \cdots & \overline{m}_{N1} \\ \vdots & & \vdots \\ \overline{m}_{1P} & \cdots & \overline{m}_{NP} \end{pmatrix},$$

i.e., M^* is the matrix obtained by transposing and complex conjugating M. We observe that, if $M \in \mathbb{R}^{N \times P}$ (that is, if M is a real matrix), then the complex conjugation may be ignored and M^* is simply called the *transpose* of M.

Let $M \in \mathbb{F}^{N \times P}$ and $L \in \mathbb{F}^{P \times Q}$. Then $(M^*)^* = M$ and $(ML)^* = L^* M^*$. The following important fact follows from a straightforward calculation:

$$\langle Mx, y \rangle = \langle x, M^* y \rangle \quad \forall x \in \mathbb{F}^P, \ \forall y \in \mathbb{F}^N. \tag{A.6}$$

If $N = P$ (that is, if M is square), then $(\exp(M))^* = \exp(M^*)$ (this follows from the definition of the matrix exponential, see Section 2.1 and Proposition A.27); moreover, if M is invertible, then M^* is invertible and $(M^{-1})^* = (M^*)^{-1}$.

Theorem A.1

Let $M \in \mathbb{F}^{N \times P}$. Then

$$(\operatorname{im} M)^{\perp} = \ker M^*.$$

Equivalently, $x \in (\operatorname{im} M)^{\perp}$ if, and only if, $x^* M = 0$.

A matrix $M \in \mathbb{C}^{N \times N}$ for which $M = M^*$ is called *Hermitian*; if $M \in \mathbb{R}^{N \times N}$, the term *symmetric* is used in place of Hermitian. The eigenvalues of a Hermitian matrix are real. A matrix $M \in \mathbb{F}^{N \times N}$ is called *positive definite*, written $M > 0$ (or *positive semi-definite*, written $M \geq 0$) if $M = M^*$ and $\langle v, Mv \rangle > 0$ (or ≥ 0) for all nonzero $v \in \mathbb{F}^N$. If $M = M^*$, then $M > 0$ ($M \geq 0$) if, and only if, the eigenvalues of M are positive (non-negative), i.e. $\sigma(M) \subset (0, \infty)$ ($\sigma(M) \subset [0, \infty)$).

Let $M \in \mathbb{F}^{N \times P}$. The *rank* of M, denoted by $\operatorname{rk} M$, is defined by $\operatorname{rk} M := \dim \operatorname{im} M$. The *row rank* (*column rank*) of M is the maximal number of linearly independent rows (columns) of M. The row and column ranks of M are denoted by $\operatorname{rk}_r M$ and $\operatorname{rk}_c M$, respectively.

Theorem A.2

For $M \in \mathbb{F}^{N \times P}$, $\operatorname{rk}_r M = \operatorname{rk}_c M = \operatorname{rk} M$.

Proof

It follows immediately from (A.4) that $\operatorname{rk}_c M = \operatorname{rk} M$. Since $\operatorname{rk}_r M = \operatorname{rk}_c M^* = \operatorname{rk} M^*$, it is sufficient to show that $\operatorname{rk} M^* = \operatorname{rk} M$. Since $\operatorname{im} M \oplus (\operatorname{im} M)^{\perp} = \mathbb{F}^N$ and, by Theorem A.1, $\ker M^* = (\operatorname{im} M)^{\perp}$, it follows that

$$\dim \ker M^* = \dim(\operatorname{im} M)^{\perp} = N - \operatorname{rk} M .$$

By the dimension formula,

$$\operatorname{rk} M^* + \dim \ker M^* = N ,$$

and therefore $\operatorname{rk} M^* = \operatorname{rk} M$. $\qquad\square$

We note that in the above proof it was shown that

$$\operatorname{rk} M = \operatorname{rk} M^* . \tag{A.7}$$

Proposition A.3

Let $M \in \mathbb{F}^{N \times P}$ and $L \in \mathbb{F}^{P \times Q}$. Then $\operatorname{rk}(ML) \leq \min\{\operatorname{rk} M, \operatorname{rk} L\}$.

Proof

By (A.3), every column of ML is a linear combination of the columns of M. Therefore, invoking Theorem A.2,

$$\operatorname{rk}(ML) = \operatorname{rk}_c(ML) \leq \operatorname{rk}_c M = \operatorname{rk} M .$$

Combining this with (A.7), we obtain

$$\mathrm{rk}(ML) = \mathrm{rk}(L^*M^*) \le \mathrm{rk}\,L^* = \mathrm{rk}\,L,$$

completing the proof.   □

As immediate consequences of Theorem A.2 we have that, for every matrix $M \in \mathbb{F}^{N \times P}$,

$$\mathrm{rk}\,M < P \text{ if, and only if, there exists } x \ne 0 \text{ such that } Mx = 0,$$

and similarly,

$$\mathrm{rk}\,M < N \text{ if, and only if, there exists } x \ne 0 \text{ such that } x^*M = 0.$$

We say that the rank of $M \in \mathbb{F}^{N \times P}$ is *full* (or that M is of *full rank*) if $\mathrm{rk}\,M = \min\{N, P\}$. We see that

$$\text{if } N \le P, \text{ then the rank of } M \text{ is full if, and only if, } \mathrm{im}\,M = \mathbb{F}^N,$$

and similarly,

$$\text{if } N \ge P, \text{ then the rank of } M \text{ is full if, and only if, } \ker M = \{0\}.$$

Furthermore, if $M \in \mathbb{F}^{N \times N}$, then $\lambda \in \sigma(M)$ if, and only if, $\lambda I - M$ is not of full rank.

Theorem A.4

Let $M \in \mathbb{F}^{N \times P}$.

(1) If $N \le P$, then there exists $M^\sharp \in \mathbb{F}^{P \times N}$ such that $MM^\sharp x = x$ for all $x \in \mathrm{im}\,M$. In particular, if M is of full rank, then $MM^\sharp = I$, that is, M has a right inverse.

(2) If $N \ge P$, then there exists $M^\sharp \in \mathbb{F}^{P \times N}$ such that $M^\sharp M x = x$ for all $x \in (\ker M)^\perp$. In particular, if M is of full rank, then $M^\sharp M = I$, that is, M has a left inverse.

Proof

To prove statement (1), set $K := \dim \mathrm{im}\,M \le N$ and let $\{x_1, x_2, \ldots, x_N\}$ be a basis of $\mathbb{F}^N$ with the property that $\{x_1, x_2, \ldots, x_K\}$ is a basis of $\mathrm{im}\,M$. Then there exist $y_1, \ldots, y_K$ in $\mathbb{F}^P$ such that $My_i = x_i$ for $i = 1, \ldots, K$. Furthermore, choose arbitrary $y_{K+1}, \ldots, y_N$ in $\mathbb{F}^P$ and set

$$X := (x_1, \ldots, x_N) \in \mathbb{F}^{N \times N}, \quad Y := (y_1, \ldots, y_N) \in \mathbb{F}^{P \times N}.$$

Obviously, X is invertible and we define $M^\sharp := YX^{-1}$. Denoting the i-th canonical basis vector in $\mathbb{F}^N$ by e_i, we have

$$M^\sharp x_i = YX^{-1}x_i = Ye_i = y_i, \quad i = 1, \ldots, N.$$

Consequently,

$$MM^\sharp x_i = My_i = x_i, \quad i = 1, \ldots, K.$$

Since $\{x_1, \ldots, x_K\}$ is a basis of $\operatorname{im} M$, it follows that $MM^\sharp x = x$ for all $x \in \operatorname{im} M$.

The proof of statement (2) is left to the reader, see Exercise A.1.   □

Exercise A.1

Prove statement (2) of Theorem A.4.

The *determinant* $\det M$ of a matrix $M = (m_{ij}) \in \mathbb{F}^{N \times N}$ is defined by

$$\det M = \sum_\varphi (-1)^{\iota(\varphi)} m_{\varphi(1)1} m_{\varphi(2)2} \cdots m_{\varphi(N)N}, \tag{A.8}$$

where the summation is over all $N!$ bijections φ mapping $\{1, 2, \ldots, N\}$ into itself (permutations of integers from 1 to N) and $\iota(\varphi)$ is the number of inversions contained in the permutation φ, that is, $\iota(\varphi)$ is the number of elements of the set $\{(i, j) : i < j \text{ and } \varphi(i) > \varphi(j)\}$.

We record two fundamental properties of the determinant: for matrices $M, L \in \mathbb{F}^{N \times N}$, (i) $\det(ML) = \det M \det L$ and (ii) M is invertible if, and only if, $\det M \neq 0$.

Theorem A.5 (Cramer's rule)

If $M \in \mathbb{F}^{N \times N}$, then

$$M \operatorname{adj} M = (\operatorname{adj} M)M = (\det M)I.$$

Here $\operatorname{adj} M = (\alpha_{ij}) \in \mathbb{F}^{N \times N}$ denotes the *adjugate* of M, that is, $\alpha_{ij} = (-1)^{i+j}\mu_{ji}$, where μ_{ji} is the determinant of the $(N-1) \times (N-1)$-matrix obtained from M by deleting its j-th row and i-th column.

In the literature, Cramer's rule appears in various forms (all of which are closely related). An important consequence of Theorem A.5 is the following formula for the inverse of an invertible matrix M:

$$M^{-1} = \frac{1}{\det M} \operatorname{adj} M.$$

The *characteristic polynomial* of a matrix $M \in \mathbb{F}^{N \times N}$ is the polynomial

$$p_M(s) := \det(sI - M) = s^N + a_{N-1}s^{N-1} + a_{N-2}s^{N-2} + \cdots + a_1 s + a_0.$$

The equation $p_M(s) = 0$ is the *characteristic equation* of M. A complex number λ satisfies $p_M(\lambda) = 0$ if, and only if, $\lambda \in \sigma(M)$, i.e. the zeros of the characteristic polynomial of M are precisely the eigenvalues of M. The *algebraic multiplicity* of an eigenvalue $\lambda \in \sigma(M)$ is its multiplicity as a zero of p_M.

Theorem A.6 (Cayley-Hamilton theorem)

A matrix $M \in \mathbb{F}^{N \times N}$ satisfies its own characteristic equation, i.e.

$$p_M(M) = M^N + a_{N-1}M^{N-1} + a_{N-2}M^{N-2} + \cdots + a_1 M + a_0 I = 0.$$

Next, we state and prove a result on the spectrum of a triangular block matrix. Recall that a *block matrix* is a matrix of the form

$$M := \begin{pmatrix} M_{11} & \cdots & M_{1Q} \\ \vdots & & \vdots \\ M_{P1} & \cdots & M_{PQ} \end{pmatrix},$$

where the M_{ij} are matrices such that, for fixed i, the matrices M_{ij} have the same number of rows for all $j = 1, \ldots, Q$ and, for fixed j, the matrices M_{ij} have the same number of columns for all $i = 1, \ldots, P$.

Theorem A.7 (Spectrum of a triangular block matrix)

If $M \in \mathbb{F}^{N \times N}$ is a matrix of the form

$$M = \begin{pmatrix} M_1 & M_3 \\ 0 & M_2 \end{pmatrix},$$

where $M_1 \in \mathbb{F}^{N_1 \times N_1}$, $M_2 \in \mathbb{F}^{N_2 \times N_2}$, $M_3 \in \mathbb{F}^{N_1 \times N_2}$ and $N_1 + N_2 = N$, then

$$\sigma(M) = \sigma(M_1) \cup \sigma(M_2).$$

Obviously, the above result extends to triangular block matrices with P diagonal blocks for arbitrary P.

Proof of Theorem A.7

To show that $\sigma(M) \subset \sigma(M_1) \cup \sigma(M_2)$, let $\lambda \in \sigma(M)$ and let $v \in \mathbb{C}^{N_1+N_2}$ be a corresponding eigenvector, that is $v \neq 0$ and $Mv = \lambda v$. Writing

$$v = \begin{pmatrix} v_1 \\ v_2 \end{pmatrix}, \quad \text{where } v_1 \in \mathbb{C}^{N_1} \text{ and } v_2 \in \mathbb{C}^{N_2},$$

we have that

$$\begin{pmatrix} \lambda v_1 \\ \lambda v_2 \end{pmatrix} = \lambda v = Mv = \begin{pmatrix} M_1 & M_3 \\ 0 & M_2 \end{pmatrix} \begin{pmatrix} v_1 \\ v_2 \end{pmatrix} = \begin{pmatrix} M_1 v_1 + M_3 v_2 \\ M_2 v_2 \end{pmatrix}.$$

Therefore,

$$\lambda v_1 = M_1 v_1 + M_3 v_2, \quad \lambda v_2 = M_2 v_2.$$

If $v_2 \neq 0$, then v_2 is an eigenvector of M_2 corresponding to the eigenvalue λ, showing that $\lambda \in \sigma(M_2)$. If $v_2 = 0$, then $v_1 \neq 0$ and $\lambda v_1 = M_1 v_1$, implying that $\lambda \in \sigma(M_1)$. Consequently, $\lambda \in \sigma(M_1) \cup \sigma(M_2)$.

Conversely, to show that $\sigma(M_1) \cup \sigma(M_2) \subset \sigma(M)$, let $\lambda \in \sigma(M_1) \cup \sigma(M_2)$. If $\lambda \in \sigma(M_1)$, then there exists $v_1 \in \mathbb{C}^{N_1}$, $v_1 \neq 0$, such that $M_1 v_1 = \lambda v_1$. Setting

$$v := \begin{pmatrix} v_1 \\ 0 \end{pmatrix},$$

it then follows that

$$Mv = \begin{pmatrix} M_1 v_1 \\ 0 \end{pmatrix} = \begin{pmatrix} \lambda v_1 \\ 0 \end{pmatrix} = \lambda \begin{pmatrix} v_1 \\ 0 \end{pmatrix} = \lambda v.$$

Thus, $\lambda \in \sigma(M)$. Finally, if $\lambda \notin \sigma(M_1)$, then $\lambda \in \sigma(M_2)$, and hence, there exists $v_2 \in \mathbb{C}^{N_2}$, $v_2 \neq 0$, such that $v_2^* M_2 = \lambda v_2^*$. Setting

$$v := \begin{pmatrix} 0 \\ v_2 \end{pmatrix},$$

we obtain

$$v^* M = (0, v_2^* M_2) = (0, \lambda v_2^*) = \lambda(0, v_2^*) = \lambda v^*,$$

implying that $\lambda \in \sigma(M)$. $\qquad\square$

If $M \in \mathbb{C}^{N \times N}$ and $\lambda \in \sigma(M) \in \mathbb{C}^{N \times N}$ with algebraic multiplicity m, then $\ker(M - \lambda I)^m$ is said to be the *generalized eigenspace* associated with λ. A *generalized eigenvector* associated with λ is a non-zero element in $\ker(M - \lambda I)^m$. Trivially, any eigenvector is also a generalized eigenvector. The generalized eigenspace associated with an eigenvalue λ of M is an M-invariant subspace of $\mathbb{C}^N$.

Theorem A.8 (Generalized eigenspace decomposition theorem)

Let $M \in \mathbb{C}^{N \times N}$ and write $\{\lambda_1, \ldots, \lambda_k\} = \sigma(M)$, that is, $\lambda_1, \ldots, \lambda_k$ are the distinct eigenvalues of M. Then $\mathbb{C}^N = \ker(M - \lambda_1 I)^{m_1} \oplus \cdots \oplus \ker(M - \lambda_k I)^{m_k}$, where m_j denotes the algebraic multiplicity of λ_j.

Finally, we recall the Jordan form theorem.

Theorem A.9 (Jordan canonical form)

Let $M \in \mathbb{C}^{N \times N}$. There exists an invertible matrix $T \in \mathbb{C}^{N \times N}$ and $k \in \mathbb{N}$ such that $J := T^{-1}MT$ has the block structure

$$
J = \begin{pmatrix}
J_1 & 0 & \cdots & 0 & 0 \\
0 & J_2 & \cdots & 0 & 0 \\
\vdots & \vdots & \ddots & \vdots & \vdots \\
0 & 0 & \cdots & J_{k-1} & 0 \\
0 & 0 & \cdots & 0 & J_k
\end{pmatrix},
$$

where the block J_j is of dimension $r_j \times r_j$ and of the form

$$
\begin{pmatrix}
\lambda & 1 & 0 & \cdots & 0 & 0 & 0 \\
0 & \lambda & 1 & \cdots & 0 & 0 & 0 \\
0 & 0 & \lambda & \cdots & 0 & 0 & 0 \\
\vdots & \vdots & \vdots & \ddots & \vdots & \vdots & \vdots \\
0 & 0 & 0 & \cdots & \lambda & 1 & 0 \\
0 & 0 & 0 & \cdots & 0 & \lambda & 1 \\
0 & 0 & 0 & \cdots & 0 & 0 & \lambda
\end{pmatrix}.
$$

Furthermore, $\sum_{j=1}^{k} r_j = N$ and if $r_j = 1$, then $J_j = \lambda$ for some $\lambda \in \sigma(M)$. Every $\lambda \in \sigma(M)$ occurs in at least one block; the same $\lambda \in \sigma(M)$ may occur in more than one block.

Let $M \in \mathbb{C}^{N \times N}$ with associated Jordan form J as in Theorem A.9. The blocks J_j are referred to as *Jordan blocks*. Let $\lambda \in \sigma(M)$ and denote the set of all indices j corresponding to Jordan blocks J_j in which λ occurs by $\eta(\lambda) \subset \{1, \ldots, k\}$. Since the characteristic polynomial p_M of M is given by

$$
p_M(s) = \det(sI - M) = \det(sI - J) = \Pi_{j=1}^{k} \det(sI - J_j)
$$

and $\det(sI - J_j) = (s - \lambda)^{r_j}$ for all $j \in \eta(\lambda)$, it follows that the algebraic multiplicity $m(\lambda)$ of λ, which by definition is its multiplicity as a zero of the characteristic polynomial p_M, is given by $m(\lambda) = \sum_{j\in\eta(\lambda)} r_j$. Defining $r(\lambda) :=$ $\max_{i\in\eta(\lambda)} r_i$, we record that

$$m(\lambda) = \dim\ker(M - \lambda I)^{r(\lambda)} = \dim\ker(M - \lambda I)^{m(\lambda)},$$

and so, the algebraic multiplicity $m(\lambda)$ of λ coincides with the dimension of its associated generalized eigenspace $\ker(M - \lambda I)^{m(\lambda)}$.

Note that, for each $j \in \eta(\lambda)$, $(J_j - \lambda I)v = 0$ if, and only if, components $v_2,\ldots,v_{r_j}$ of $v \in \mathbb{C}^{r_j}$ are zero. Therefore,

$$\ker\left(J_j - \lambda I\right) = \mathrm{span} \begin{pmatrix} 1 \\ 0 \\ \vdots \\ 0 \end{pmatrix} \quad \forall\, j \in \eta(\lambda)$$

whilst, for $j \in \{1,\ldots,k\}$ with $j \notin \eta(\lambda)$, $\ker\left(J_j - \lambda I\right) = \{0\}$. Thus, we may infer that the geometric multiplicity of λ, which by definition is equal to $\dim\ker(M - \lambda I) = \dim\ker(J - \lambda I)$, coincides with the number of elements in the set $\eta(\lambda)$. In particular, the geometric multiplicity of λ is less than or equal to its algebraic multiplicity $m(\lambda)$.

If the geometric and algebraic multiplicities of an eigenvalue coincide, then the eigenvalue is said to be *semisimple*. Observe that $\lambda \in \sigma(M)$ is semisimple if, and only if, the dimension of each of its associated Jordan blocks is 1, that is, $r_j = 1$ for all $j \in \eta(\lambda)$. An immediate consequence of this observation is the fact that λ is semisimple if, and only if, its eigenspace coincides with its generalized eigenspace. Thus, we have the following equivalence.

Proposition A.10

$\lambda \in \sigma(M)$ is semisimple $\Leftrightarrow \ker(M - \lambda I) = \ker(M - \lambda I)^{m(\lambda)}$.

A.2 Metric and normed spaces

We start by recalling some concepts and results from the theory of metric spaces. Let X be a metric space with metric μ, that is, X is non-empty set and $\mu : X \times X \to \mathbb{R}_+$ is a map with the following properties: (i) $\mu(x,y) = 0$ if, and only if, $x = y$, (ii) $\mu(x,y) = \mu(y,x)$ for all $x,y \in X$, and (iii) $\mu(x,y) \leq \mu(x,z) + \mu(z,y)$ for all $x,y,z \in X$ (triangle inequality).

A sequence (x_n) in X is said to *converge* to a *limit* $x \in X$ if $\mu(x_n,x) \to 0$ as $n \to \infty$ (equivalently, for all $\varepsilon > 0$, there exists $N \in \mathbb{N}$ such that $\mu(x_n,x) \leq \varepsilon$

for all $n \geq N$). The limit of a convergent sequence is unique (as is readily verified using the triangle inequality). We say that a sequence (x_n) in X is a *Cauchy sequence* if for every ε there exists $N \in \mathbb{N}$ such that $\mu(x_n, x_m) \leq \varepsilon$ for all $n, m \geq N$. It follows easily from the triangle inequality that a convergent sequence is a Cauchy sequence. The metric space X is said to be *complete* if every Cauchy sequence in X converges in X.

For $x \in X$ and $r > 0$, we define $\mathbb{B}(x, r) := \{y \in X \colon \mu(x, y) < r\}$. Let $S \subset X$. We say that S is *open* if, for every $x \in S$, there exists $\varepsilon > 0$ such that $\mathbb{B}(x, \varepsilon) \subset S$. Note that, for every $x \in X$ and $r > 0$, the set $\mathbb{B}(x, r)$ is open and is referred to as the *open ball* centred at x with radius r. The set S is said to be a *neighbourhood* of a point $x \in X$ if there exists an open set $U \subset X$ such that $x \in U \subset S$. The set S is said to be *closed* if its complement $X \backslash S$ is open. Any union of open sets is open and any intersection of closed sets is closed. Furthermore, intersections of finitely many open sets are open and unions of finitely many closed sets are closed. A point $x \in S$ is *an interior point* of S if x has a neighbourhood contained in S: the set of all interior points is the *interior* of S. The *closure* of S, written $\mathrm{cl}(S)$ or $\bar{S}$ is the intersection of all closed sets containing S. In an obvious sense, $\mathrm{cl}(S)$ is the "smallest" closed set containing S. A point $x \in X$ is a *boundary point* of S if, for all $\varepsilon > 0$, $\mathbb{B}(x, \varepsilon)$ has non-empty intersection with both S and its complement $X \backslash S$. The *boundary* of S is the set of all its boundary points and is denoted by ∂S. The set S is closed if, and only if, it contains its boundary ∂S. The set S is *dense* in X if $\mathrm{cl}(S) = X$.

Exercise A.2

Let S be a subset of a metric space X.

(a) Show S is closed if, and only if, for every convergent sequence (x_n) with $x_n \in S$ for all $n \in \mathbb{N}$, the limit is also in S.

(b) Show that $x \in \mathrm{cl}(S)$ if, and only if, for every $\varepsilon > 0$, $\mathbb{B}(x, \varepsilon)$ has non-empty intersection with S.

We say that S is *bounded* if $\sup_{x,y \in S} \mu(x, y) < \infty$. The set S is said to be *compact* if every sequence (x_n) in S has a convergent subsequence with limit in S. Furthermore, S is said to be *pre-compact* if the closure of S is a compact subset of X.

Proposition A.11

Let $S \subset X$.

(1) If S is compact, then S is bounded and closed.

(2) If S is pre-compact, then S is bounded.

Let $S \subset T \subset X$. The set S is said to be *relatively open* in T if $S = U \cap T$ for some open set U. It is readily shown that S is relatively open in T if, and only if, for every $x \in S$ there exists $\varepsilon > 0$ such that $T \cap \mathbb{B}(x, \varepsilon) \subset S$. Note that, if S is an open set, then it is relatively open in T. Note further that, if T is an open set, then $S \subset T$ is relatively open in T if, and only if, S is open. We also record that, if S is relatively open in T, then $x \in S$ is an interior point of S if, and only if, x is an interior point of T.

Example A.12

Let $I, J \subset \mathbb{R}$ be intervals with $I \subset J$. By the previous observation, we know that, if J is open, then I is relatively open in J if, and only if, I is open. We proceed to consider the three possible cases in which J is not open.

Case 1. Let $J = [a, b)$ with $-\infty < a < b \le \infty$. Then I is relatively open in J if, and only if, I takes one of the following two forms: $I = [a, d)$ with $a < d \le b$, or $I = (c, d)$ with $a \le c < d \le b$.

Case 2. Let $J = (a, b]$ with $-\infty \le a < b < \infty$. Then I is relatively open in J if, and only if, I takes one of the following two forms: $I = (c, b]$ with $a \le c < b$, or $I = (c, d)$ with $a \le c < d \le b$.

Case 3. Let $J = [a, b]$ with $-\infty < a < b < \infty$. Then I is relatively open in J if, and only if, I takes one of the following four forms: $I = J$, or $I = [a, d)$ with $a < d \le b$, or $I = (c, b]$ with $a \le c < b$, or $I = (c, d)$ with $a \le c < d \le b$. $\triangle$

A set $S \subset T \subset X$ is said to be *relatively closed* in T if its complement $T \backslash S$ in T is relatively open in T; equivalently, S is relatively closed in T if $S = V \cap T$ for some closed set V.

A set $S \subset X$ is *disconnected* if $S = S_1 \cup S_2$, where the sets S_1 and S_2 are non-empty, disjoint and relatively open in S, in which case $S_1 \cup S_2$ is said to be a *disconnection* of S. Observe that, if $S_1 \cup S_2$ is a disconnection of S, then S_2 (respectively, S_1) is the complement in S of S_1 (respectively, S_2) and so the sets S_1 and S_2 (each relatively open in S) are also relatively closed in S. As a consequence, in the above definition of disconnectedness, if the phrase "relatively open" is replaced by "relatively closed", then we obtain an equivalent characterization of disconnectedness. A set $S \subset X$ is said to be *connected* if there does not exist a disconnection of S.

Consider a function $f : X \to Y$, where X and Y are metric spaces with metrics μ_X and μ_Y, respectively. The function f is said be *continuous at a point* $x \in X$ if, for every $\varepsilon > 0$, there exists $\delta > 0$ such that, for every $w \in X$ with $\mu_X(w, x) \le \delta$, we have $\mu_Y(f(w), f(x)) \le \varepsilon$. Equivalently, f is continuous at $x \in X$ if, for every sequence (x_n) converging to x (in X), the sequence $(f(x_n))$ converges to $f(x)$ in Y. The function f is said to be *continuous* if f

is continuous at x for every $x \in X$. Furthermore, f is said to be *uniformly continuous* if, for every $\varepsilon > 0$, there exists a $\delta > 0$ such that, for all $v, w \in X$ with $\mu_X(v, w) \leq \delta$, we have $\mu_Y(f(v), f(w)) \leq \varepsilon$.

Let X be a vector space over the field $\mathbb{F}$, where $\mathbb{F} = \mathbb{R}$ or $\mathbb{F} = \mathbb{C}$. A map $\nu : X \to [0, \infty)$ is called a *norm* (on X) if ν has the following three properties: $\nu(x) = 0$ if and only if $x = 0$, $\nu(sx) = |s|\nu(x)$ for all $(s, x) \in \mathbb{F} \times X$ and

$$\nu(x + y) \leq \nu(x) + \nu(y), \quad \forall\, x, y \in X \qquad \text{(triangle inequality)}$$

A norm ν is usually written in the form $\nu(x) = \|x\|$. A vector space endowed with a norm is called a *normed vector space* or a *normed space*. As an immediate consequence of the triangle inequality we have the "reverse triangle inequality":

$$\big| \|x\| - \|y\| \big| \leq \|x - y\| \quad \forall\, x, y \in X.$$

Let X be a normed space with norm $\| \cdot \|$. Then, trivially, X is also a metric space, with metric $\mu : X \times X \to [0, \infty)$ given by $\mu(x, y) = \|x - y\|$. Consequently, normed spaces are special cases of metric spaces and hence every concept introduced in the context of metric spaces (such as boundedness, compactness, completeness, convergence, continuity and uniform continuity) applies to normed spaces. In particular, it is easy to show that a subset S of a normed space X is bounded if, and only if, $\sup_{x \in S} \|x\| < \infty$. Furthermore, the reverse triangle inequality shows that a norm is continuous as a function from X to $\mathbb{R}$. We add that a complete normed space is also referred to as a *Banach space*[1].

Example A.13

For $x \in \mathbb{F}^N$, let x_i denote the components of x.

(1) Let $p \in [1, \infty)$. Setting $\|x\|_p := (\Sigma_{i=1}^N |x_i|^p)^{1/p}$, defines a norm on $\mathbb{F}^N$. Note that, for $p = 2$, the norm can be expressed in terms of the inner product (A.2): $\|x\|_2 = \sqrt{\langle x, x \rangle}$.

(2) If we define $\|x\|_\infty := \max_{1 \leq i \leq N} |x_i|$, then $\| \cdot \|_\infty$ is a norm on $\mathbb{F}^N$. $\qquad \triangle$

Using the completeness of $\mathbb{F}$, it is not difficult to see that $\mathbb{F}^N$, equipped with any of the norms $\| \cdot \|_p$ $(1 \leq p \leq \infty)$ is complete and so is a Banach space.

Example A.14

Let $C(I, \mathbb{R})$ denote the set of all continuous functions $I \to \mathbb{R}$, where $I \subset \mathbb{R}$ is a compact interval. With the obvious notions of addition and scalar multiplication, $C(I, \mathbb{R})$ is a vector space. In view of the fact that continuous functions

---

[1] Stefan Banach (1892-1945), Polish.

$I \to \mathbb{R}$ are bounded, we see that the following is well defined for all $f \in C(I, \mathbb{R})$

$$\|f\|_\infty := \sup_{t \in I} |f(t)| < \infty.$$

It is straightforward to show that $\| \cdot \|_\infty$ is a norm in $C(I, \mathbb{R})$. These ideas will shortly be subsumed in Proposition A.22, a particular consequence of which is completeness of $C(I, \mathbb{R})$ and so $C(I, \mathbb{R})$ is a Banach space. $\qquad \triangle$

The next two results relate to compactness and pre-compactness in $\mathbb{F}^N$. In particular, Proposition A.11 can be strengthened considerably if $X = \mathbb{F}^N$.

Theorem A.15

Let $S \subset \mathbb{F}^N$.

(1) S is compact if, and only if, S is bounded and closed.

(2) S is pre-compact if, and only if, S is bounded.

As a consequence, we obtain the following result.

Theorem A.16 (Bolzano-Weierstrass theorem)

Every bounded sequence in $\mathbb{F}^N$ contains a convergent subsequence.

Let (x_n) be a sequence in a normed space X and consider the associated infinite series $\sum_{n=1}^\infty x_n$. We say that the series converges if the sequence (s_n) of partial sums $s_n := \sum_{k=1}^n x_k$ converges as $n \to \infty$. A series $\sum_{n=1}^\infty x_n$ is said to *converge absolutely* if the infinite series $\sum_{n=1}^\infty \|x_n\|$ (a series in $\mathbb{R}$) converges. If X is a Banach space, then, as in the $\mathbb{F}$-valued case, it can be shown that $\sum_{n=1}^\infty x_n$ converges if, and only if, for every $\varepsilon > 0$, there exists $N \in \mathbb{N}$ such that $\|\sum_{k=n}^m x_k\| \le \varepsilon$ for all $m \ge n \ge N$ (Cauchy criterion). As a consequence we have the following result.

Proposition A.17

In a Banach space, every absolutely convergent series converges.

Exercise A.3

Give a detailed proof of Proposition A.17.

The next proposition is used freely throughout the book.

Proposition A.18

Let X and Y be normed spaces with norms $\|\cdot\|_X$ and $\|\cdot\|_Y$, respectively. If $K \subset X$ is compact and $f : X \to Y$ is continuous, then

(1) the restriction of f to K is uniformly continuous;

(2) there exist $x_{\max}$ and $x_{\min}$ in K such that

$$\|f(x_{\max})\|_Y = \sup_{x \in K} \|f(x)\|_Y \quad \text{and} \quad \|f(x_{\min})\|_Y = \inf_{x \in K} \|f(x)\|_Y.$$

Two norms $\|\cdot\|$ and $\|\cdot\|_*$ in the (same) vector space X are said to be *equivalent* if there exist positive constants α and β such that

$$\alpha\|x\|_* \leq \|x\| \leq \beta\|x\|_* \quad \forall\, x \in X.$$

Theorem A.19

Any two norms on $\mathbb{F}^N$ are equivalent.

Proof

Let $\|\cdot\|$ be a norm in $\mathbb{F}^N$. It is sufficient to prove that there exist positive constants α and β such that

$$\alpha\|x\|_1 \leq \|x\| \leq \beta\|x\|_1 \quad \forall\, x \in \mathbb{F}^N, \tag{A.9}$$

where $\|\cdot\|_1$ is the 1-norm defined in part (1) of Example A.13. Let $e_1, \ldots, e_N$ be the canonical basis of $\mathbb{F}^N$ and write $x \in \mathbb{F}^N$ in the form $x = \sum_{j=1}^N x_j e_j$. Then, noting that, for all $j = 1, 2, \ldots, N$, $|x_j| \leq \|x\|_1$ and using the triangle inequality, we obtain

$$\|x\| \leq \sum_{j=1}^N |x_j| \|e_j\| \leq \|x\|_1 \sum_{j=1}^N \|e_j\|.$$

The second inequality in (A.9) now follows with $\beta := \sum_{j=1}^N \|e_j\|$.

To prove the first inequality in (A.9), consider the function $\nu : \mathbb{F}^N \to \mathbb{R}_+$, $x \mapsto \|x\|$. Then, by the reverse triangle inequality,

$$\left| \nu(x) - \nu(y) \right| = \left| \|x\| - \|y\| \right| \leq \|x - y\| \leq \beta\|x - y\|_1 \quad \forall\, x, y \in \mathbb{F}^N.$$

This estimate shows that if $\mathbb{F}^N$ is equipped with the norm $\|\cdot\|_1$, then ν is continuous. By Proposition A.18, ν attains its infimum on the compact set $S = \{x \in \mathbb{F}^N : \|x\|_1 = 1\}$, that is, there exists $x_0 \in S$ such that $\nu(x_0) = \min_{x \in S} \nu(x) =: \alpha$. Note that $\alpha > 0$ (because otherwise $\alpha = \nu(x_0) = \|x_0\| = 0$, implying $x_0 = 0$ which is impossible since $\|x_0\|_1 = 1$). For $x \in \mathbb{F}^N$, $x \neq 0$, set $u_x := (1/\|x\|_1)x$. Then $u_x \in S$ and

$$\frac{\|x\|}{\|x\|_1} = \frac{\|u_x\|}{\|u_x\|_1} = \|u_x\| = \nu(u_x) \geq \alpha,$$

from which the first inequality in (A.9) follows.   □

As we have already pointed out, the vector space $\mathbb{F}^N$, equipped with any of the norms $\|\cdot\|_p$ (see Example A.13) is complete. Combining this with Theorem A.19, we obtain the following corollary.

Corollary A.20

$\mathbb{F}^N$ is complete whatever the norm is.

It is clear that a sequence (x_n) in $\mathbb{F}^N$ converges to x with respect to the norm $\|\cdot\|_1$, that is, $\|x_n - x\|_1 \to 0$ as $n \to \infty$, if, and only if, the components of x_n converge to the corresponding components of x as $n \to \infty$. An application of Theorem A.19 yields the following corollary.

Corollary A.21

A sequence (x_n) converges to x with respect to an arbitrary norm $\|\cdot\|$ in $\mathbb{F}^N$ if, and only if, the components of x_n converge to the corresponding components of x.

Let X and Y be normed spaces and let $S \subset X$. We say that a sequence (f_n) of functions $f_n : S \to Y$ converges uniformly to a function $f : S \to Y$ if, for every $\varepsilon > 0$, there exists $N \in \mathbb{N}$ such that $\|f_n(x) - f(x)\| \leq \varepsilon$ for all $x \in S$ and all $n \geq N$. In the following, we denote the set of continuous functions $S \to Y$ by $C(S, Y)$. We remark that, with the usual operations of addition and scalar multiplication, $C(S, Y)$ is a vector space.

Proposition A.22

Let X be a normed space, let Y be a Banach space and let $K \subset X$ be a compact set. Let $f \in C(K, Y)$ and define

$$\|f\|_\infty := \sup_{x \in K} \|f(x)\| \quad \text{(sup norm)}$$

where $\|\cdot\|$ is the norm in Y. Then $\|\cdot\|_\infty$ is a norm in $C(K, Y)$ and, equipped with this norm, $C(K, Y)$ is complete.

We note that convergence in $C(K,Y)$, equipped with the norm $\|\cdot\|_\infty$, is equivalent to uniform convergence.

Proof of Proposition A.22

By Proposition A.18, $\|f\|_\infty < \infty$ for every $f \in C(K,Y)$. It is straightforward to show that $\|\cdot\|_\infty$ is a norm. To prove completeness, let (f_n) be a Cauchy sequence. Then, for every $x \in K$, $(f_n(x))$ is a Cauchy sequence in Y. Hence, by the completeness of Y, for every $x \in X$, $f_n(x)$ converges to a limit $f(x) \in Y$ as $n \to \infty$. Let $\varepsilon > 0$. Then there exists $N \in \mathbb{N}$ such that $\|f_n(x) - f_m(x)\| \le \varepsilon$ for all $x \in K$ and all $n, m \ge N$. Letting $n \to \infty$, it follows that $\|f(x) - f_m(x)\| \le \varepsilon$ for all $x \in K$ and all $m \ge N$, showing that $\|f - f_n\|_\infty \to 0$ as $n \to \infty$. It remains to show that the function $f : K \to Y$ is continuous. To this end, let $y \in X$ be fixed, but arbitrary. Choose $m \in \mathbb{N}$ such that $\|f - f_m\|_\infty \le \varepsilon/3$. Furthermore, choose $\delta > 0$ such that $\|f_m(y) - f_m(x)\| \le \varepsilon/3$ whenever $\|y - x\| \le \delta$. Applying the triangle inequality twice, we obtain, for all $x \in K$,

$$\|f(y) - f(x)\| \le \|f(y) - f_m(y)\| + \|f_m(y) - f_m(x)\| + \|f_m(x) - f(x)\|.$$

Consequently, $\|f(y) - f(x)\| \le \varepsilon$ whenever $\|y - x\| \le \delta$. $\qquad\square$

Exercise A.4

Let X be a normed space, let Y be a Banach space and let $S \subset X$ be a subset (not necessarily compact). Define $C_b(S,Y) := \{f \in C(S,Y) : f \text{ bounded}\}$, where a function $f : S \to Y$ is said to be bounded if

$$\|f\|_\infty := \sup_{x \in S} \|f(x)\| < \infty.$$

Show that $\|\cdot\|_\infty$ is a norm in $C_b(S,Y)$ and that $C_b(S,Y)$, equipped with this norm, is complete.

Let $K \subset X$ be compact. For a non-empty set $S \subset Y$, the set $C(K,S)$ of all continuous functions $K \to S$ is a subset of $C(K,Y)$. Note that $C(K,S)$ is a subspace of $C(K,Y)$ (and hence a normed space with norm $\|\cdot\|_\infty$) if, and only if, S is a linear subspace of Y. Endowed with the metric $\mu(f,g) = \|f - g\|_\infty$, $C(K,S)$ becomes a metric space. This space is complete if, and only if, S is closed.

As an application of Proposition A.17 and Proposition A.22, we prove the Weierstrass criterion for uniform convergence of a series of functions.

Corollary A.23 (Weierstrass criterion)

Let X be a normed space, Y be a Banach space, $K \subset X$ be compact, (f_n) be a sequence in $C(K, Y)$ and let (m_n) be a real sequence such that $\|f_n(x)\| \leq m_n$ for all $x \in K$ and all $n \in \mathbb{N}$, where $\| \cdot \|$ is the norm in Y. If $\sum_{n=1}^{\infty} m_n$ is convergent, then $\sum_{n=1}^{\infty} f_n$ is uniformly convergent.

Proof

By Proposition A.22, $C(K, Y)$, equipped with the norm $\|f\|_\infty = \sup_{x \in K} \|f(x)\|$, is a Banach space. By hypothesis, $\sum_{n=1}^{\infty} \|f_n\|_\infty \leq \sum_{n=1}^{\infty} m_n < \infty$, that is, the series $\sum_{n=1}^{\infty} f_n$ is absolutely convergent. Therefore, by Proposition A.17, $\sum_{n=1}^{\infty} f_n$ converges in the Banach space $C(K, Y)$, or equivalently, $\sum_{n=1}^{\infty} f_n$ converges uniformly. $\qquad\square$

Whilst Proposition A.11 provides necessary conditions for compactness and pre-compactness in the context of a metric space X, these conditions are not sufficient in general (see, however, Proposition A.15). The Arzelà-Ascoli theorem gives a necessary and sufficient condition for pre-compactness in the normed space $X = C(I, \mathbb{F}^N)$, where I is a compact interval. To formulate the theorem, the concept of equicontinuity is needed: a set $\mathcal{F} \subset C(I, \mathbb{F}^N)$ is said to be *equicontinuous* if, for each $\varepsilon > 0$, there exists $\delta > 0$ such that, for all $f \in \mathcal{F}$, $\|f(s) - f(t)\| \leq \varepsilon$ for all $s, t \in I$ with $|s - t| \leq \delta$. The essence of this definition is that, given $\varepsilon > 0$, the same $\delta > 0$ will "work" for all $f \in \mathcal{F}$. We are now in a position to state the Arzelà-Ascoli theorem.

Theorem A.24 (Arzelà-Ascoli theorem)

A set $\mathcal{F} \subset C(I, \mathbb{F}^N)$ is pre-compact if, and only if, $\mathcal{F}$ is bounded and equicontinuous.

Let S be a non-empty subset of a metric space X and $F \colon S \to S$. An element $x \in S$ is a *fixed point* of F if $x = F(x)$. Next we record two results which provide sufficient conditions for the existence of a fixed point, namely, *Banach's fixed-point theorem* (also known as the *contraction-mapping theorem*) and *Brouwer's fixed-point theorem*. Proofs of these theorems can be found in, for example, Walter [21])

Theorem A.25 (Contraction-mapping theorem)

Let X be a complete metric space with metric μ. Let $F \colon X \to X$ be a contraction, that is, there exists a constant $0 \le q < 1$ such that

$$\mu(F(x), F(y)) \le q\mu(x, y), \quad \forall x, y \in X.$$

Then F has a unique fixed point $x^* \in X$, that is, the equation $F(x) = x$ has exactly one solution $x = x^*$ in X. Moreover, for arbitrary $x \in X$, $F^n(x) \to x^*$ as $n \to \infty$ and

$$\mu(F^n(x), x^*) \le \frac{1}{1-q}\mu(F^{n+1}(x), F^n(x)) \le \frac{q^n}{1-q}\mu(F(x), x),$$

where F^n denotes the n-fold composition $F \circ \cdots \circ F$.

Note that the contraction-mapping theorem holds in the general context of a complete metric space, but under the restrictive assumption that the map F is a contraction. By contrast, the next result, Brouwer's fixed-point theorem, applies to maps that are assumed only to be continuous but under the restrictive assumption that the domain is a convex compact subset of $\mathbb{R}^N$. Before stating the result, we recall the concept of a convex subset of $\mathbb{R}^N$. A set $C \subset \mathbb{R}^N$ is *convex* if it contains the line segment joining any two of its points, that is, if $a, b \in C$, then $(1 - \mu)a + \mu b \in C$ for all $\mu \in [0, 1]$.

Theorem A.26 (Brouwer's fixed-point theorem)

Let $C \subset \mathbb{R}^N$ be a non-empty convex and compact set, and $F : C \to C$ continuous. Then F has a fixed point, that is, there exists at least one point $x \in C$ such that $F(x) = x$.

We conclude this section with some basic concepts and results on matrix norms and matrix-valued functions. The space of matrices $\mathbb{F}^{P \times Q}$ (under the usual operations of addition and multiplication by scalars) is a vector space over $\mathbb{F}$. A norm on this vector space is also called a *matrix norm*. Let $M \in \mathbb{F}^{P \times Q}$ and $1 \le p \le \infty$. Defining

$$\|M\| := \sup_{z \ne 0} \frac{\|Mz\|_p}{\|z\|_p}, \tag{A.10}$$

where $\|\cdot\|_p$ is the p-norm on $\mathbb{F}^N$ as defined in Example A.13, it is straightforward to show that $\|\cdot\|$ is a norm on $\mathbb{F}^{P \times Q}$. Moreover, as an immediate consequence of the definition, we have that

$$\|Mz\|_p \le \|M\|\|z\|_p \quad \forall z \in \mathbb{F}^Q.$$

Sometimes it is said that the matrix norm defined by (A.10) is the matrix norm *induced* by $\|\cdot\|_p$. It is straightforward to show that

$$\|M\| = \sup_{\|z\|_p = 1} \|Mz\|_p = \sup_{\|z\|_p \le 1} \|Mz\|_p, \tag{A.11}$$

and, furthermore,

$$\|M\| = \inf\{\gamma \geq 0 \colon \|Mz\|_p \leq \gamma\|z\|_p \ \forall z \in \mathbb{F}^Q\}. \qquad (A.12)$$

Exercise A.5

Prove (A.11) and (A.12).

For $M \in \mathbb{F}^{P \times Q}$ and $N \in \mathbb{F}^{Q \times R}$, we have for all $z \in \mathbb{F}^R$,

$$\|MNz\|_p \leq \|M\|\|Nz\|_p \leq \|M\|\|N\|\|z\|_p, \quad \forall z \in \mathbb{F}^R$$

and therefore, by (A.12),

$$\|MN\| \leq \|M\|\|N\|. \qquad (A.13)$$

The matrix space $\mathbb{F}^{P \times Q}$ is closely related to the space $\mathbb{F}^{PQ}$. Indeed, for $x \in \mathbb{F}^{PQ}$, define $x^i \in \mathbb{F}^P$, for $i = 1, \ldots, Q$, by $x_j^i := x_{(i-1)P+j}$ for $j = 1, \ldots, P$, where x_j^i and x_k denote the components of x^i and x, respectively. It is obvious that the map

$$\varphi : \mathbb{F}^{PQ} \to \mathbb{F}^{P \times Q}, \ x \mapsto (x^1, \ldots, x^Q),$$

is linear and bijective, that is, φ is an isomorphism. Note that φ "acts" by partitioning a vector in $\mathbb{F}^{PQ}$ into Q columns of length P and arranging them in form of a $P \times Q$-matrix.

Let $\|\cdot\|$ be a norm on $\mathbb{F}^{P \times Q}$. It is a routine exercise to check that $\|\cdot\|_\varphi$ defined by

$$\|x\|_\varphi := \|\varphi(x)\| \quad \forall x \in \mathbb{F}^{PQ}$$

is a norm on $\mathbb{F}^{PQ}$. Since, by Corollary A.20, the space $\mathbb{F}^{PQ}$ is complete whatever the norm is, it is complete when equipped with the norm $\|\cdot\|_\varphi$. It follows that, endowed with the norm $\|\cdot\|$, the matrix space $\mathbb{F}^{P \times Q}$ is complete. Similarly, Corollary A.21 can be used to show that a sequence (M_n) in $\mathbb{F}^{P \times Q}$ converges to M, that is, $\|M_n - M\| \to 0$ as $n \to \infty$, if, and only if, the entries of M_n converge to the corresponding entries of M.

Unless said otherwise, the matrix norm used in this book is the norm defined by (A.10) with $p = 2$. For notational convenience, the norm $\|\cdot\|_2$ will usually also be denoted by $\|\cdot\|$ and we write $\|Mz\| \leq \|M\|\|z\|$, where $M \in \mathbb{F}^{P \times Q}$ and $z \in \mathbb{F}^Q$.

Consider the power series

$$f(z) = \sum_{n=0}^{\infty} a_n z^n. \qquad (A.14)$$

The coefficients a_n are complex numbers and z is a complex variable. We recall that the power series (A.14) has a *radius of convergence* r, where $0 \leq r \leq \infty$.

The series converges absolutely for $z \in \mathbb{C}$ such that $|z| < r$ (for $z = 0$ if $r = 0$) and diverges for every $z \in \mathbb{C}$ such that $|z| > r$.

The following result will enable us to make sense of $f(M)$, where $M \in \mathbb{C}^{N \times N}$.

Proposition A.27

Assume that the radius of convergence of the power series (A.14) is infinite, that is, the power series converges for all $z \in \mathbb{C}$. Then, for every $M \in \mathbb{C}^{N \times N}$, the series $f(M) := \sum_{n=0}^{\infty} a_n M^n$ converges in $\mathbb{C}^{N \times N}$.

Proof

Let $M \in \mathbb{C}^{N \times N}$. By (A.13), $\|M^n\| \leq \|M\|^n$, and hence

$$\sum_{n=0}^{\infty} \|a_n M^n\| \leq \sum_{n=0}^{\infty} |a_n| \|M\|^n \tag{A.15}$$

By hypothesis, the radius of convergence of the power series (A.14) is equal to ∞, and hence (A.14) converges absolutely for every $z \in \mathbb{C}$. Consequently, the right-hand side of (A.15) converges and thus, $\sum_{n=0}^{\infty} a_n M^n$ converges absolutely. The claim follows now from Proposition A.17 and the completeness of $\mathbb{C}^{N \times N}$.

$\square$

Proposition A.27 enables us, for example, to define $\exp(M)$ and $\sin(M)$, for every $M \in \mathbb{C}^{N \times N}$, by setting

$$\exp(M) := \sum_{n=0}^{\infty} \frac{1}{n!} M^n, \quad \sin(M) := \sum_{n=0}^{\infty} \frac{(-1)^n}{(2n+1)!} M^{2n+1}.$$

Obviously, the definition of the matrix exponential is consistent with that given in Section 2.1.

A.3 Differentiation and integration

Throughout this section, $I \subset \mathbb{R}$ denotes an interval. Let $f : I \to \mathbb{C}$ be a complex-valued function. Then there exist real-valued functions $g : I \to \mathbb{R}$ and $h : I \to \mathbb{R}$ (the real and imaginary parts of f, respectively) such that $f(t) = g(t) + ih(t)$ for all $t \in I$. We say that f is differentiable at a point $a \in I$

if g and h are differentiable at a and we set

$$f'(a) = \dot{f}(a) = \frac{\mathrm{d}f}{\mathrm{d}t}(a) = g'(a) + ih'(a).$$

If f is differentiable at very point in I, then f is said to be differentiable. Right and left differentiability of a complex-valued function is defined in an analogous way. Moreover, if $a < b$ and $a, b \in I$, then f is said to be integrable (in the sense of Riemann) on $[a, b]$ if g and h are integrable on $[a, b]$ and we set

$$\int_a^b f(t)\mathrm{d}t = \int_a^b g(t)\mathrm{d}t + i \int_a^b h(t)\mathrm{d}t.$$

Throughout this section, let $\mathbb{M}_{\mathbb{F}}$ denote the space of matrices $\mathbb{F}^{P \times Q}$ (in particular, if $Q = 1$, then $\mathbb{M}_{\mathbb{F}} = \mathbb{F}^P$). Derivatives and integrals (when they exist) of functions $f \colon I \to \mathbb{M}_{\mathbb{F}}$ should be interpreted in the natural entrywise manner, that is,

$$f'(t) = \dot{f}(t) = \frac{\mathrm{d}f}{\mathrm{d}t}(t) = \begin{pmatrix} f'_{11}(t) & \cdots & f'_{1Q}(t) \\ \vdots & \ddots & \vdots \\ f'_{P1}(t) & \cdots & f'_{PQ}(t) \end{pmatrix} \quad t \in I,$$

and

$$\int_a^b f(t)\mathrm{d}t = \begin{pmatrix} \int_a^b f_{11}(t)\mathrm{d}t & \cdots & \int_a^b f_{1Q}(t)\mathrm{d}t \\ \vdots & \ddots & \vdots \\ \int_a^b f_{P1}(t)\mathrm{d}t & \cdots & \int_a^b f_{PQ}(t)\mathrm{d}t \end{pmatrix}, \quad a, b \in I, \ a < b,$$

where the functions $f_{ij} \colon I \to \mathbb{F}$ denote the entries of f.

Piecewise continuity and piecewise continuous differentiability. We deem a function $f \colon I \to \mathbb{M}_{\mathbb{F}}$ to be *piecewise continuous* if the following hold: for every $a, b \in I$ with $a < b$, the interval $[a, b]$ admits a finite partition $a = t_1 < t_2 < \cdots < t_{n-1} < t_n = b$ such that f (i) is continuous on every subinterval (t_i, t_{i+1}), $i = 1, \ldots, n-1$, (ii) has right limit at t_1, (iii) has left limit at t_n, and (iv) has both left and right limits at every t_i, $i = 2, \ldots, n-1$. Left and right limits of f at t are denoted as follows:

$$f(t^-) := \lim_{s \uparrow t} f(s) \quad \text{and} \quad f(t^+) := \lim_{s \downarrow t} f(s).$$

The vector space of piecewise continuous functions $I \to \mathbb{M}_{\mathbb{F}}$ is denoted by $PC(I, \mathbb{M}_{\mathbb{F}})$. Let $\| \cdot \|$ denote a norm on $\mathbb{M}_{\mathbb{F}}$. A function $f \in PC(I, \mathbb{M}_{\mathbb{F}})$ is (Riemann) integrable on $[a, b]$ for all $a, b \in I$ with $a < b$. The following result is invoked freely throughout the book.

Proposition A.28 (Triangle inequality for integrals)

Let $f \in PC([a, b], \mathbb{M}_{\mathbb{F}})$. Then

$$\left\| \int_a^b f(t)\mathrm{d}t \right\| \leq \int_a^b \|f(t)\|\mathrm{d}t.$$

Proof

For $n \in \mathbb{N}$, we set $\delta_n := (b - a)/n$ and define the Riemann sums

$$R_n := \delta_n \sum_{j=0}^{n-1} f(a + j\delta_n), \quad S_n := \delta_n \sum_{j=0}^{n-1} \|f(a + j\delta_n)\|.$$

By Riemann integration theory (see, for example, [6]),

$$\lim_{n \to \infty} R_n = \int_a^b f(t)\mathrm{d}t, \quad \lim_{n \to \infty} S_n = \int_a^b \|f(t)\|\mathrm{d}t.$$

The triangle inequality for finite sums implies that $\|R_n\| \leq S_n$ for all $n \in \mathbb{N}$, and thus,

$$\left\| \int_a^b f(t)\mathrm{d}t \right\| = \lim_{n \to \infty} \|R_n\| \leq \lim_{n \to \infty} S_n = \int_a^b \|f(t)\|\mathrm{d}t,$$

completing the proof. □

Let $f \in PC([a, \infty), \mathbb{M}_{\mathbb{F}})$. We define the improper integral $\int_a^\infty f(t)\mathrm{d}t$ by

$$\int_a^\infty f(t)\mathrm{d}t := \lim_{\tau \to \infty} \int_a^\tau f(t)\mathrm{d}t,$$

provided the limit exists.

Exercise A.6

Let $f \in PC([a, \infty), \mathbb{M}_{\mathbb{F}})$. Show that the improper integral $\int_a^\infty f(t)\mathrm{d}t$ exists if, and only if, for every $\varepsilon > 0$, there exists $r \geq a$ such that $\left\| \int_\sigma^\tau f(t)\mathrm{d}t \right\| \leq \varepsilon$ for all $\tau, \sigma \in [r, \infty)$ with $\tau \geq \sigma$ (Cauchy criterion for improper integrals).

Next we recall the mean-value theorem for integrals. We state this result in a slightly more general form than is usually given in most undergraduate text books on analysis.

Theorem A.29 (Mean-value theorem for integrals)

Let $f \in C([a, b], \mathbb{R})$, $g \in PC([a, b], \mathbb{R})$ and assume that $g(t) \geq 0$ for all $t \in [a, b]$. Then there exists $\tau \in [a, b]$ such that

$$\int_a^b f(t)g(t)\mathrm{d}t = f(\tau) \int_a^b g(t)\mathrm{d}t.$$

Proof

Set $\mu_1 := \min_{[a,b]} f(t)$ and $\mu_2 := \max_{[a,b]} f(t)$. Then, since g is non-negative,

$$\mu_1 \int_a^b g(t)\mathrm{d}t \leq \int_a^b f(t)g(t)\mathrm{d}t \leq \mu_2 \int_a^b g(t)\mathrm{d}t.$$

Consequently, there exists $\mu \in [\mu_1, \mu_2]$ such that $\int_a^b f(t)g(t)\mathrm{d}t = \mu \int_a^b g(t)\mathrm{d}t$. By the intermediate-value theorem for continuous functions we have that $\mu = f(\tau)$ for some $\tau \in [a, b]$ and therefore, $\int_a^b f(t)g(t)\mathrm{d}t = f(\tau) \int_a^b g(t)\mathrm{d}t$. □

Let $f: I \to \mathbb{M}_\mathbb{F}$ and $t \in I$. The function f is *left differentiable* at t with *left derivative* $(\mathrm{d}^- f/\mathrm{d}t)(t)$ if the following limit exists

$$\lim_{h\uparrow 0} \frac{f(t + h) - f(t)}{h} =: \frac{\mathrm{d}^- f}{\mathrm{d}t}(t)$$

and is *right differentiable* at t with *right derivative* $(\mathrm{d}^+ f/\mathrm{d}t)(t)$ if the following limit exists

$$\lim_{h\downarrow 0} \frac{f(t + h) - f(t)}{h} =: \frac{\mathrm{d}^+ f}{\mathrm{d}t}(t).$$

The function f is deemed to be *piecewise continuously differentiable* if the following holds: for every $a, b \in \mathbb{R}$ with $a < b$, the interval $[a, b]$ admits a finite partition $a = t_1 < t_2 < \cdots < t_{n-1} < t_n = b$ such that f is continuously differentiable on every subinterval $[t_i, t_{i+1}]$, $i = 1, \ldots, n-1$, where the derivative at the subinterval endpoints is interpreted as the right derivative at t_i and the left derivative at t_{i+1}. The points t_i, $i = 2, \ldots, n - 1$, are points at which f may fail to be differentiable.

The vector space of piecewise continuously differentiable functions $I \to \mathbb{M}_\mathbb{F}$ is denoted by $PC^1(I, \mathbb{M}_\mathbb{F})$. Note that a piecewise continuously differentiable function is *a fortiori* continuous. Let $f \in PC^1(I, \mathbb{M}_\mathbb{F})$ and let $E \subset I$ be the set of points in I at which f fails to be differentiable. Then any function $f^\nabla : I \to \mathbb{M}_\mathbb{F}$ with the property that $f^\nabla(t) = f'(t)$ for all $t \in I \backslash E$ is in $PC(I, \mathbb{M}_\mathbb{F})$.

Theorem A.30 (Generalized fundamental theorem of calculus I)

Let $f \in PC(I, \mathbb{M}_{\mathbb{F}})$, $t_0 \in I$, define $F \colon I \to \mathbb{M}_{\mathbb{F}}$ by $F(t) := \int_{t_0}^{t} f(\tau) \mathrm{d}\tau$ for all $t \in I$ and let $E \subset I$ be the set of points at which f fails to be continuous. Then $F \in PC^1(I, \mathbb{M}_{\mathbb{F}})$ and

$$\frac{\mathrm{d}F}{\mathrm{d}t}(t) = f(t) \ \forall t \in I \backslash E \quad \text{and} \quad \frac{\mathrm{d}F^{\pm}}{\mathrm{d}t}(t) = f(t^{\pm}) \ \forall t \in E.$$

Proof

For simplicity, we assume that I is open. The arguments given below can be suitably modified to allow for boundary points contained in I. Let $t \in I$. Then there exist $t_1, t_2 \in I$ with $t_1 < t < t_2$ and such that t is the only potential discontinuity of f in $[t_1, t_2]$.

Assume that $t \notin E$. Then f is continuous on $[t_1, t_2]$, and, since $F(s) = F(t_1) + \int_{t_1}^{s} f(\tau) \mathrm{d}\tau$, it follows from the fundamental theorem of calculus for continuous f that F is continuously differentiable on $[t_1, t_2]$ and $F'(s) = f(s)$ for all $s \in [t_1, t_2]$. In particular, $F'(t) = f(t)$.

Now assume that $t \in E$. Define $\tilde{f} \colon [t_1, t] \to \mathbb{M}_{\mathbb{F}}$ by $\tilde{f}(\tau) = f(\tau)$ for all $\tau \in [t_1, t)$ and $\tilde{f}(t) = f(t^-)$. Then $\tilde{f}$ is continuous and

$$F(t) = F(t_1) + \int_{t_1}^{t} \tilde{f}(\tau) \mathrm{d}\tau.$$

Let $h < 0$ be such that $t + h \geq t_1$. Since $\tilde{f}$ is continuous, we may apply the mean-value theorem for integrals (see Theorem A.29) to conclude the existence of a number $\tau_h \in [t + h, t]$ such that

$$F(t + h) - F(t) = -\int_{t+h}^{t} \tilde{f}(\tau) \mathrm{d}\tau = h \tilde{f}(\tau_h).$$

As $h \to 0$, $\tau_h \to t$, and consequently, $(\mathrm{d}F^-/\mathrm{d}t)(t) = \tilde{f}(t) = f(t^-)$. Moreover, *mutatis mutandis*, the same argument can be used to prove that $(\mathrm{d}F^+/\mathrm{d}t)(t) = f(t^+)$.

Finally, note that the above argument shows that $F \in PC^1(I, \mathbb{M}_{\mathbb{F}})$. $\qquad \square$

Theorem A.31 (Generalized fundamental theorem of calculus II)

Let $f \in PC^1(I, \mathbb{M}_{\mathbb{F}})$, $t_0 \in I$ and let $E \subset I$ be the set of points at which f fails to be differentiable. Then

$$f(t) = f(t_0) + \int_{t_0}^{t} f^{\nabla}(\tau) \mathrm{d}\tau \quad \forall t \in I,$$

where f^∇ is any function $I \to \mathbb{M}_\mathbb{F}$ with the property that $f^\nabla(t) = f'(t)$ for all $t \in I \backslash E$.

Proof

Without loss of generality we may assume that that $t > t_0$. There exists a finite partition $t_0 < t_1 < \cdots < t_k = t$ of the interval $[t_0, t]$ such that f is continuously differentiable on every interval $[t_{j-1}, t_j]$. Therefore, by the fundamental theorem of calculus for continuously differentiable functions, we have that, for $j = 1, \ldots, k$,

$$f(t_j) - f(t_{j-1}) = \int_{t_{j-1}}^{t_j} f'(\tau)\mathrm{d}\tau.$$

Now $\int_{t_0}^t f^\nabla(\tau)\mathrm{d}\tau = \sum_{j=1}^k \int_{t_{j-1}}^{t_j} f'(\tau)\mathrm{d}\tau$ and we conclude that

$$\int_{t_0}^t f^\nabla(\tau)\mathrm{d}\tau = \sum_{j=1}^k \left(f(t_j) - f(t_{j-1}) \right) = f(t_k) - f(t_0) = f(t) - f(t_0),$$

completing the proof. $\square$

Uniform convergence and integrals. Assume that the interval I is compact. Then, see Proposition A.22, $C(I, \mathbb{M}_\mathbb{F})$ is a Banach space when endowed with the norm $\|f\|_\infty = \sup_{t \in I} \|f(t)\|$, where $\|\cdot\|$ is the norm in $\mathbb{M}_\mathbb{F}$. Convergence of a sequence (f_n) in $C(I, \mathbb{M}_\mathbb{F})$ is equivalent to uniform convergence. The following basic result shows that integration and uniform limits can be interchanged.

Theorem A.32

Let $a < b$, let (f_n) be a uniformly convergent sequence in $C([a, b], \mathbb{F}^{P \times Q})$ with limit f and let $g \in PC([a, b], \mathbb{F}^{N \times P})$. Then

$$\lim_{n \to \infty} \int_a^b g(\tau) f_n(\tau)\mathrm{d}\tau = \int_a^b g(\tau) f(\tau)\mathrm{d}\tau.$$

Proof

Set $\gamma := \sup_{\tau \in [a,b]} \|g(\tau)\| < \infty$. Invoking the triangle inequality for integrals, we obtain

$$\left\| \int_a^b g(\tau) f(\tau)\mathrm{d}\tau - \int_a^b g(\tau) f_n(\tau)\mathrm{d}\tau \right\| \leq \gamma \int_a^b \|f(\tau) - f_n(\tau)\|\mathrm{d}\tau.$$

Let $\varepsilon > 0$. Then there exists $N \in \mathbb{N}$ such that

$$\sup_{\tau \in [a,b]} \|f(\tau) - f_n(\tau)\| = \|f - f_n\|_\infty \leq \frac{\varepsilon}{\gamma(b-a)} \quad \forall n \geq N.$$

Consequently,

$$\left\| \int_a^b g(\tau)f(\tau)\mathrm{d}\tau - \int_a^b g(\tau)f_n(\tau)\mathrm{d}\tau \right\| \leq \gamma(b-a)\|f - f_n\|_\infty \leq \varepsilon \quad \forall n \geq N,$$

completing the proof. $\qquad\qquad\qquad\qquad\qquad\qquad\qquad\qquad\qquad\qquad\qquad\quad$ $\square$

Differential calculus in several variables. Throughout, in the context of a function of several variables, differentiability should be interpreted in the following sense (Fréchet[2] differentiability). Let $X \subset \mathbb{R}^N$ be a non-empty open set. A function $f\colon X \to \mathbb{R}^M$ is *differentiable* at $x \in X$ if there exists a real $M \times N$ matrix, which we denote by $(Df)(x) \in \mathbb{R}^{M \times N}$, such that

$$\lim_{z \to 0} \frac{\|f(x+z) - f(x) - ((Df)(x))z\|}{\|z\|} = 0.$$

We record two basic facts: if f is differentiable at $x \in X$, then
(a) f is continuous at x;
(b) all partial derivatives of components f_i of f with respect to components of its argument exist at x and $(Df)(x)$ coincides with the Jacobian matrix of f at x, that is,

$$(Df)(x) = \big((\partial_j f_i)(x)\big).$$

Specifically, the entry in row i and column j of the matrix $(Df)(x) \in \mathbb{R}^{M \times N}$ is $(\partial_j f_i)(x)$, the partial derivative, at x, of the i-th component of f with respect to the j-th component of its argument. Note that, implicit in (b) is the fact that, if f is differentiable at $x \in X$, then the derivative $(Df)(x)$ at x is unique.

The converse of (b) does not hold: the existence, at $x \in X$, of all partial derivatives is not a sufficient condition for differentiability of f at x (or indeed for continuity of f at x).

The function f is *differentiable* if $(Df)(x)$ exists for every $x \in X$, in which case, the map $Df\colon X \to \mathbb{R}^{M \times N}$, $x \mapsto (Df)(x)$, is the *derivative* of f (and $(Df)(x)$ is the *derivative of f at $x \in X$*). If the map Df is continuous, then f is said to be *continuously differentiable*. The space of continuously differentiable maps $X \to \mathbb{R}^M$ is denoted by $C^1(X, \mathbb{R}^M)$. Observe that, if $f \in C^1(X, \mathbb{R}^M)$, then all partial derivatives $\partial_j f_i$ (where $j = 1, \ldots, N$ and $i = 1, \ldots, M$) exist and are continuous. The converse is also true: if all partial derivatives exist and are continuous, then f is continuously differentiable.

---

[2] Maurice René Fréchet (1878-1973), French.

The particular case wherein $M = 1$ features frequently. In this case, f is differentiable at $x \in X$, then

$$(Df)(x) = \big(\partial_1 f(x), \ldots \partial_N(x)\big) =: (\nabla f)(x),$$

where $(\nabla f)(x)$ is said to be the *gradient* of f and x.

Let $X \subset \mathbb{R}^N$ and $Y \subset \mathbb{R}^M$ be a non-empty open sets, and $g: X \times Y \to \mathbb{R}^P$. If g is differentiable in its first argument, that is, if, for each $y \in Y$, the function $g_y: X \to \mathbb{R}^P$, $x \mapsto g(x,y)$, is differentiable, then we define $D_1 g: X \times Y \to \mathbb{R}^{P \times N}$ by

$$(D_1 g)(x,y) := (Dg_y)(x) \ \ \forall\, (x,y) \in X \times Y.$$

If $N = 1$, then $D_1 g = \partial_1 g$.

Similarly, if g is differentiable in its second argument, that is, if, for each $x \in X$, the function $g_x: Y \to \mathbb{R}^P$, $y \mapsto g(x,y)$, is differentiable, then we define $D_2 g: X \times Y \to \mathbb{R}^{P \times M}$ by

$$(D_2 g)(x,y) := (Dg_x)(y) \ \ \forall\, (x,y) \in X \times Y.$$

If g is differentiable then its derivative $Dg: X \times Y \to \mathbb{R}^{P \times (N+M)}$ can be expressed as the block-matrix-valued function

$$Dg = \big(D_1 g\,, D_2 g\big).$$

Next, we record the *multivariable chain rule* for differentiation of compositions.

Theorem A.33 (Chain rule)

Let $X \subset \mathbb{R}^N$ and $Y \subset \mathbb{R}^M$ be non-empty open sets. If $g: X \to Y$ is differentiable at $x \in X$ and $f: Y \to \mathbb{R}^P$ is differentiable at $g(x) \in Y$, then $f \circ g$ is differentiable at x with derivative

$$\big(D(f \circ g)\big)(x) = (Df)(g(x))(Dg)(x).$$

In the case $N = 1$, the requirement that $X \subset \mathbb{R}$ be open can be relaxed: the following special case of the chain rule is used freely throughout the book.

Proposition A.34

Let I be an interval, and let $Y \subset \mathbb{R}^M$ be non-empty and open. Assume that the functions $g: I \to Y$ and $f: Y \to \mathbb{R}$ are differentiable. Then the composition $f \circ g$ is differentiable and

$$\frac{d}{dt}(f \circ g)(t) = (f \circ g)'(t) = \sum_{i=1}^{M} (\partial_i f)(g(t))g_i'(t) = \langle (\nabla f)(g(t)), g'(t) \rangle \ \ \forall\, t \in I,$$

where g_i denotes the i-th component of g.

We proceed to state a special case of *the implicit function theorem* which is used in Sections 4.6. Given an equation of the form $f(y, z) = 0$, then to each value of z there may correspond one or more values of y which satisfy the equation - or indeed there may be no values of y for which the equation holds. If there exists a non-empty set Z and a function u defined on Z such that, for each $z \in Z$, the equation holds with $y = u(z)$, that is, $f(u(z), z) = 0$ for all $z \in Z$, then we say that the equation implicitly defines the function $z \mapsto u(z)$ on Z.

Implicit function theorems seek to determine conditions under which an equation implicitly defines a suitably regular function. One such theorem follows.

Theorem A.35 (Implicit function theorem)

Let $J \subset \mathbb{R}$ be an open interval and $G \subset \mathbb{R}^P$ be a non-empty open set. Let $f \in C(J \times G, \mathbb{R})$ be differentiable in its first variable (with derivative $\partial_1 f$). Assume that $(y^0, z^0) \in J \times G$ is such that $f(y^0, z^0) = 0$, $\partial_1 f$ is continuous at (y^0, z^0) and $(\partial_1 f)(y^0, z^0) \neq 0$. Then there exists an open neighbourhood $Z \subset G$ of z^0 and a continuous function $u \colon Z \to J$ such that $u(z^0) = y^0$ and $f(u(z), z) = 0$ for all $z \in Z$.

Proof

We may assume that $(\partial_1 f)(y^0, z^0) > 0$ (otherwise, consider $-f$). By continuity of $\partial_1 f$ at (y^0, z^0), there exist $\beta > 0$, $\delta > 0$ and a neighbourhood $Z_0 \subset G$ of z^0 such that

$$(\partial_1 f)(y, z) \geq \beta \quad \forall\, (y, z) \in [y^0 - \delta,\, y^0 + \delta] \times Z_0. \tag{A.16}$$

Consequently, $f(y^0 - \delta, z^0) < 0 < f(y^0 + \delta, z^0)$. By continuity of f, there exists an open neighbourhood $Z \subset Z_0$ of z^0 such that

$$f(y^0 - \delta, z) < 0 < f(y^0 + \delta, z) \quad \forall\, z \in Z.$$

Set $I := [y^0 - \delta, y^0 + \delta]$. For each $z \in Z$, an application of the intermediate value theorem to the continuous function $f_z \colon I \to \mathbb{R}$, $y \mapsto f(y, z)$, yields the existence of $y_z \in I$ such that $f_z(y_z) = f(y_z, z) = 0$, Defining $u \colon Z \to I$ by $u(z) := y_z$, we have $f(u(z), z) = 0$ for all $z \in Z$. It remains to show that u is continuous. Seeking a contradiction, suppose that u is not continuous. Then there exist $z \in Z$, (z_n) in Z, with $z_n \to z$ as $n \to \infty$, and $w \in I$, with $w \neq u(z)$, such that $\lim_{n \to \infty} u(z_n) = w$. By continuity of f,

$$f(w, z) = \lim_{n \to \infty} f(u(z_n), z_n) = 0$$

and so, by the mean value theorem for differentiation, there exists $y \in I$ such

$$(\partial_1 f)(y, z) = \frac{f(w, z) - f(u(z), z)}{w - u(z)} = 0,$$

which contradicts (A.16). $\qquad\qquad\qquad\qquad\qquad\qquad\qquad\qquad\qquad\qquad\square$

A.4 Elements of the Laplace transform

A function $f : \mathbb{R}_+ \to X$, where X is an arbitrary normed space, is said to be *exponentially bounded* if there exists $\beta \in \mathbb{R}$ such that

$$\sup_{t \geq 0} \left(\|f(t)\| e^{-\beta t} \right) < \infty. \tag{A.17}$$

We say that f has *exponential growth* γ_f if f is exponentially bounded and γ_f is the infimum of all β such that (A.17) holds. Note that $-\infty \leq \gamma_f < \infty$. A function f is said to be of *class* $\mathcal{E}_\gamma$, $-\infty \leq \gamma < \infty$, if f has exponential growth not greater than γ.

In this section, let $\mathbb{M}_{\mathbb{C}}$ denote the space $\mathbb{C}^{P \times Q}$ (with $\mathbb{M}_{\mathbb{C}} = \mathbb{C}^P$ if $Q = 1$). Let $f \in PC(\mathbb{R}_+, \mathbb{M}_{\mathbb{C}})$ be of class $\mathcal{E}_\gamma$ and let $s \in \mathbb{C}$ with $\operatorname{Re} s > \gamma$. Choose $\beta \in \mathbb{R}$ such that $\operatorname{Re} s > \beta > \gamma$. Then there exists $\Gamma_\beta > 0$ such that $\|f(t)\| \leq \Gamma_\beta e^{\beta t}$ for all $t \in \mathbb{R}_+$ and thus, invoking the triangle inequality for integrals, we obtain for all $\tau \geq \sigma \geq 0$,

$$\left\| \int_\sigma^\tau e^{-st} f(t) \mathrm{d}t \right\| \leq \int_\sigma^\tau e^{-t \operatorname{Re} s} \|f(t)\| \mathrm{d}t \leq \Gamma_\beta \int_\sigma^\tau e^{(\beta - \operatorname{Re} s)t} \mathrm{d}t.$$

Therefore, since $\beta - \operatorname{Re} s < 0$, it follows from Exercise A.6 that the improper integral $\int_0^\infty e^{-st} f(t) \mathrm{d}t$ exists.

Let $-\infty \leq \gamma < \infty$ and set $\mathbb{C}_\gamma := \{s \in \mathbb{C} \colon \operatorname{Re} s > \gamma\}$ (obviously, $\mathbb{C}_{-\infty} = \mathbb{C}$). The *Laplace transform* $\mathcal{L}\{f\}$ of the function $f \in PC(\mathbb{R}_+, \mathbb{M}_{\mathbb{C}})$ of class $\mathcal{E}_\gamma$ is the function $\mathbb{C}_\gamma \to \mathbb{M}_{\mathbb{C}}$ given by

$$\mathcal{L}\{f\}(s) := \int_0^\infty e^{-st} f(t) \, \mathrm{d}t \quad \forall s \in \mathbb{C}_\gamma.$$

Laplace transformation is a linear operation: for $f, g \in PC(\mathbb{R}_+, \mathbb{M}_{\mathbb{C}})$ of class $\mathcal{E}_\gamma$ and scalars $\lambda, \mu \in \mathbb{C}$, we have

$$\mathcal{L}\{\lambda f + \mu g\}(s) = \lambda \mathcal{L}\{f\}(s) + \mu \mathcal{L}\{g\}(s) \quad \forall s \in \mathbb{C}_\gamma.$$

As an alternative notation, which is sometimes more convenient to use, we introduce $\hat{f} := \mathcal{L}\{f\}$.

Example A.36

Let $a, b \in \mathbb{R}$ and write $c := a + ib \in \mathbb{C}$. Consider the scalar-valued functions $\exp_c$, $\sin_c$ and $\cos_c$ from $\mathbb{R}_+$ to $\mathbb{C}$ defined by

$$\exp_c(t) = e^{ct}, \quad \sin_c(t) = \sin(ct), \quad \cos_c(t) = \cos(ct); \quad \forall t \in \mathbb{R}_+.$$

Then $\exp_c$ is of class $\mathcal{E}_a$, with Laplace transform given by

$$\mathcal{L}\{\exp_c\}(s) = \int_0^\infty e^{-(s-c)t} \mathrm{d}t = \frac{1}{s-c} - \lim_{t\to\infty} \frac{e^{-(s-c)t}}{s-c}.$$

Since the limit on the right is zero for all $s \in \mathbb{C}$ with $\operatorname{Re} s > \operatorname{Re} c = a$, it follows that

$$\mathcal{L}\{\exp_c\}(s) = \frac{1}{s-c} \quad \forall s \in \mathbb{C}_a. \tag{A.18}$$

Moreover, since $2i \sin z = \exp(iz) - \exp(-iz)$ and $2 \cos z = \exp(iz) + \exp(-iz)$, $\sin_c$ and $\cos_c$ are of class $\mathcal{E}_{|b|}$, and (A.18), together with linearity of the Laplace transform, yields

$$\mathcal{L}\{\sin_c\}(s) = \frac{c}{s^2+c^2}, \quad \mathcal{L}\{\cos_c\}(s) = \frac{s}{s^2+c^2}; \quad \forall s \in \mathbb{C}_{|b|}.$$

$\triangle$

A key feature of the Laplace transform is that it transforms analytic operations (such as differentiation, integration and convolution) into algebraic operations, as the following theorem shows.

Theorem A.37 (Key properties of the Laplace transform)

(1) Let $f \in PC^1(\mathbb{R}_+, \mathbb{M}_\mathbb{C})$ and let $E \subset \mathbb{R}_+$ be the set of points in $\mathbb{R}_+$ at which f fails to be differentiable. Let $f^\nabla : \mathbb{R}_+ \to \mathbb{M}_\mathbb{C}$ be such that $f^\nabla(t) = f'(t)$ for all $t \in I \backslash E$. Assume that f and f^∇ are of class $\mathcal{E}_\gamma$ for some $\gamma \in \mathbb{R}$. Then $\mathcal{L}\{f^\nabla\}(s) = s\mathcal{L}\{f\}(s) - f(0)$ for all $s \in \mathbb{C}_\gamma$.

(2) Let $f \in PC(\mathbb{R}_+, \mathbb{M}_\mathbb{C})$ be of class $\mathcal{E}_\gamma$ for some $\gamma \in \mathbb{R}$, set $\beta := \max(\gamma, 0)$ and define $F : \mathbb{R}_+ \to \mathbb{M}_\mathbb{C}$ by $F(t) := \int_0^t f(\tau)\mathrm{d}\tau$. Then F is of class $\mathcal{E}_\beta$ and $\mathcal{L}\{F\}(s) = (1/s)\mathcal{L}\{f\}(s)$ for all $s \in \mathbb{C}_\beta$.

(3) Let $f \in PC(\mathbb{R}_+, \mathbb{C}^{P\times Q})$ and $g \in PC(\mathbb{R}_+, \mathbb{C}^{Q\times R})$ be of class $\mathcal{E}_\gamma$ for some $\gamma \in \mathbb{R}$. Then the convolution $f \star g$ of f and g, defined by $(f \star g)(t) = \int_0^t f(t-\tau)g(\tau)\mathrm{d}\tau$ for $t \in \mathbb{R}_+$, is of class $\mathcal{E}_\gamma$ and $\mathcal{L}\{f \star g\}(s) = \mathcal{L}\{f\}(s)\mathcal{L}\{g\}(s)$ for all $s \in \mathbb{C}_\gamma$.

Proof

To prove statement (1), note that

$$(f(t)e^{-st})' = f^\nabla(t)e^{-st} - sf(t)e^{-st} \quad \forall t \in \mathbb{R}_+ \backslash E.$$

Integrating the above identity, invoking Theorem A.31 and noting that, for all $s \in \mathbb{C}_\gamma$, $\lim_{t\to\infty}(f(t)e^{-st}) = 0$, establishes the claim.

The proof of statement (2) is left to the reader (see Exercise A.7). We proceed to prove statement (3). Let $\beta > \gamma$ and choose $\alpha \in (\gamma, \beta)$. Since f and g are of class $\mathcal{E}_\gamma$, there exists $\Gamma > 0$ such that, for all $t, \tau \in \mathbb{R}_+$,

$$\|f(t-\tau)g(\tau)\| \leq \Gamma e^{\alpha(t-\tau)} e^{\alpha\tau} = \Gamma e^{\alpha t}.$$

Hence $\|(f \star g)(t)\| \leq \Gamma t e^{\alpha t}$ for all $t \geq 0$. Consequently, there exists $\Gamma_\beta > 0$ such that $\|(f \star g)(t)\| \leq \Gamma_\beta e^{\beta t}$ for all $t \geq 0$. This holds for every $\beta > \gamma$ and so $f \star g$ is of class $\mathcal{E}_\gamma$. Therefore, the Laplace transform $\mathcal{L}\{f \star g\}(s)$ exists for all $s \in \mathbb{C}_\gamma$ and, for all such s, is given by

$$\mathcal{L}\{f \star g\}(s) = \int_0^\infty e^{-st} \int_0^t f(t-\tau)g(\tau)\mathrm{d}\tau \, \mathrm{d}t$$

$$= \int_0^\infty \int_\tau^\infty e^{-st} f(t-\tau)g(\tau)\mathrm{d}t \, \mathrm{d}\tau = \int_0^\infty \int_0^\infty e^{-s(\sigma+\tau)} f(\sigma)g(\tau)\mathrm{d}\sigma \, \mathrm{d}\tau$$

$$= \left(\int_0^\infty e^{-s\sigma} f(\sigma)\mathrm{d}\sigma \right) \left(\int_0^\infty e^{-s\tau} g(\tau)\mathrm{d}\tau \right) = \mathcal{L}\{f\}(s)\mathcal{L}\{g\}(s).$$

$\square$

Exercise A.7

Prove statement (2) of Theorem A.37.

Finally, we compute the Laplace transform of the function $t \mapsto \exp(tM)$, where M is a square matrix.

Proposition A.38

Let $M \in \mathbb{C}^{P\times P}$, set $\gamma := \max\{\mathrm{Re}\,\lambda \colon \lambda \in \sigma(M)\}$ and define $\exp_M \colon \mathbb{R}_+ \to \mathbb{C}^{P\times P}$ by $\exp_M(t) = \exp(tM)$ for $t \geq 0$. Then $\exp_M$ is of class $\mathcal{E}_\gamma$ and $\mathcal{L}\{\exp_M\}(s) = (sI - M)^{-1}$ for all $s \in \mathbb{C}_\gamma$.

Proof

Setting $f := \exp_M$, we have that $f(0) = I$ and f is continuously differentiable with derivative $f'(t) = Mf(t)$ for all $t \geq 0$ (recall Lemma 2.10). Moreover, by Theorem 2.11, f is of class $\mathcal{E}_\gamma$. Hence, by statement (1) of Theorem A.37,

$$M\mathcal{L}\{f\}(s) = \mathcal{L}\{f'\}(s) = s\mathcal{L}\{f\}(s) - I \quad \forall s \in \mathbb{C}_\gamma.$$

Consequently, $\mathcal{L}\{f\}(s) = (sI - M)^{-1}$ for all $s \in \mathbb{C}_\gamma$, completing the proof. $\square$

A.5 Zorn's lemma

Let S be a non-empty set. A relation $\preceq$ is said to be a *partial ordering* on S if it has the following properties:

(a) *reflexivity*: for all $x \in S$, $x \preceq x$;

(b) *antisymmetry*: for all $x, y \in S$, if $x \preceq y$ and $y \preceq x$, then $x = y$;

(c) *transitivity*: for all $x, y, z \in S$, if $x \preceq y$ and $y \preceq z$, then $x \preceq z$.

A *partially ordered set* is a set S on which a partial ordering is defined. The term "partial" is intended to convey the idea that S may contain elements x and y for which neither $x \preceq y$ nor $y \preceq x$: such elements are said to be incomparable; conversely, two elements x and y of S are said to be comparable if $x \preceq y$ or $y \preceq x$.

A *maximal element* of a partially ordered non-empty set S is an element $m \in S$ (which may or may not exist) such that

$$s \in S \text{ and } m \preceq s \quad \Longrightarrow \quad m = s.$$

Let T be a non-empty subset of a partially ordered set S. An *upper bound* for T (which may or may not exist) is an element u of S such that

$$t \preceq u \quad \forall t \in T.$$

If every two elements of T are comparable, then T is said to be a *totally ordered* subset.

Lemma A.39 (Zorn's lemma)

Let $S \neq \emptyset$ be a partially ordered set. If every totally ordered subset T of S has an upper bound, then S has at least one maximal element.

Zorn's lemma is equivalent to the *axiom of choice*, which states the following: given any non-empty set S, there exists a function

$$F \colon \{R \subset S \colon R \neq \emptyset\} \to S$$

such that $F(R) \in R$ for every non-empty subset R of S (the function F is called a *choice function*). In words, the axiom of choice simply states that, given a non-empty set S, there exists a map F which assigns to every non-empty subset $R \subset S$ an element in R, or, in more colloquial term, which "chooses" an element from every non-empty subset. The axiom of choice appears intuitively obvious: the axiom itself, or some principle equivalent to it, is usually postulated in an axiomatic treatment of set theory. Whilst the axiom of choice (or equivalently, Zorn's lemma) is useful in many existence proofs, it does not lead to "explicit" constructions or formulae. There is a (very small) minority of mathematicians who do not accept the axiom of choice because of its non-constructive character: they object to postulating the existence of a choice function when no indication is given of how this function is constructed. For more details on the axiom of choice, Zorn's lemma and equivalent principles see, for example, [7, 9].

Solutions to selected exercises from
Chapters 1 – 6

Chapter 1

Exercise 1.2
(a) Let $x\colon I \to \mathbb{R}^N$ be a solution. By the chain rule (Proposition A.34), the derivative $(E \circ x)'$ of the composition $E \circ x$ satisfies

$$(E \circ x)'(t) = \langle (\nabla E)(x(t)), \dot{x}(t) \rangle = \langle (\nabla E)(x(t)), f(x(t)) \rangle = 0, \quad \forall t \in I.$$

Consequently, there exists $\gamma \in \mathbb{R}$ such that $E(x(t)) = (E \circ x)(t) = \gamma$ for all $t \in I$.
(b) $\langle (\nabla E)(z), f(z) \rangle = -g(z_1)z_2 + z_2 g(z_1) = 0$ for all $z \in \mathbb{R}^2$.
(c) Applying part (b) with g given by $g(s) = -b \sin s$ shows that

$$E(z) = E(z_1, z_2) = b \int_0^{z_1} \sin s \, ds = b(1 - \cos z_1) + z_2^2/2$$

is a first integral.
(d) $(\nabla E)(z) = (\nabla E)(z_1, z_2) = \big(d - c/z_1, \, b - a/z_2\big)$ and so

$$
\begin{aligned}
\langle (\nabla E)(z_1, z_2), \big(z_1(-a + bz_2), z_2(c - dz_1)\big) \rangle \\
= ac - bcz_2 - adz_1 + bdz_1 z_2 - ac + adz_1 + bcz_2 - bdz_1 z_2 \\
= 0, \quad \forall (z_1, z_2) \in (0, \infty) \times (0, \infty).
\end{aligned}
$$

(e) Assume $E\colon G \to \mathbb{R}$ is a first integral for (1.12). We have seen that the image of any solution of (1.12) is contained in some level set of E. Therefore, in principle, a study of the level sets of a first integral can provide insight into the qualitative behaviour of solutions of (1.12). For any *constant* function E, trivially we have $\langle (\nabla E)(z), f(z) \rangle = 0$ for all $z \in G$ and, moreover, G is the only non-empty level set E. Therefore, if non-constancy is removed from the definition of a first integral, then every constant function is a first integral and the result in (a) above does not provide any useful information.

Exercise 1.3
In parts (a)-(d), it is assumed that $k(\xi) \neq 0$.

(a) $K'(z) = 1/k(z) \neq 0$ for all $z \in U$. Therefore, $K\colon U \to K(U)$ is strictly monotone and so has an inverse function $K^{-1}\colon K(U) \to U$. Moreover, $K(U)$ is an open interval containing 0 and $K^{-1}(0) = \xi$.

(b) Since H is continuous with $H(\tau) = 0$, there exists $\varepsilon > 0$ such that $I := (\tau - \varepsilon, \tau + \varepsilon)$ is contained in J and $H(I)$ is contained in $K(U)$.

(c) Differentiating the relation $K(x(t)) = H(t)$ for all $t \in I$ gives $K'(x(t))\dot{x}(t) = h(t)$ for all $t \in I$. Since $K' = 1/k$, we have $\dot{x}(t) = k(x(t))h(t)$ for all $t \in I$. Moreover, $x(\tau) = K^{-1}(H(\tau)) = K^{-1}(0) = \xi$ and so $x\colon I \to G, \, t \mapsto K^{-1}(H(t))$ is a solution of the initial-value problem. Assume $x_1, x_2\colon I \to G$ are two solutions of the initial-value problem. Then $K(x_1(t)) = H(t) = K(x_2(t))$ for all $t \in I$ and so $x_1(t) = K^{-1}(K(x_2(t))) = x_2(t)$ for all $t \in I$.

H. Logemann and E. P. Ryan, *Ordinary Differential Equations*,
Springer Undergraduate Mathematics Series,
DOI: 10.1007/978-1-4471-6398-5, © Springer-Verlag London 2014

(d) Set $J := \mathbb{R}$, $I := (-1,1)$, $U := (0,\infty)$, $k(x) = z^3$ for all $z \in U$ and $h(t) := t$ for all $t \in J$. Define $K \colon U \to K(U)$ by $K(z) := \int_1^z ds/k(s) = (1 - z^{-2})/2$ for all $z \in U$ and define $H \colon J \to \mathbb{R}$ by $H(t) := \int_0^t h(s)ds = t^2/2$ for all $t \in J$. Then $K(U) = (-\infty, 1/2)$, $H(I) = (-1/2, 1/2) \subset K(U)$ and $K^{-1} \colon K(U) \to U$ is given by $K^{-1}(z) = 1/\sqrt{1-2z}$. By parts (a)-(c), it follows that $x \colon I \to \mathbb{R}$, $t \mapsto K^{-1}(H(t)) = 1/\sqrt{1-t^2}$, solves the initial-value problem. Moreover, since $x(t) \to \infty$ as $t \to \pm 1$, the solution x is maximal.

In parts (e) and (f) below, it is assumed that $k(\xi) = 0$.

(e) First, we prove that $x(t) = \xi$ for all $t \in I$ with $t \geq \tau$. Suppose that this claim is false. Then there exists $I^* = (\sigma, \rho) \subset I$ such that $\sigma \geq \tau$, $x(\sigma) = \xi$, $x(\rho) \neq \xi$ and $x(t) \in (\xi - \delta, \xi) \cup (\xi, \xi + \delta)$ for all $t \in I^*$. Set $c := (\rho - \sigma) \max_{t \in [\sigma, \rho]} |h(t)|$. Since, for all $t \in I^*$, $\dot{x}(t) = k(x(t))h(t)$ and $k(x(t)) \neq 0$, we have

$$c \geq \int_r^\rho |h(t)|dt \geq \left| \int_r^\rho h(t)dt \right| = \left| \int_r^\rho \frac{\dot{x}(t)}{k(x(t))}dt \right| = \left| \int_{x(r)}^{x(\rho)} \frac{ds}{k(s)} \right| \quad \forall r \in I^*.$$

Observe that either $x(t) \in (\xi, \xi + \delta)$ for all $t \in I^*$ or $x(t) \in (\xi - \delta, \xi)$ for all $t \in I^*$. If the former is the case, then $x(\rho) > \xi$ and passing to the limit $r \to \sigma$ (and so $x(r) \downarrow x(\sigma) = \xi$) yields a contradiction to the second of properties (1.16). If the latter is the case, then $x(\rho) < \xi$ and passing to the limit $r \to \sigma$ (and so $x(r) \uparrow x(\sigma) = \xi$) yields a contradiction to the first of properties (1.16). We may now conclude that $x(t) = \xi$ for all $t \in I$ with $t \geq \tau$.

The above argument applies *mutatis mutandis* to conclude that $x(t) = \xi$ for all $t \in I$ with $t \leq \tau$.

(f) The function k fails to satisfy properties (1.16).

Exercise 1.4

(a) Let $x \colon J \to \mathbb{R}$ be a solution of (1.18). We first show that $x(t) = 0$ for all $t \in J$ with $t \geq \tau$. Suppose otherwise, then there exists $I = (\sigma, \rho) \subset J$ with $\sigma \geq \tau$, $x(\sigma) = 0$ and $x(t) \neq 0$ for all $t \in I$. Define $\alpha := (\rho - \sigma) \max_{t \in [\sigma, \rho]} |a(t)|$. Observe that $(d/dt)(\ln|x(t)|) = \dot{x}(t)/x(t) = a(t)$ for all points $t \in I$ at which a is continuous. Therefore,

$$\left| \ln|x(\rho)/x(s)| \right| = \left| \int_s^\rho a(t)dt \right| \leq \int_s^\rho |a(t)|dt \leq \alpha \quad \forall s \in (\sigma, \rho)$$

which is impossible since, by choosing s sufficiently close to σ, $x(s)$ can be made arbitrarily close to $x(\sigma) = 0$ and so the term on the left can be made arbitrarily large. Therefore, $x(t) = 0$ for all $t \in J$ with $t \geq \tau$.

The above argument applies *mutatis mutandis* to conclude that $x(t) = 0$ for all $t \in J$ with $t \leq \tau$.

(b) Clearly, $x(\tau) = \xi$ and, invoking Theorem A.30, we have $\dot{x}(t) = a(t)x(t)$ at all points t of continuity of a. Therefore, x is a solution. Suppose $y \colon J \to \mathbb{R}$ is also a solution, and write $z := x - y$. Then $z(\tau) = \xi - \xi = 0$ and $\dot{z}(t) = \dot{x}(t) - \dot{y}(t) = a(t)(x(t) - y(t)) = a(t)z(t)$. By the result in (a), the zero function is the only solution on J of the initial-value problem: $\dot{z}(t) = a(t)z(t)$, $z(0) = 0$. Therefore, $y(t) = x(t)$ for all $t \in J$ and so x is the unique maximal solution.

(c) By properties of the exponential function, sufficiency of the condition is clear. We proceed to prove necessity and argue by contraposition. Assume that $\int_\tau^t a(s)ds \not\to -\infty$ as $t \to \infty$. Then there exist $\alpha \in \mathbb{R}$ and a sequence (t_n) in $\mathbb{R}$, with $t_n \to \infty$ as $n \to \infty$, such that $\int_\tau^{t_n} a(s)ds \geq \alpha$ for all $n \in \mathbb{N}$. Therefore, $|x(t_n)| \geq e^\alpha |\xi| > 0$ for all $n \in \mathbb{N}$ and so, for $\xi \neq 0$, $x(t) \not\to 0$ as $t \to \infty$.

(d) Define $A := \int_0^T a(s)ds$, $B := \int_0^T |a(s)|ds$ and $C := \left|\int_0^\tau a(s)ds\right|$. Observe that, for every integer m,

$$\int_0^{mT} a(s)ds = mA \quad \text{and} \quad \int_{mT}^{(m+1)T} |a(s)|ds = B.$$

Let (t_n) be any sequence in $\mathbb{R}$ with $t_n \to \infty$ as $n \to \infty$. For each $n \in \mathbb{N}$, there exists a unique integer m_n such that $m_n T \le t_n < (m_n+1)T$. Clearly, $m_n \to \infty$ as $n \to \infty$. Now

$$\int_\tau^{t_n} a(s)ds = \left(\int_\tau^0 + \int_0^{m_n T} + \int_{m_n T}^{t_n}\right) a(s)ds$$

Therefore, $m_n A - B - C \le \int_\tau^{t_n} a(s)ds \le m_n A + B + C$ for all $n \in \mathbb{N}$ and so $\int_\tau^{t_n} a(s)ds \to -\infty$ as $n \to \infty$ if, and only if, $A < 0$. We may now infer that $\int_\tau^t a(s)ds \to -\infty$, as $t \to \infty$ if, and only if, $A < 0$. Invoking the result in (c) completes the proof.

Chapter 2

Exercise 2.2
Let $J = \mathbb{R}$ and $N = 2$. Let $a\colon \mathbb{R} \to \mathbb{F}$ be any piecewise continuous function with the property that the set E of points at which it fails to be continuous is non-empty. Define $A\colon \mathbb{R} \to \mathbb{F}^{2\times 2}$ and $\xi \in \mathbb{F}^2$ by

$$A(t) := \begin{pmatrix} 0 & 1 \\ 0 & a(t) \end{pmatrix}, \quad \xi := \begin{pmatrix} 1 \\ 0 \end{pmatrix}.$$

Then the initial-value problem $\dot{x}(t) = A(t)x(t)$, $x(0) = \xi$ has constant solution $x\colon \mathbb{R} \to \mathbb{F}^2$, $t \mapsto x(t) = \xi$, whilst A fails to be continuous at each $\sigma \in E$.

Exercise 2.4
Observe that $M_2(t, s) - M_1(t, s) = \int_s^t A(\sigma)d\sigma$ for all $(t, s) \in J \times J$ and, for all $n \in \mathbb{N}$,

$$M_{n+2}(t, s) - M_{n+1}(t, s) = \int_s^t A(\sigma)\big[M_{n+1}(\sigma, s) - M_n(\sigma, s)\big]d\sigma \quad \forall (t, s) \in J \times J.$$

The result (2.3) follows by induction.
Assume that for, some $n \in \mathbb{N}$, the equality in (2.4) holds for all $(t, s) \in J \times J$. Then

$$\int_s^t \int_s^{\sigma_1} \cdots \int_s^{\sigma_n} d\sigma_{n+1} \cdots d\sigma_2 d\sigma_1 = \int_s^t \frac{(\sigma_1 - s)^n}{n!} d\sigma_1$$

$$= \frac{1}{n!} \int_0^{t-s} \sigma^n d\sigma = \frac{(t-s)^{n+1}}{(n+1)!} \quad \forall (t, s) \in J \times J.$$

Since $\int_s^t d\sigma_1 = t - s$ for all $(t, s) \in J \times J$, (2.4) follows by induction.

Exercise 2.5
The result follows from the Peano-Baker series (2.6) if it can be shown that, for all $n \in \mathbb{N}$,

$$\frac{1}{n!}\left(\int_\tau^t A(\sigma)d\sigma\right)^n =$$
$$\int_\tau^t A(\sigma_1) \int_\tau^{\sigma_1} A(\sigma_2) \cdots \int_\tau^{\sigma_{n-1}} A(\sigma_n)d\sigma_n \cdots d\sigma_2 d\sigma_1 \quad \forall t, \tau \in \mathbb{R}. \quad (*)$$

Clearly, $(*)$ holds for $n = 1$. Let $n \in \mathbb{N}$ and assume that $(*)$ holds. Observe that commutativity of $A(t)$ and $A(\sigma)$ for all t, σ implies commutativity of $A(t)$ and $\int_\tau^t A(\sigma) \mathrm{d}\sigma$ which, in conjunction with the product rule for differentiation and Theorem A.30, gives

$$\frac{\mathrm{d}}{\mathrm{d}t} \left(\int_\tau^t A(\sigma) \mathrm{d}\sigma \right)^{n+1} = (n+1) A(t) \left(\int_\tau^t A(\sigma) \mathrm{d}\sigma \right)^n$$

at all points t of continuity of A. Integrating and dividing by $(n+1)!$, we have

$$\frac{1}{(n+1)!} \left(\int_\tau^t A(\sigma) \mathrm{d}\sigma \right)^{n+1} = \int_\tau^t A(\sigma) \frac{1}{n!} \left(\int_\tau^\sigma A(\rho) \mathrm{d}\rho \right)^n \mathrm{d}\sigma$$

$$= \int_\tau^t A(\sigma) \int_\tau^\sigma A(\sigma_1) \cdots \int_\tau^{\sigma_{n-1}} A(\sigma_n) \mathrm{d}\sigma_n \cdots \mathrm{d}\sigma_1 \mathrm{d}\sigma.$$

By induction, it follows that $(*)$ holds for all $n \in \mathbb{N}$.

Exercise 2.7
Let $\tau \in J$ be arbitrary. Consider the initial-value problems $\dot{x}(t) = A(t)x(t)$, $x(\tau) = \xi$, and $\dot{\tilde{x}}(t) = \tilde{A}(t)\tilde{x}(t)$, $\tilde{x}(\tau) = \tilde{\xi}$. The unique solutions on J are given, respectively, by $x(t) = \Phi(t, \tau)\xi$ and $\tilde{x}(t) = \tilde{\Phi}(t, \tau)\tilde{\xi}$ for all $t \in J$. Now,

$$\frac{\mathrm{d}}{\mathrm{d}t} \langle \tilde{x}(t), x(t) \rangle = \langle \tilde{A}(t)\tilde{x}(t), x(t) \rangle + \langle \tilde{x}(t), A(t)x(t) \rangle = \langle \big(\tilde{A}(t) + A^*(t) \big) \tilde{x}(t), x(t) \rangle = 0$$

for all points $t \in J$ at which A is continuous. Therefore, $\langle \tilde{x}(t), x(t) \rangle = \langle \tilde{\xi}, \xi \rangle$ for all $t \in J$ and so

$$\langle \tilde{\xi}, \xi \rangle = \langle \tilde{\Phi}(t, \tau)\tilde{\xi}, \Phi(t, \tau)\xi \rangle = \langle \Phi^*(t, \tau)\tilde{\Phi}(t, \tau)\tilde{\xi}, \xi \rangle \quad \forall t \in J.$$

Since $\tilde{\xi}, \xi \in \mathbb{F}^N$ and $\tau \in J$ are arbitrary, we may now infer that $\Phi^*(t, \tau)\tilde{\Phi}(t, \tau) = I$ for all $(t, \tau) \in J \times J$. Therefore, $\Phi^*(t, \tau) = \tilde{\Phi}^{-1}(t, \tau) = \tilde{\Phi}(\tau, t)$ for all $(t, \tau) \in J \times J$, whence the required result.

Exercise 2.10
(1) For all $k \in \mathbb{N}$, $P^k = \mathrm{diag}(p_1^k, \ldots, p_N^k)$ and so

$$\exp(P) = \sum_{k=0}^\infty P^k / k! = \mathrm{diag}\left(\sum_{k=0}^\infty p_1^k / k!, \ldots, \sum_{k=0}^\infty p_N^k / k! \right) = \mathrm{diag}\left(e^{p_1}, \ldots, e^{p_N} \right).$$

(2) $\qquad (\exp(P))^* = \left(\sum_{k=0}^\infty P^k / k! \right)^* = \sum_{k=0}^\infty (P^*)^k / k! = \exp(P^*).$

(3) By Corollary 2.3, $(\mathrm{d}/\mathrm{d}t)\exp(Pt) = P\exp(Pt)$. Moreover,

$$P\exp(Pt) = P \sum_{k=0}^\infty (Pt)^k / k! = \left(\sum_{k=0}^\infty (Pt)^k / k! \right) P = \exp(Pt)P.$$

Exercise 2.12
Let $\{v_1, \ldots, v_K\}$ be a basis of V. Since V is closed under complex conjugation, it follows that $\{\bar{v}_1, \ldots, \bar{v}_K\}$ is also a basis of V. Therefore,

$$V = \mathrm{span}\{v_1, \ldots, v_K, \bar{v}_1, \ldots, \bar{v}_K\} = \mathrm{span}\{\mathrm{Re}\, v_1, \ldots, \mathrm{Re}\, v_K, \mathrm{Im}\, v_1, \ldots, \mathrm{Im}\, v_K\}$$

and so the family $\{\mathrm{Re}\, v_1, \ldots, \mathrm{Re}\, v_K, \mathrm{Im}\, v_1, \ldots, \mathrm{Im}\, v_K\}$ of vectors in $\mathbb{R}^N$ contains a basis.

Exercise 2.14

Let S_{ih} denote the set of all solutions of $\dot{x}(t) = A(t)x(t) + b(t)$ and let $y \in S_{\text{ih}}$. Assume $z \in S_{\text{ih}}$ and write $x := z - y$. Then $\dot{x}(t) = \dot{z}(t) - \dot{y}(t) = A(t)z(t) + b(t) - A(t)y(t) - b(t) = A(t)(z(t) - y(t)) = A(t)x(t)$ at every $t \in J$ which is not a point of discontinuity of A or b. Therefore, $x \in S_{\text{hom}}$ and so $z \in y + S_{\text{hom}}$. This establishes the inclusion $S_{\text{ih}} \subset y + S_{\text{hom}}$. To establish the reverse inclusion, assume $z \in y + S_{\text{hom}}$. Then $z = y + x$ for some $x \in S_{\text{hom}}$ and so $\dot{z}(t) = A(t)y(t) + b(t) + A(t)x(t) = A(t)z(t) + b(t)$ at every $t \in J$ which is not a point of discontinuity of A or b. Therefore, $z \in S_{\text{ih}}$.

Exercise 2.16

Note initially that, since $\Phi(p, 0)$ is invertible, $0 \notin \sigma(\Phi(p, 0))$ and so, for the function $f \colon z \mapsto z^n$, we have $f'(\mu) \neq 0$ for all $\mu \in \sigma(\Phi(p, 0))$. Therefore, by the spectral mapping theorem (Theorem 2.19),

$$\ker(\Phi^n(p, 0) - I) = \ker(\Phi(p, 0) - \lambda I).$$

(a) Let $x \colon \mathbb{R} \to \mathbb{F}^N$ be a non-zero solution (and so, in particular, $x(0) \neq 0$). Assume $x(0) \in \ker(\Phi(p, 0) - \lambda I)$. Then $\Phi^n(p, 0)x(0) = \lambda^n x(0) = x(0)$ and so, invoking (2.32), we have, for all $t \in \mathbb{R}$,

$$x(t + np) = \Phi(t + np, 0)x(0) = \Phi(t, 0)\Phi^n(p, 0)x(0) = \Phi(t, 0)x(0) = x(t).$$

Therefore, x is np-periodic.
Conversely, assume that x is np-periodic. Then $x(np) = \Phi(np, 0)x(0) = x(0)$. By (2.32), we have $\Phi(np, 0) = \Phi^n(p, 0)$. Therefore, $(\Phi^n(p, 0) - I)x(0) = 0$ and so $x(0) \in \ker(\Phi^n(p, 0) - I) = \ker(\Phi(p, 0) - \lambda I)$.

(b) That S_{np} is a vector space is clear. Let $\mathcal{B}$ be a basis of $\ker(\Phi(p, 0) - I)$. For $z \in \mathcal{B}$, let x_z denote the np-periodic solution $t \mapsto \Phi(t, 0)z$. By part (a), the set $\{x_z \colon z \in \mathcal{B}\}$ is a basis for S_{np}, whence the result.

Exercise 2.17

Sufficiency. Assume that λ is an eigenvalue of $\Phi(p, 0)$ and $\lambda^n = \mu$. Let $v \in \mathbb{C}^N$ be an associated eigenvector and so $\Phi^n(p, 0)v = \lambda^n v = \mu v$. Define x by $x(t) := \Phi(t, 0)v$ for all $t \in \mathbb{R}$. Invoking (2.32), with $\tau = 0$, gives

$$x(t + np) = \Phi(t + np, 0)v = \Phi(t, 0)\Phi^n(p, 0)v = \mu\Phi(t, 0)v = \mu x(t) \quad \forall t \in \mathbb{R}.$$

Necessity. Assume that x is a non-zero solution of (2.30), with the property $x(t + np) = \mu x(t)$ for all $t \in \mathbb{R}$. Write $v := x(0) \neq 0$. Invoking (2.32), with $\tau = 0$, we have

$$\mu\Phi(t, 0)v = \mu x(t) = x(t + np) = \Phi(t + np, 0)v = \Phi(t, 0)\Phi^n(p, 0)v,$$

and thus, $\Phi(t, 0)(\Phi^n(p, 0) - \mu I)v = 0$. Consequently $(\Phi^n(p, 0) - \mu I)v = 0$ and so μ is an eigenvalue of $\Phi^n(p, 0)$. By Theorem 2.19 (with $f(z) = z^n$),

$$\sigma(\Phi^n(p, 0)) = \{\lambda^n \colon \lambda \in \sigma(\Phi(p, 0))\}.$$

Therefore, $\Phi(p, 0)$ has an eigenvalue λ with the property that $\lambda^n = \mu$.

Exercise 2.19

Necessity. Assume that (2.33) has a p-periodic solution x. Write $\xi := x(0)$. Then

$$\xi = x(p) = \Phi(p, 0)\xi + \int_0^p \Phi(p, s)b(s)\mathrm{d}s = \Phi(p, 0)\xi + \eta$$

whence $\eta = \big(I - \Phi(p,0)\big)\xi$ and so $\eta \in \mathrm{im}\,\big(I - \Phi(p,0)\big)$.

Sufficiency. Assume that $\eta \in \mathrm{im}\,\big(I - \Phi(p,0)\big)$. Let $\xi \in \mathbb{F}^N$ be such that $\eta = \big(I - \Phi(p,0)\big)\xi$ and define $x\colon \mathbb{R} \to \mathbb{F}^N$ by

$$x(t) = \Phi(t,0)\xi + \int_0^t \Phi(t,s)b(s)\mathrm{d}s \quad \forall\, t \in \mathbb{R}.$$

Clearly, x is a solution of (2.33). We will show that x is p-periodic. Invoking (2.31), (2.32) and periodicity of b, we have $\Phi(t+p,s) = \Phi(t,0)\Phi(p,s)$ and $\Phi(t+p,s+p)b(s+p) = \Phi(t,s)b(s)$ for all $t,s \in \mathbb{R}$. Therefore,

$$
\begin{aligned}
x(t+p) &= \Phi(t+p,0)\xi + \int_0^{t+p} \Phi(t+p,s)b(s)\mathrm{d}s \\
&= \Phi(t,0)\left(\Phi(p,0)\xi + \int_0^p \Phi(p,s)b(s)\mathrm{d}s\right) + \int_p^{t+p} \Phi(t+p,s)b(s)\mathrm{d}s \\
&= \Phi(t,0)\big(\Phi(p,0)\xi + \eta\big) + \int_0^t \Phi(t+p,s+p)b(s+p)\mathrm{d}s \\
&= \Phi(t,0)\xi + \int_0^t \Phi(t,s)b(s)\mathrm{d}s = x(t) \quad \forall\, t \in \mathbb{R}
\end{aligned}
$$

and so x is p-periodic.

Exercise 2.23

$$\int_0^{2\pi} \mathrm{tr}\,A(s)\mathrm{d}s = \int_0^{2\pi} \big(2 + \sin s - \cos s\big)\mathrm{d}s = 4\pi > 0.$$

An application of Corollary 2.33 (with $p = 2\pi$) shows that there exists a solution which is unbounded on $\mathbb{R}_+$.

Exercise 2.25

The identity (2.49) clearly holds for $k = 1$. Assume that (2.49) holds for some $k \in \mathbb{N}$. Then

$$
\begin{aligned}
X^{k+1} - Y^{k+1} &= (X+Y)(X^k - Y^k) + XY^k - YX^k = X(X^k - Y^k) + (X-Y)Y^k \\
&= (X-Y)Y^k + \sum_{j=1}^k X^{k+1-j}(X-Y)Y^{j-1} = \sum_{j=1}^{k+1} X^{k+1-j}(X-Y)Y^{j-1}
\end{aligned}
$$

and so the identity holds for $k+1$. The result follows by induction.

Chapter 3

Exercise 3.3

For notational convenience, write $\alpha := -(M+m)g/Ml$ and $\beta = mg/M$. The reachability matrix is

$$
\mathcal{C}(A,B) := (B, AB, A^2B, A^3B) = \frac{1}{Ml}\begin{pmatrix} 0 & -1 & 0 & \alpha \\ -1 & 0 & \alpha & 0 \\ 0 & l & 0 & \beta \\ l & 0 & \beta & 0 \end{pmatrix}
$$

with determinant $(\alpha l + \beta)^2/(Ml)^4 = 1/(Ml)^4 > 0$. Therefore, $\operatorname{rk} \mathcal{C}(A, B) = 4 = N$ and so the system is controllable.

Exercise 3.4

By Proposition 3.8, $\operatorname{im} \mathcal{C}(A, B)$ is A-invariant. It immediately follows that $\operatorname{im} \mathcal{C}(A, B)$ is A^k-invariant for all $k \in \mathbb{N}$. Let $v \in \operatorname{im} \mathcal{C}(A, B)$ and $t \in \mathbb{R}$. Then $(t^k/k!)A^k v \in \operatorname{im} \mathcal{C}(A, B)$ for all $k \in \mathbb{N}$. Since $\operatorname{im} \mathcal{C}(A, B)$ is a subspace of $\mathbb{R}^N$, $\operatorname{im} \mathcal{C}(A, B)$ is closed and so $\exp(At)v = \sum_{k=0}^{\infty}(t^k/k!)A^k v$ is in $\operatorname{im} \mathcal{C}(A, B)$.

Exercise 3.7

Recall that $\tilde{A} = \begin{pmatrix} A_1 & A_2 \\ 0 & A_3 \end{pmatrix}$ and $\tilde{B} = \begin{pmatrix} B_1 \\ 0 \end{pmatrix}$. Hence, $\tilde{A}^k \tilde{B} = \begin{pmatrix} A_1^k B_1 \\ 0 \end{pmatrix}$ for all $k \in \mathbb{N}$.

By the Cayley-Hamilton theorem,

$$\operatorname{rk} \mathcal{C}(A_1, B_1) = \operatorname{rk}(B_1, A_1 B_1, \ldots, A_1^{K-1} B_1) = \operatorname{rk}(B_1, A_1 B_1, \ldots, A_1^{N-1} B_1),$$

and thus,

$$\operatorname{rk} \mathcal{C}(A_1, B_1) = \operatorname{rk}\left(\begin{pmatrix} B_1 \\ 0 \end{pmatrix}, \begin{pmatrix} A_1 B_1 \\ 0 \end{pmatrix}, \ldots, \begin{pmatrix} A_1^{N-1} B_1 \\ 0 \end{pmatrix}\right)$$

$$= \operatorname{rk}(\tilde{B}, \tilde{A}\tilde{B}, \ldots, \tilde{A}^{N-1}\tilde{B}) = \operatorname{rk}(S^{-1}(B, AB, \ldots, A^{N-1} B))$$

$$= \operatorname{rk}(B, AB, \ldots, A^{N-1} B) = \operatorname{rk} \mathcal{C}(A, B) = K.$$

Therefore, by Theorem 3.6, (A_1, B_1) is controllable.

Exercise 3.10

$$\begin{pmatrix} sI - A \\ C \end{pmatrix} = \begin{pmatrix} s & -1 & 0 & 0 \\ -3\omega^2 & s & 0 & -2\omega \\ 0 & 0 & s & -1 \\ 0 & 2\omega & 0 & s \\ 1 & 0 & 0 & 0 \\ 0 & 0 & 1 & 0 \end{pmatrix}$$

Rows 1, 3, 4 and 5 are linearly independent for all $s \in \mathbb{C}$. Therefore, the system is observable.

Assume that only the radial measurement y_1 is available, in which case C is replaced by its first row $C_1 = (1\ 0\ 0\ 0)$. Then we have

$$\begin{pmatrix} sI - A \\ C_1 \end{pmatrix} = \begin{pmatrix} s & -1 & 0 & 0 \\ -3\omega^2 & s & 0 & -2\omega \\ 0 & 0 & s & -1 \\ 0 & 2\omega & 0 & s \\ 1 & 0 & 0 & 0 \end{pmatrix}$$

Noting column 3, it is clear that this matrix fails to have full rank for $s = 0$. Therefore, the system with radial measurement only is not observable.

Now, assume that only the angular measurement y_2 is available, in which case C is replaced by its second row $C_2 = (0\ 0\ 1\ 0)$. Then we have

$$\begin{pmatrix} sI - A \\ C_2 \end{pmatrix} = \begin{pmatrix} s & -1 & 0 & 0 \\ -3\omega^2 & s & 0 & -2\omega \\ 0 & 0 & s & -1 \\ 0 & 2\omega & 0 & s \\ 0 & 0 & 1 & 0 \end{pmatrix}$$

If $s \neq 0$, then it is readily verified that rows 1, 3, 4 and 5 are linearly independent, whilst, if $s = 0$, then rows 1, 2, 3 and 5 are linearly independent. Therefore, the system with angular measurement only is observable.

Exercise 3.11
Noting that $\left(\mathcal{O}(C, A)\right)^* = \mathcal{C}(A^*, C^*)$ and applying the Kalman controllability decomposition lemma (Lemma 3.10) to the pair (A^*, C^*), we may infer the existence of $T \in GL(N, \mathbb{R})$ such that

$$T^{-1}A^*T = \begin{pmatrix} M_1 & M_2 \\ 0 & M_3 \end{pmatrix}, \quad T^{-1}C^* = \begin{pmatrix} M_4 \\ 0 \end{pmatrix}$$

with $M_1 \in \mathbb{R}^{K \times K}$, $M_4 \in \mathbb{R}^{K \times P}$ and (M_1, M_4) a controllable pair.
Writing $S := (T^*)^{-1}$, $A_1 := M_1^*$, $A_2 := M_2^*$, $A_3 := M_3^*$ and $C_1 := M_4^*$, we have

$$S^{-1}AS = \tilde{A} = \left(T^{-1}A^*T\right)^* = \begin{pmatrix} A_1 & 0 \\ A_2 & A_3 \end{pmatrix}, \quad CS = \left(T^{-1}C^*\right)^* = (C_1, 0)$$

and, by controllability of (M_1, M_4), we have observability of $(M_4^*, M_1^*) = (C_1, A_1)$.

Exercise 3.12
(a) By the Kalman controllability decomposition lemma (Lemma 3.10), there exists $S \in GL(N, \mathbb{R})$ such that the matrices $\tilde{A}$, $\tilde{B}$ and $\tilde{C}$ have the requisite structure and the pair (A_1, B_1) is controllable. It remains to show that (C_1, A_1) is observable. Suppose otherwise. Then there exists $v \neq 0$ such that $C_1 A_1^k v = 0$ for all $k \in \mathbb{N}_0$. A straightforward computation shows that $\tilde{C}\tilde{A}^k$ has the structure $\tilde{C}\tilde{A}^k = (C_1 A_1^k, *)$. Writing $\tilde{v} := \begin{pmatrix} v \\ 0 \end{pmatrix} \neq 0$, we have

$$\tilde{C}\tilde{A}^k\tilde{v} = C_1 A_1^k v = 0 \quad \forall k \in \mathbb{N}_0$$

which contradicts observability of the pair $(\tilde{C}, \tilde{A})$.

(b) By the Kalman observability decomposition lemma (Lemma 3.22), there exists $S \in GL(N, \mathbb{R})$ such that the matrices $\tilde{A}$, $\tilde{B}$ and $\tilde{C}$ have the requisite structure and the pair (C_1, A_1) is observable. It remains to show that (A_1, B_1) is controllable. Suppose otherwise. Then there exists $v \neq 0$ such that $v^* B_1 A_1^k = 0$ for all $k \in \mathbb{N}_0$. A straightforward computation shows that $\tilde{A}^k\tilde{B}$ has the structure $\tilde{A}^k\tilde{B} = \begin{pmatrix} A_1^k B_1 \\ * \end{pmatrix}$. Writing $\tilde{v} := \begin{pmatrix} v \\ 0 \end{pmatrix} \neq 0$, we have

$$\tilde{v}^*\tilde{A}^k\tilde{B} = v^* A_1^k B_1 = 0 \quad \forall k \in \mathbb{N}_0$$

which contradicts controllability of the pair $(\tilde{A}, \tilde{B})$.

Exercise 3.15
Obviously, the transfer function $\hat{G}_K$ is given by $\hat{G}_K(s) = C(sI - (A - BKC))^{-1}B$.
Now $sI - (A - BKC) = (sI - A)(I + (sI - A)^{-1}BKC) = (I + BKC(sI - A)^{-1})(sI - A)$, and so

$$\hat{G}_K(s) = C(sI - A)^{-1}\left(I + BKC(sI - A)^{-1}\right)^{-1}B$$
$$= C\left(I + (sI - A)^{-1}BKC\right)^{-1}(sI - A)^{-1}B.$$

Therefore,

$$
\begin{aligned}
\hat{G}_K(s)(I + K\hat{G}(s)) &= C(sI - A)^{-1}\big(I + BKC(sI - A)^{-1}\big)^{-1}B\big(I + KC(sI - A)^{-1}B\big) \\
&= C(sI - A)^{-1}\big(I + BKC(sI - A)^{-1}\big)^{-1}\big(I + BKC(sI - A)^{-1}\big)B \\
&= C(sI - A)^{-1}B = \hat{G}(s)
\end{aligned}
$$

and

$$
\begin{aligned}
(I + \hat{G}(s)K)\hat{G}_K(s) &= \big(I + C(sI - A)^{-1}BK\big)C\big(I + (sI - A)^{-1}BKC\big)^{-1}(sI - A)^{-1}B \\
&= C\big(I + (sI - A)^{-1}BKC\big)\big(I + (sI - A)^{-1}BKC\big)^{-1}(sI - A)^{-1}B \\
&= C(sI - A)^{-1}B = \hat{G}(s).
\end{aligned}
$$

Exercise 3.17
(a) Since $R(s)$ is not identically equal to the zero matrix, it follows that $B \neq 0$ and $C \neq 0$.
First consider the case that (A, B) is controllable and (C, A) is observable. Then there is nothing to show: the claim follows with $T = I$. (To identify the triple (A, B, C) with the block structure given in Exercise 3.17, in the latter simply disregard the last two block rows and the last two block columns in $\tilde{A}$ and the last two blocks in $\tilde{B}$ and $\tilde{C}$.)
Now consider the case wherein (C, A) is observable and (A, B) is not controllable. By the result in part (a) of Exercise 3.12, there exists $T \in GL(N, \mathbb{R})$ such that

$$
T^{-1}AT = \begin{pmatrix} A_1 & A_2 \\ 0 & A_3 \end{pmatrix}, \quad T^{-1}B = \begin{pmatrix} B_1 \\ 0 \end{pmatrix}, \quad CT = (C_1, C_2),
$$

with (A_1, B_1) controllable and (C_1, A_1) observable, proving the claim in this case.(To identify the above structure with the block structure given in Exercise 3.17, in the latter simply disregard the third block row and third block column in $\tilde{A}$ and the third blocks in $\tilde{B}$ and $\tilde{C}$.)
Finally, consider the case wherein (C, A) is not observable. By the observability decomposition lemma (Lemma 3.22), there exists $S \in GL(N, \mathbb{R})$ such that

$$
S^{-1}AS = \begin{pmatrix} A_1 & 0 \\ A_2 & A_3 \end{pmatrix}, \quad S^{-1}B = \begin{pmatrix} B_1 \\ B_2 \end{pmatrix}, \quad CS = (C_1, 0),
$$

with (C_1, A_1) observable. If the pair (A_1, B_1) is controllable, then the claim follows with $T = S$. If (A_1, B_1) is not controllable, then, by the result in part (a) of Exercise 3.12 applied in the context of the triple (A_1, B_1, C_1), there exists an invertible matrix S_1 such that

$$
S_1^{-1}A_1S_1 = \begin{pmatrix} A_{11} & A_{12} \\ 0 & A_{22} \end{pmatrix}, \quad S_1^{-1}B_1 = \begin{pmatrix} B_{11} \\ 0 \end{pmatrix}, \quad C_1S_1 = (C_{11}, C_{12})
$$

where (A_{11}, B_{11}) is controllable and (C_{11}, A_{11}) is observable. Defining

$$
T := S\tilde{S}, \quad \text{where } \tilde{S} := \begin{pmatrix} S_1 & 0 \\ 0 & I \end{pmatrix},
$$

and setting $(A_{31}, A_{32}) = A_2 S_1$, $A_{33} = A_3$ and $B_{31} = B_2$, we have

$$T^{-1}AT = \begin{pmatrix} S_1^{-1} & 0 \\ 0 & I \end{pmatrix} \begin{pmatrix} A_1 & 0 \\ A_2 & A_3 \end{pmatrix} \begin{pmatrix} S_1 & 0 \\ 0 & I \end{pmatrix} = \begin{pmatrix} A_{11} & A_{12} & 0 \\ 0 & A_{22} & 0 \\ A_{31} & A_{32} & A_{33} \end{pmatrix}$$

and

$$T^{-1}B = \begin{pmatrix} S_1^{-1} & 0 \\ 0 & I \end{pmatrix} \begin{pmatrix} B_1 \\ B_2 \end{pmatrix} = \begin{pmatrix} B_{11} \\ 0 \\ B_{31} \end{pmatrix}, \quad CT = (C_1, 0) \begin{pmatrix} S_1 & 0 \\ 0 & I \end{pmatrix} = (C_{11}, C_{12}, 0)$$

with (A_{11}, B_{11}) controllable and (C_{11}, A_{11}) observable.

(b) A straightforward calculation reveals that

$$CA^k B = (CT)(T^{-1}A^k T)(T^{-1}B) = C_{11}A_{11}^k B_{11} \quad \forall k \in \mathbb{N}_0.$$

Therefore,

$$C \exp(At)B = C_{11} \exp(A_{11}t)B_{11} \quad \forall t \in \mathbb{R},$$

and applying Laplace transform gives

$$R(s) = C(sI - A)^{-1} B = C_{11}(sI - A_{11})^{-1} B_{11}.$$

Therefore, (A_{11}, B_{11}, C_{11}) is a realization of R.

Exercise 3.19
(a) The claim follows immediately from the relations

$$\dot{x}_1 = A_1 x_1 + B_1 C_2 x_2, \quad \dot{x}_2 = A_2 x_2 + B_2 u, \quad y = C_1 x_1.$$

(b) Note that the inverse of

$$sI - A = \begin{pmatrix} sI - A_1 & -B_1 C_2 \\ 0 & sI - A_2 \end{pmatrix}$$

is given by

$$(sI - A)^{-1} = \begin{pmatrix} (sI - A_1)^{-1} & (sI - A_1)^{-1} B_1 C_2 (sI - A_2)^{-1} \\ 0 & (sI - A_2)^{-1} \end{pmatrix}.$$

Therefore,

$$\begin{aligned}
\hat{G}(s) &= C(sI - A)^{-1} B \\
&= (C_1, 0) \begin{pmatrix} (sI - A_1)^{-1} & (sI - A_1)^{-1} B_1 C_2 (sI - A_2)^{-1} \\ 0 & (sI - A_2)^{-1} \end{pmatrix} \begin{pmatrix} 0 \\ B_2 \end{pmatrix} \\
&= C_1 (sI - A_1)^{-1} B_1 C_2 (sI - A_2)^{-1} B_2 = \hat{G}_1(s)\hat{G}_2(s).
\end{aligned}$$

(c) For $j = 1, 2$, write $\hat{G}_j = n_j/d_j$, where n_j and d_j are coprime polynomials. It follows from Proposition 3.29 and Theorem 3.30 that the degree of d_j is equal to N_j. Moreover, note that the dimension of the realization (A, B, C) of $\hat{G}_1\hat{G}_2$ is equal to $N_1 + N_2$ and $\hat{G}_1\hat{G}_2 = n_1 n_2/(d_1 d_2)$.
If the realization (A, B, C) is minimal, then, by Proposition 3.29, n_1 and d_2 are coprime and, furthermore, n_2 and d_1 are coprime, or, equivalently, there is no pole/zero

cancellation in the product $\hat{G}_1\hat{G}_2$.

Conversely, assume that there is no pole/zero cancellation in the product $\hat{G}_1\hat{G}_2$. Then, the polynomials n_1n_2 and d_1d_2 are coprime. Since the degree of d_1d_2 is equal to $N_1 + N_2$, another application of Proposition 3.29 shows that the realization (A, B, C) is minimal.

Chapter 4

Exercise 4.1
Let $a, b \in I_z$ be arbitrary and, without loss of generality, assume $a \leq b$. To conclude that I_z is an interval it suffices to show that $[a, b] \subset I_z$. Since $I_z := \cup_{y \in T} I_y$, there exist $y_a, y_b \in T$ such that $a \in I_{y_a}$ and $b \in I_{y_b}$. Since T is totally ordered, either $y_a \preceq y_b$ or $y_b \preceq y_a$. In the former case, $I_{y_a} \subset I_{y_b}$ and so $[a, b] \subset I_{y_b} \subset I_z$. In the latter case, $I_{y_b} \subset I_{y_a}$ and so $[a, b] \subset I_{y_a} \subset I_z$.

We proceed to show that z is well defined. Let $t \in I_z$ be arbitrary. Then $t \in I_y$ for some $y \in T$. Define $v := y(t)$. Assume $\hat{y} \in T$ is such that $t \in I_{\hat{y}}$ and define $\hat{v} := \hat{y}(t)$. Since T is totally ordered, either $y \preceq \hat{y}$ or $\hat{y} \preceq y$. In each case, $y(t) = \hat{y}(t)$. Therefore, with each $t \in I_z$, we may associate a unique element $z(t)$ of G given by $z(t) = y(t)$, where y is any element of T such that $t \in I_y$. The function $z \colon I_z \to G$, so defined, has the property

$$z|_{I_y} = y \ \forall y \in T$$

and is the only function with that property.

Exercise 4.3
For $\xi \neq 0$, separation of variables gives

$$\int_\xi^{x(t)} \frac{ds}{s^2} = \int_\tau^t s^3 ds \quad \Longrightarrow \quad \left[-\frac{1}{s}\right]_\xi^{x(t)} = \left[\frac{1}{4}s^4\right]_\tau^t \quad \Longrightarrow \quad \frac{1}{x(t)} - \frac{1}{\xi} = \frac{1}{4}(\tau^4 - t^4).$$

Consequently,

$$\frac{1}{x(t)} = \frac{4 + \xi(\tau^4 - t^4)}{4\xi} \quad \Longrightarrow \quad x(t) = \frac{4\xi}{4 + \xi(\tau^4 - t^4)}.$$

(a) For $(\tau, \xi) \in \mathbb{R} \times (0, \infty)$, the maximal interval of existence is bounded and is given by

$$\left(-(\tau^4 + 4/\xi)^{1/4}, (\tau^4 + 4/\xi)^{1/4}\right).$$

(b) For (τ, ξ) such that $\xi \in (-4/\tau^4, 0)$, the maximal interval of existence is $\mathbb{R}$.

Exercise 4.4
Seeking a contradiction, suppose that $f(x^\infty) \neq 0$. Setting $\lambda := f(x^\infty)$, it follows that λ has at least one component, λ_j say, which is not equal to zero: $\lambda_j \neq 0$. Since

$$\lim_{t \to \infty} \dot{x}(t) = \lim_{t \to \infty} f(x(t)) = \lambda,$$

we have $\lim_{t \to \infty} \dot{x}_j(t) = \lambda_j \neq 0$. Hence

$$\lim_{t \to \infty} x_j(t) = \begin{cases} \infty, & \text{if } \lambda_j > 0 \\ -\infty, & \text{if } \lambda_j < 0, \end{cases}$$

contradicting the assumption that $\lim_{t \to \infty} x(t) = x^\infty$.

Exercise 4.6
Observe that, if $\{\inf I, \sup I\} \cap (J\backslash I) \neq \emptyset$, then $J\backslash I \neq \emptyset$ and so $I \neq J$. Conversely, assume $I \neq J$. Then, $J\backslash I \neq \emptyset$ and so there exists $\gamma \in J$ with $\gamma \notin I$. Write $\alpha := \inf I$ and $\omega := \sup I$. Then, either (i) $\gamma \geq \omega$ or (ii) $\gamma \leq \alpha$. If (i) holds with $\gamma = \omega$, then $\omega \notin I$ and so $\omega \in J\backslash I$. If (i) holds with $\gamma > \omega$, then, since I is relatively open in J, we again have $\omega \notin I$ and so $\omega \in J\backslash I$. If (ii) holds, then analogous reasoning shows that $\alpha \in J\backslash I$. Therefore, $\{\alpha, \omega\} \cap (J\backslash I) \neq \emptyset$.

Exercise 4.7
Let $x, y \in \mathbb{R}^N$. Let $\varepsilon > 0$ be arbitrary. Then there exists $v \in V$ such that $\mathrm{dist}(y, V) \geq \|y - v\| - \varepsilon$. Therefore,

$$\mathrm{dist}(x, V) \leq \|x - v\| \leq \|x - y\| + \|y - v\| \leq \|x - y\| + \mathrm{dist}(y, V) + \varepsilon$$

and so, since $\varepsilon > 0$ is arbitrary, $\mathrm{dist}(x, V) - \mathrm{dist}(y, V) \leq \|x - y\|$. Repeating this argument, with the roles of x and y interchanged, yields the second requisite inequality

$$\mathrm{dist}(y, V) - \mathrm{dist}(x, V) \leq \|x - y\|.$$

Exercise 4.8
(a) Let $\tau \in I$. Then, by the variation of parameters formula,

$$x(t) = e^{A(t-\tau)}x(\tau) + \int_\tau^t e^{A(t-s)}b(s, x(s))\mathrm{d}s, \quad \forall\, t \in I.$$

Let $\alpha, \beta \in \mathbb{R}$ be such that $\tau \in (\alpha, \beta) \subset I$ (since $\tau \in I$ and I, as a maximal interval of existence, is open, such α and β exist). Then, setting

$$K := \max\{\|e^{A\sigma}\| : \alpha - \beta \leq \sigma \leq \beta - \alpha\} < \infty,$$

we obtain

$$\|x(t)\| \leq \|e^{A(t-\tau)}\|\|x(\tau)\| + \left|\int_\tau^t \|e^{A(t-s)}\|\|b(s, x(s))\|\mathrm{d}s\right|$$

$$\leq K\|x(\tau)\| + \left|\int_\tau^t K\gamma(s)\|x(s)\|\mathrm{d}s\right|, \quad \forall\, t \in (\alpha, \beta).$$

Setting $c := K\|x(\tau)\|$, an application of Gronwall's lemma yields

$$\|x(t)\| \leq c\exp\left(K\left|\int_\tau^t \gamma(s)\mathrm{d}s\right|\right) \leq c\exp\left(K\int_\alpha^\beta \gamma(s)\mathrm{d}s\right) < \infty, \quad \forall\, t \in (\alpha, \beta).$$

Setting $\alpha^* := \inf I$, $\beta^* := \sup I$, it follows that $\alpha^* = -\infty$ and $\beta^* = \infty$, because otherwise, if, for example, $\beta^* < \infty$, the above argument would apply with $\beta = \beta^*$ and so x would be bounded on (τ, β^*), which, by Theorem 4.11, is impossible.
(b) By the variation of parameters formula,

$$x(t) = e^{At}x(0) + \int_0^t e^{A(t-s)}b(s, x(s))\mathrm{d}s, \quad \forall\, t \geq 0,$$

and so

$$\|x(t)\| \leq Me^{\mu t}\|x(0)\| + \int_0^t Me^{\mu(t-s)}\gamma(s)\|x(s)\|\mathrm{d}s \quad \forall\, t \geq 0.$$

Therefore,

$$\|x(t)\|e^{-\mu t} \le M\|x(0)\| + \int_0^t M\gamma(s)\|x(s)\|e^{-\mu s}\mathrm{d}s \quad \forall t \ge 0.$$

By Gronwall's lemma,

$$\|x(t)\|e^{-\mu t} \le M\|x(0)\| \exp\left(M\int_0^t \gamma(s)\mathrm{d}s\right), \quad \forall t \ge 0,$$

and so,

$$\|x(t)\| \le M\|x(0)\| \exp\left(\mu t + M\int_0^t \gamma(s)\mathrm{d}s\right), \quad \forall t \ge 0. \tag{$*$}$$

(c) If $\mu < 0$ and there exists $T > 0$ such that

$$\sup_{t \ge T}\left(\frac{1}{t}\int_0^t \gamma(s)\mathrm{d}s\right) < \frac{|\mu|}{M}, \tag{$**$}$$

then

$$\mu t + M\int_0^t \gamma(s)ds = t\left(\mu + M\frac{1}{t}\int_0^t \gamma(s)ds\right) \to -\infty \quad \text{as} \quad t \to \infty,$$

and thus $x(t) \to 0$ as $t \to \infty$, by $(*)$. The existence of a number $T > 0$ such that $(**)$ holds, is guaranteed, for example, if the improper Riemann integral $\int_0^\infty \gamma(s)ds$ of γ converges or if $\sup_{t \ge t^*} \gamma(t) < |\mu|/M$ for some $t^* > 0$.

Exercise 4.9
Let $\varepsilon > 0$.
(a) For $z \in (0, \varepsilon)$,

$$\frac{|g(z) - g(0)|}{|z - 0|} = \frac{\sqrt{z}}{z} = \frac{1}{\sqrt{z}} \to \infty \quad \text{as} \quad z \downarrow 0.$$

It follows that the function g is not Lipschitz on $\mathbb{R}$.
(b) For $z \in (0, \varepsilon)$,

$$\frac{|g(z) - g(0)|}{|z - 0|} = \left|\frac{z \ln z}{z}\right| = |\ln z| \to \infty \quad \text{as} \quad z \downarrow 0.$$

It follows that g is not Lipschitz on $\mathbb{R}$.

Exercise 4.10
Let $z \in V$ and choose $\varepsilon > 0$ such that $U := \{w \in \mathbb{R}^Q : \|w - z\| \le \varepsilon\} \subset V$. It follows from the continuity of the first order partial derivatives of g and compactness of U that

$$\gamma := \max_{1 \le i, j \le N}\left(\sup_{w \in U}|\partial_i g_j(w)|\right) < \infty,$$

wherein $\partial_i g_j$ denotes the partial derivative of component j of g with respect to argument i.
Let $z_1, z_2 \in U$ and define $h_j : [0, 1] \to \mathbb{R}$ by

$$h_j(t) = g_j((1 - t)z_1 + tz_2), \quad \forall t \in [0, 1].$$

Note that

$$\dot{h}_j(t) = \langle(\nabla g_j)((1 - t)z_1 + tz_2), z_2 - z_1\rangle, \quad \forall t \in [0, 1],$$

and so, by the Cauchy-Schwarz inequality,

$$|\dot{h}_j(t)| \leq \|(\nabla g_j)((1-t)z_1 + tz_2)\|\|z_2 - z_1\| \lesssim \gamma\sqrt{Q}\|z_2 - z_1\|, \quad \forall t \in [0,1].$$

By the mean-value theorem of differentiation, there exists $\tau \in [0,1]$ such that

$$|g_j(z_2) - g_j(z_1)| = |h_j(1) - h_j(0)| = |\dot{h}_j(\tau)|.$$

Hence, $|g_j(z_2) - g_j(z_1)| \leq \gamma\sqrt{Q}\|z_2 - z_1\|$, and thus,

$$\|g(z_2) - g(z_1)\| \leq \gamma\sqrt{MQ}\|z_2 - z_1\|.$$

This holds for all $z_1, z_2 \in U$ and the claim follows.

Exercise 4.12
(a) Let $(t_0, z_0) \in J \times G$ be arbitrary. The hypotheses ensure that there exist neighbourhoods J_0 and $G_0 \subset G$ of t_0 and z_0, respectively, and a constant $L_2 \geq 0$ such that $J_0 \cap J$ and G_0 are compact, $C := \text{cl}\{(f_1(t), f_2(z)) : (t, z) \in (J_0 \cap J) \times G_0\} \subset D$ and

$$\|f_2(x) - f_2(y)\| \leq L_2\|x - y\| \quad \forall x, y \in G_0.$$

By compactness of $J_0 \cap J$ and G_0, piecewise continuity of f_1 and continuity of f_2, there exists $K > 0$ such that $\|(f_1(t), f_2(z))\| \leq K$ for all $(t, z) \in (J_0 \cap J) \times G_0$. Therefore, the set C is compact. By Corollary 4.16, there exists $L_3 \geq 0$ such that

$$\|f_3(s, u) - f_3(s, v)\| \leq L_3\|u - v\| \quad \forall (s, u), (s, v) \in C.$$

Defining $L := L_3 L_2$, we have, for all $(t, x), (t, y) \in (J_0 \cap J) \times G_0$,

$$\|f(t, x) - f(t, y)\| = \|f_3(f_1(t), f_2(x)) - f_3(f_1(t), f_2(y))\|$$
$$\leq L_3\|f_2(x) - f_2(y)\| \leq L\|x - y\|,$$

and so f is locally Lipschitz with respect to its second argument.
Finally, let $y : J \to G$ be continuous. By piecewise continuity of f_1 and continuity of f_2 and f_3, it immediately follows that the function $t \mapsto f(t, y(t)) = f_3(f_1(t), f_2(y(t)))$ is piecewise continuous. Therefore, f satisfies Assumption **A**.
(b) Let f be given by $f(t, z) := g(z) + k(t)h(z)$, where $g, h : \mathbb{R}^N \to \mathbb{R}^N$ are locally Lipschitz and $k : \mathbb{R} \to \mathbb{R}$ is piecewise continuous. Defining the piecewise continuous function $f_1 := k$, the locally Lipschitz function $f_2 := (g, h) : \mathbb{R}^N \to \mathbb{R}^{2N}$ and the continuous function $f_3 : \mathbb{R} \times \mathbb{R}^{2N} \to \mathbb{R}^N$ by $f_3(r, s) = f_3(r, (s_1, s_2)) := s_1 + rs_2$, we see that f_3 is locally Lipschitz in its second argument $s = (s_1, s_2)$ and $f(t, z) = f_3(f_1(t), f_2(z))$. By Proposition 4.20, f satisfies Assumption **A**.

Exercise 4.13
Let $(\tau, \xi), (\sigma, \eta), (\rho, \theta) \in J \times G$ be arbitrary. Since $\psi(\tau, \tau, \xi) = \xi$, it follows that $(\tau, \xi) \sim (\tau, \xi)$ and so the relation $\sim$ is reflexive. Next, assume $(\tau, \xi) \sim (\sigma, \eta)$ and so $\psi(\sigma, \tau, \xi) = \eta$. By Theorem 4.26, we have $I(\tau, \xi) = I(\sigma, \eta)$ and $\psi(\tau, \sigma, \eta) = \psi(\tau, \sigma, \psi(\sigma, \tau, \xi)) = \psi(\tau, \tau, \xi) = \xi$. Therefore, $(\sigma, \eta) \sim (\tau, \xi)$ and so the relation $\sim$ is symmetric. Finally, assume $(\tau, \xi) \sim (\sigma, \eta)$ and $(\sigma, \eta) \sim (\rho, \theta)$. Then $\psi(\sigma, \tau, \xi) = \eta$ and $\psi(\rho, \sigma, \eta) = \theta$. Hence,

$$\psi(\rho, \tau, \xi) = \psi(\rho, \sigma, \psi(\sigma, \tau, \xi)) = \psi(\rho, \sigma, \eta) = \theta$$

and so $(\tau, \xi) \sim (\rho, \theta)$. Therefore, the reflexive and symmetric relation $\sim$ is also transitive and so is an equivalence relation.
Let $\mathcal{G}$ denote the graph of the maximal solution $\psi(\cdot, \tau, \xi)$, that is, $\mathcal{G} := \{(t, \psi(t, \tau, \xi)) : t \in I(\tau, \xi)\}$. Observe that

$$(\tau, \xi) \sim (\sigma, \eta) \Leftrightarrow \psi(\sigma, \tau, \xi) = \eta \Leftrightarrow (\sigma, \eta) \in \mathcal{G}$$

and so the equivalence class of (τ, ξ) coincides with $\mathcal{G}$.

Exercise 4.16

For $(\xi_1, \xi_2) = \xi$, consider the initial-value problem $\dot{x} = f(x)$, $x(0) = \xi$. If $\xi = 0$, then it is clear that $\varphi(t, \xi) = \varphi(t, 0) = 0$ for all $t \in \mathbb{R}$. Assume $\xi \neq 0$. A straightforward calculation gives the polar form of the initial-value problem

$$\dot{r} = r(1 - r^2), \quad \dot{\theta} = -1, \quad (r(0), \theta(0)) = (\rho, \sigma),$$

where $\rho = \|\xi\|$, $\xi_1 = \rho \cos \sigma$, and $\xi_2 = \rho \sin \sigma$. Clearly, $\theta(t) = \sigma - t$. Assume $\rho = 1$, then, from the first of the differential equations, it is clear that $r(t) = 1$ for all t and so the solution of the original initial-value problem is given componentwise by

$$x_1(t) = \rho \cos(t - \sigma) = \rho \big(\cos \sigma \cos t + \sin \sigma \sin t \big) = \xi_1 \cos t + \xi_2 \sin t$$
$$x_2(t) = \rho \sin(\sigma - t) = \rho \big(\sin \sigma \cos t - \cos \sigma \sin t \big) = -\xi_1 \sin t + \xi_2 \cos t$$

for all $t \in \mathbb{R}$. Writing $R(t) = \begin{pmatrix} \cos t & \sin t \\ -\sin t & \cos t \end{pmatrix}$, it follows that

$$\varphi(t, \xi) = R(t)\xi \quad \forall t \in \mathbb{R}, \ \forall \xi \in \mathbb{R}^2, \ \|\xi\| = 1.$$

We proceed to resolve the cases of $\rho > 1$ and $\rho < 1$. Observe that, if $\rho > 1$, then $r(t) > 1$ for all t and, if $\rho < 1$, then $r(t) < 1$ for all t. Therefore, in each case $(1 - r(t))/(1 - \rho) > 0$ for all t. Separating variables in the differential equation for r, we have

$$t = \int_0^t ds = \int_\rho^{r(t)} \frac{ds}{s(1 - s^2)} = \int_\rho^{r(t)} \left(\frac{1}{s} + \frac{1}{2(1 - s)} - \frac{1}{2(1 + s)} \right) ds,$$

and so

$$t = \ln \frac{r(t)}{\rho} - \frac{1}{2} \ln \frac{1 - r(t)}{1 - \rho} - \frac{1}{2} \ln \frac{1 + r(t)}{1 + \rho} \quad \Longrightarrow \quad r(t) = \rho \big(\rho^2 + (1 - \rho^2) e^{-2t} \big)^{-1/2}.$$

Consequently,

$$\varphi(t, \xi) = \big(\|\xi\|^2 + (1 - \|\xi\|^2) e^{-2t} \big)^{-1/2} R(t)\xi \quad \forall t \in I_\xi.$$

If $\rho = \|\xi\| < 1$, then we may infer that $I_\xi = \mathbb{R}$. Furthermore, if $\rho = \|\xi\| > 1$, then the above expression for $\varphi(t, \xi)$ has a singularity: the maximal interval of existence of the solution is given by $I_\xi = (\alpha_\xi, \infty)$ with

$$\alpha_\xi := -\ln \big(\|\xi\| / \sqrt{\|\xi\|^2 - 1} \big).$$

Assembling the four cases (viz. $\xi = 0$, $0 < \|\xi\| < 1$, $\|\xi\| = 1$ and $\|\xi\| > 1$) treated above, we may infer that the local flow $\varphi \colon D \to \mathbb{R}^2$ has domain

$$D := \{ (t, \xi) \in \mathbb{R} \times \mathbb{R}^2 \colon \|\xi\|^2 + (1 - \|\xi\|^2) e^{-2t} > 0 \}$$

and is given by

$$\varphi(t, \xi) := \big(\|\xi\|^2 + (1 - \|\xi\|^2) \big)^{-1/2} R(t)\xi \quad \forall (t, \xi) \in D.$$

Exercise 4.18

Let $\xi, \eta, \theta \in G$ be arbitrary. Since $\xi \in O(\xi)$, it follows that $\xi \sim \xi$ and so the relation $\sim$

is reflexive. Next, assume $\xi \sim \eta$ and so $\varphi(\tau, \xi) = \eta$ for some $\tau \in I_\xi$. Invoking Theorem 4.35, we have $-\tau \in I_\xi - \tau = I_\eta$ and $\varphi(-\tau, \eta) = \varphi(-\tau, \varphi(\tau, \xi)) = \xi$ and so $\eta \sim \xi$. Therefore, the relation $\sim$ is symmetric. Assume $\xi \sim \eta$ and $\eta \sim \theta$. Then $\varphi(\tau, \xi) = \eta$ for some $\tau \in I_\xi$ and $\varphi(\sigma, \eta) = \theta$ for some $\sigma \in I_\eta = I_\xi - \tau$. Then $\sigma + \tau \in I_\xi$ and $\varphi(\sigma + \tau, \xi) = \varphi(\sigma, \varphi(\tau, \xi)) = \varphi(\sigma, \eta) = \theta$ and so $\xi \sim \theta$. Therefore, the reflexive and symmetric relation $\sim$ is also transitive and so is an equivalence relation. Finally, observe that

$$\xi \sim \eta \iff (\exists \tau \in I_\xi \colon \varphi(\tau, \xi) = \eta) \iff \eta \in O(\xi),$$

and so the equivalence class of ξ coincides with $O(\xi)$.

Exercise 4.20
Let $\xi \in G$. Since the hypotheses of Theorem 4.38 hold, $\Omega(\xi)$ is non-empty, compact and is approached by $\varphi(t, \xi)$ as $t \to \infty$. Assume that $S \subset \mathbb{R}^N$ is non-empty and closed, and is approached by $\varphi(t, \xi)$ as $t \to \infty$. Seeking a contradiction, suppose that $\Omega(\xi) \not\subset S$. Then there exists $z \in \Omega(\xi)$ with $z \notin S$. Since S is closed, it follows that $\varepsilon := \mathrm{dist}(z, S) > 0$. Since $z \in \Omega(\xi)$, there exists (t_n), with $t_n \to \infty$ as $n \to \infty$, such that $\varphi(t_n, \xi) \to z$ as $n \to \infty$. By continuity of the map $u \mapsto \mathrm{dist}(u, S)$ (recall Exercise 4.7), we have $\mathrm{dist}(\varphi(t_n, \xi), S) \geq \varepsilon/2$ for all sufficiently large n. This contradicts the fact that $\mathrm{dist}(\varphi(t, \xi), S) \to 0$ as $t \to \infty$. Therefore, $\Omega(\xi) \subset S$.

Exercise 4.22
(a) By the fundamental theorem of calculus, D is continuously differentiable. Moreover,

$$D(-u) = \int_0^{-u} d(v)\mathrm{d}v = -\int_0^u d(v)\mathrm{d}v = -D(u) \quad \forall u \in \mathbb{R}$$

and so D is an odd function. Since $D'(0) = d(0) < 0$, there exists $\varepsilon > 0$ such that $D(u) < 0$ for all $u \in (0, \varepsilon)$ and, since $D(u) \to \infty$ as $u \to \infty$, there exists $E > \varepsilon$ such that $D(u) > 0$ for all $u > E$. By continuity of D, the set $Z := \{u \in [\varepsilon, E]\colon D(u) = 0\}$ is non-empty, and the requisite properties hold for $a := \inf Z$ and $b := \sup Z$.
(b) By direct calculation we obtain $\dot{x}_1(t) = \dot{y}(t) = x_2(t) - D(y(t)) = x_2(t) - D(x_1(t))$ and $\dot{x}_2(t) = \ddot{y}(t) + d(y(t))\dot{y}(t) = -y(t) = -x_1(t)$.
(c) Note that, if $(z_1, z_2) = z \in \mathbb{R}^2$ is such that $\|z\| = a$, then $|z_1| \leq a$ and so $z_1 D(z_1) \leq 0$. Therefore,

$$\|z\| = a \implies \langle z, f(z) \rangle = z_1 z_2 - z_1 D(z_1) - z_1 z_2 = -z_1 D(z_1) \geq 0$$

and so the vector $f(z)$ does not point into the disc of radius a centred at 0 at any point z of its boundary. Therefore the exterior of the open disc of radius a centred at 0 is positively invariant under the (local) flow.
(d) For $(z_1, z_2) = (0, \gamma)$, we have $z_1^2 + 2z_2^2 = 2(b^2 + c^2 + 4m^2) = 2(c^2 + 4m^2 + 3b^2/2) - b^2 = 2r_1^2 - b^2$ and so $(0, \gamma) \in E_1 \subset \Gamma$. For $(z_1, z_2) = (0, -\gamma)$, we have $2z_1^2 + z_2^2 = b^2 + c^2 + 4m^2 = r_2^2 + b^2$ and so $(0, -\gamma) \in E_2 \subset \Gamma$.
(e) We first investigate the nature of f on $\Gamma = E_1 \cup C_1 \cup L \cup C_2 \cup E_2$. Let $z = (z_1, z_2) \in E_1$. Then the vector $n = (z_1, 2z_2)$ is an outward pointing normal to Γ^*. Moreover, $\langle n, f(z) \rangle = z_1 z_2 - z_1 D(z_1) - 2z_2 z_2 = -z_1 z_2 - z_1 D(z_1) \leq -|z_1|(|z_2| - m) \leq 0$. Now, let $z = (z_1, z_2) \in C_1 \cup C_2$. Then z is an outward pointing normal to Γ^* and $\langle z, f(z) \rangle = -z_1 D(z_1) \leq 0$. Next, let $z = (z_1, z_2) \in L$. Then the vector $n = (1, 0)$ is an outward pointing normal to Γ^* and $\langle n, f(z) \rangle = z_2 - D(z_1) = z_2 - D(c) \leq 0$. Finally, let $z = (z_1, z_2) \in E_2$. Then the vector $n = (2z_1, z_2)$ is an outward pointing normal to Γ^* and $\langle n, f(z) \rangle = 2z_1 z_2 - 2z_1 D(z_1) - z_1 z_2 \leq -|z_1|(|z_2| - m) \leq 0$. By symmetry, the above analysis may be extended to the entire closed curve Γ^* to conclude that, at all points $z \in \Gamma^*$ the vector $f(z)$ is not outward pointing. This fact, in conjunction

with the result in part (c), implies that the annular region $\mathcal{A}$ is positively invariant under the (local) flow. Moreover, $\mathcal{A}$ contains no equilibrium point. By the Poincaré-Bendixson theorem, we may infer the existence of a periodic solution $x = (x_1, x_2)$ of (4.38) with orbit in $\mathcal{A}$. Therefore, $y = x_1$ is a periodic solution of the Liénard equation (4.37).

(f) By part (e), the system (4.38) has a periodic solution (x_1, x_2) in $\mathcal{A}$. Since $\dot{x}_2 = -x_1$, there exists $\tau > 0$ such that $x_1(\tau) = 0$. Setting $v := x_2(\tau) \in [-\gamma, -a] \cup [a, \gamma]$, consider the solution y of the Liénard equation (4.37) satisfying $y(0) = 0$ and $\dot{y}(0) = v$. Then (y_1, y_2) given by

$$y_1 = y, \quad y_2 = \dot{y} + D(y),$$

solves system (4.38) and satisfies $y_1(0) = 0 = x_1(\tau)$ and $y_2(0) = v = x_2(\tau)$. Consequently, $(y_1(t), y_2(t)) = (x_1(t+\tau), x_2(t+\tau))$ for all $t \in \mathbb{R}$ and so, the function $y = y_1$ is periodic.

Chapter 5

Exercise 5.1
In this case, $G := (-1, \infty) \times (-1, \infty)$ and $f: G \to \mathbb{R}^2$ is given by

$$f(z) = f(z_1, z_2) := \big((z_1 + 1)z_2, \, -z_1(z_2 + 1)\big).$$

Set $U := \{(z_1, z_2) \in \mathbb{R}^2 : z_1^2 + z_2^2 < 1\}$ and define $V: U \to \mathbb{R}$ by

$$V(z) = V(z_1, z_2) := z_1 + z_2 - \ln(z_1 + 1) - \ln(z_2 + 1).$$

Clearly, $V(0) = 0$. Moreover, since $\ln(s + 1) < s$ for all $s \in (-1, 1) \setminus \{0\}$, we have $V(z) > 0$ for all $z \in U \setminus \{0\}$. Furthermore,

$$(\nabla V)(z) = (\nabla V)(z_1, z_2) = \left(\frac{z_1}{z_1 + 1}, \frac{z_2}{z_2 + 1}\right),$$

and so $V_f(z) = V_f(z_1, z_2) = z_1 z_2 - z_1 z_2 = 0$ for all $z \in U$. It now follows from Theorem 5.2 that the equilibrium 0 is stable.

Exercise 5.3
Define $f: \mathbb{R}^2 \to \mathbb{R}^2$ by $f(z) = f(z_1, z_2) := (z_2, b \sin z_1)$. Set $U := (-\pi, \pi) \times \mathbb{R}$ and define $V: U \to \mathbb{R}$ by $V(z) = V(z_1, z_2) := z_1 z_2$. Let $z = (z_1, z_2) \in U$ be such that $V(z) = z_1 z_2 > 0$. Then, $z_1 \neq 0$, $z_2 \neq 0$ and so $V_f(z) = z_2^2 + b z_1 \sin z_1 > 0$. Therefore, hypothesis (1) of Theorem 5.7 holds. Let $\delta > 0$ be arbitrary and set $\theta := \min\{\delta, \pi\}/2$. For $\xi := (\theta, \theta)$, we have $\xi \in U$, $\|\xi\| < \delta$ and $V(\xi) = \theta^2 > 0$. Therefore, hypothesis (2) of Theorem 5.7 also holds and so $(0, 0)$ is an unstable equilibrium.

Exercise 5.5
Let $\xi = (\xi_1, \xi_2) \in G = \mathbb{R}^2$ and write $\varphi(t, \xi) := (x(t), y(t))$ for all $t \in [0, \omega_\xi) := I_\xi \cap \mathbb{R}_+$. Then

$$x\dot{x} = x^3 \tanh(x)(1 - y) = \dot{y}(1 - y) = \dot{y} - y\dot{y}.$$

Integration yields $x^2(t) - \xi_1^2 = 2y(t) - 2\xi_2 - y^2(t) + \xi_2^2$ for all $t \in [0, \omega_\xi)$. Rearranging, we have

$$0 \leq x^2(t) = \|\xi\|^2 - 2\xi_2 + 2y(t) - y^2(t) \; \forall\, t \in [0, \omega_\xi),$$

whence boundedness of y and x. Therefore, by Theorem 4.11, $\omega_\xi = \infty$ and $O^+(\xi)$ is bounded. Moreover, since $\dot{y}(t) = x^3(t) \tanh(x(t)) \geq 0$ for all $t \in [0, \omega_\xi)$, y is non-decreasing. Combining this with the fact that y is bounded shows that $\lim_{t \to \infty} y(t) =: \lambda$ exists and is finite. Consequently,

$$\lim_{t \to \infty} \int_0^t x^3(s) \tanh(x(s)) \mathrm{d}s = \lim_{t \to \infty} (y(t) - \xi_2) = \lambda - \xi_2.$$

By the integral-invariance principle (Theorem 5.10) with $U = \mathbb{R}^2$ and g given by $g(z) = g(z_1, z_2) = z_1^3 \tanh(z_1)$ for all $z \in \mathbb{R}^2$, it follows that $\lim_{t \to \infty} x(t) = 0$. Note that any point of the form $(0, z_2)$ is an equilibrium point and thus, $g^{-1}(0) = \{(0, z_2) : z_2 \in \mathbb{R}\}$ is an invariant set.

Exercise 5.7

As in Example 5.3, introducing the function $f \colon \mathbb{R}^2 \to \mathbb{R}^2$ given by $f(z) = f(z_1, z_2) := (z_2, -b \sin z_1 - a z_2)$, the system may be expressed in the form $\dot{x} = f(x)$. Let φ denote the local flow generated by f. Define the vertical strip $S := (-\pi, \pi) \times \mathbb{R}$. By Example 5.3, the function $V \colon S \to \mathbb{R}$ given by

$$V(z) = V(z_1, z_2) := z_2^2 + 2b(1 - \cos z_1).$$

is a Lyapunov function with $V_f(z_1, z_2) = -2a z_2^2 \le 0$ for all $(z_1, z_2) \in S$, and so, the equilibrium 0 is stable. Consequently, there exists a neighbourhood $U \subset S$ of 0 such that, for every $\xi \in U$, the the closure of the semi-orbit $O^+(\xi)$ is contained in S. By Theorem 5.12, it follows that, for every $\xi \in U$, $\mathbb{R}_+ \subset I_\xi$, and moreover, as $t \to \infty$, $\varphi(t, \xi)$ approaches the largest invariant set M in $V_f^{-1}(0) = \{z = (z_1, z_2) \in S : z_2 = 0\}$. Let $z = (z_1, 0)$ be an arbitrary point of M and write $(x_1(t), x_2(t)) = \varphi(t, z)$ for all $t \in I_z$. Obviously, $x_2(t) = 0$ for all $t \in \mathbb{R}$. Therefore, $0 = \dot{x}_2(t) = -a x_2(t) - b \sin x_1(t) = -b \sin x_1(t)$ for all $t \in \mathbb{R}$. Since $x_1(t) \in (-\pi, \pi)$ for all $t \in \mathbb{R}$, it follows that $x_1(t) = 0$ for all $t \in \mathbb{R}$. In particular, $0 = x_1(0) = z_1$ and so, $z = 0$. Therefore $M = \{0\}$ and thus, $\varphi(t, \xi) \to 0$ as $t \to \infty$.

Exercise 5.9

Set $x(\cdot) := \varphi(\cdot, \xi)$. By continuity of V and compactness of $\mathrm{cl}(O^+(\xi))$, V is bounded on $O^+(\xi)$ and so the function $V \circ x$ is bounded. Since $(d/dt)(V \circ x))(t) = V_f(x(t)) \le 0$ for all $t \in \mathbb{R}_+$, $V \circ x$ is non-increasing. We conclude that the limit $\lim_{t \to \infty} V(x(t)) =: \lambda$ exists and is finite. Let $z \in \Omega(\xi)$ be arbitrary. Then there exists a sequence (t_n) in $\mathbb{R}_+$ such that $t_n \to \infty$ and $x(t_n) \to z$ as $n \to \infty$. By continuity of V, it follows that $V(z) = \lambda$. Consequently,

$$V(z) = \lambda \quad \forall z \in \Omega(\xi). \tag{$*$}$$

By invariance of $\Omega(\xi)$, if $z \in \Omega(\xi)$, then $\varphi(t, z) \in \Omega(\xi)$ for all $t \in \mathbb{R}$ and so $V(\varphi(t, z)) = \lambda$ for all $t \in \mathbb{R}$. Therefore, $V_f(\varphi(t, z)) = 0$ for all $t \in \mathbb{R}$. Since $\varphi(0, z) = z$ and z is an arbitrary point of $\Omega(\xi)$, it follows that

$$V_f(z) = 0 \quad \forall z \in \Omega(\xi), \tag{$**$}$$

and so $\Omega(\xi) \subset V_f^{-1}(0)$. The claim now follows because, by Theorem 4.38, $\Omega(\xi)$ is invariant and $x(t)$ approaches $\Omega(\xi)$ as $t \to \infty$.

    Comment. It might be tempting to conclude from $(*)$ that $(\nabla V)(z) = 0$ for all $z \in \Omega(\xi)$, which then immediately would yield $(**)$. However, this conclusion is not correct: the set $\Omega(\xi)$ is not open and therefore $(*)$ does not imply that $(\nabla V)(z) = 0$ for all $z \in \Omega(\xi)$. (The invalidity of the conclusion is illustrated by the following simple example: if $V(z) = \|z\|^2$ and $\Omega(\xi) = \{z \in \mathbb{R}^N : \|z\| = 1\}$, then $V(z) = 1$ for all $z \in \Omega(\xi)$, but $(\nabla V)(z) = 2z \ne 0$ for all $z \in \Omega(\xi)$.)

Exercise 5.10

(a) For $r^0, \theta^0 \in (0, \infty) \times [0, 2\pi)$, let $r(\cdot; r^0)$ and $\theta(\cdot; \theta^0)$ denote the unique maximal solutions of the initial-value problems

$$\dot{r} = r(1 - r), \ r(0) = r^0 \quad \text{and} \quad \dot{\theta} = \sin^2(\theta/2), \ \theta(0) = \theta^0,$$

respectively. Invoking separation of variables, a routine calculation shows that

$$r(t; r^0) = \frac{r^0}{r^0 + (1 - r^0)e^{-t}} \quad \forall t \ge 0,$$

and hence, $\lim_{t\to\infty} r(t; r^0) = 1$.

If $\theta^0 = 0$, then $\theta(t; \theta^0) = \theta(t; 0) = 0$ for all $t \in \mathbb{R}$ and the claim in part (i) follows. Assume now that $\theta^0 \in (0, 2\pi)$. Then, $\theta(t; \theta^0) < 2\pi$ for all $t \geq 0$, because otherwise there would exist $\tau > 0$ such that $\theta(\tau; \theta^0) = 2\pi$, in which case the initial-value problem

$$\dot\theta(t) = \sin^2(\theta(t)/2), \quad \theta(t_0) = 2\pi$$

would have two solutions on $\mathbb{R}$, namely, $\theta(\cdot\,; \theta^0)$ and $\theta(\cdot) = 2\pi$, contradicting uniqueness. Since $\theta(\cdot\,; \theta^0)$ is strictly increasing and $\theta(t; \theta) \in [\theta^0, 2\pi)$ for all $t \geq 0$, it follows that $\theta^* := \lim_{t\to\infty}\theta(t; \theta^0)$ exists and is contained in $(\theta^0, 2\pi]$. Suppose $\theta^* < 2\pi$. Then, $c := \sin^2(\theta^*/2) > 0$ and, for all $t > 0$ sufficiently large, $(d/dt)\theta(t; \theta^0) \geq c/2 > 0$ which contradicts the fact that $\theta(t; \theta^0) \in [\theta^0, 2\pi)$ for all $t \geq 0$. Therefore, $\lim_{t\to\infty}\theta(t; \theta^0) = 2\pi$. Since

$$\psi(t; (r^0, \theta^0)) = (r(t; r^0), \theta(t; \theta^0)) \ \forall\, t \geq 0,$$

the claim in part (ii) now follows.

(b) Writing $x = r\cos\theta$ and $y = r\sin\theta$, a straightforward calculation gives the system

$$\dot x = g(x,y)x - h(x,y)y, \quad \dot y = g(x,y)y + h(x,y)x$$

on $\mathbb{R}^2\backslash\{(0,0)\}$. The point $(1,0)$ is an equilibrium of this system. Denoting the corresponding local flow by ψ_c, it follows from (a) that
- $\lim_{t\to\infty}\psi_c(t, (x^0, y^0)) = (1, 0)$ for all $(x^0, y^0) \in \mathbb{R}^2 \setminus \{(0,0)\}$;
- $\|\psi_c(t, (\cos\theta^0, \sin\theta^0))\| = 1$ for all $t \geq 0$ and all $\theta^0 \in [0, 2\pi)$;
- for each $n \in \mathbb{N}$, there exists $t_n > 0$ such that $\psi_c(t_n, (\cos(1/n), \sin(1/n))) = (-1, 0)$.

(c) Applying the coordinate transformation $x \mapsto x + 1$ to the system in (b) yields the equivalent system

$$\dot x = g(x+1, y)(x+1) - h(x+1, y)y, \quad \dot y = g(x+1, y)y + h(x+1, y)(x+1)$$

on $G := \mathbb{R}^2\backslash\{(-1, 0)\}$, with equilibrium $(0, 0)$. Let φ denote the local flow generated by this system. Then, for all $(x^0, y^0) \in G$, $\varphi(t, (x^0, y^0)) = \psi_c(t, (x^0 + 1, y^0)) - (1, 0)$ and $\lim_{t\to\infty}\varphi(t, (x^0, y^0)) = 0$. Therefore, the equilibrium is globally attractive. To see that the equilibrium is not stable, define $\xi_n := (\cos(1/n) - 1, \sin(1/n))$. Then there exists $\delta > 0$ such that, for all $\xi \in G$ with $\|\xi\| \leq \delta$, $\|\varphi(t, \xi)\| \leq 1$ for all $t \geq 0$. For $n \in \mathbb{N}$, define $\xi_n := (\cos(1/n) - 1, \sin(1/n))$ and observe that, by the result in the third bullet item in (b),

$$\|\varphi(t_n, \xi_n)\| = \|\psi_c(t_n, (\cos(1/n), \sin(1/n))) - (1, 0)\| = \|(-2, 0)\| = 2.$$

Since $\xi_n \to (0, 0)$ as $n \to \infty$, it follows that the equilibrium $(0, 0)$ is not stable.

Exercise 5.12
(a) The Liénard system is of the form (5.8) with $g\colon \mathbb{R}^2 \to \mathbb{R}$ given by $g(z) = g(z_1, z_2) = k(z_1) + d(z_1)z_2$. By assumption, there exists $\varepsilon > 0$ such that $z_1 k(z_1) > 0$ and $d(z_1) > 0$ for all $z_1 \in (-\varepsilon, \varepsilon) \setminus \{0\}$. Define $U := (-\varepsilon, \varepsilon) \times (-\varepsilon, \varepsilon)$. Then $z_1 g(z_1, 0) = z_1 k(z_1) > 0$ for all $z_1 \in (-\varepsilon, \varepsilon) \setminus \{0\}$ and $\partial_2 g(z_1, z_2) = d(z_1) > 0$ for all $(z_1, z_2) \in U$ with $z_1 z_2 \neq 0$. Therefore, by the result in Example 5.16, $0 \in \mathbb{R}^2$ is an asymptotically stable equilibrium.

(b) Define $K\colon (-\varepsilon, \varepsilon) \to \mathbb{R}$ by $K(z_1) := \int_0^{z_1} k(s)\mathrm{d}s$. Observe that, by hypothesis (a), $k(s) \geq 0$ for all $s \in [0, \varepsilon)$ and $k(s) \leq 0$ for all $s \in (-\varepsilon, 0)$ which, together with continuity of k and hypothesis (b), ensures that $K(z_1) > 0$ for all $z_1 \in (-\varepsilon, \varepsilon) \setminus \{0\}$. Set $U := (-\varepsilon, \varepsilon) \times (-\varepsilon, \varepsilon)$ and define $V\colon U \to \mathbb{R}$ by $V(z) = V(z_1, z_2) := K(z_1) + z_2^2/2$. Define $f\colon \mathbb{R}^2 \to \mathbb{R}^2$ by $f(z) = f(z_1, z_2) := (z_2, -g(z)) = (z_2, -k(z_2) - d(z_1)z_2)$, in which case, the Liénard system may be expressed in the form $\dot{x} = f(x)$. We may now infer that $V(0) = 0$, $V(z) > 0$ for all $z \in U \setminus \{0\}$ and

$$V_f(z) = V_f(z_2, z_2) = k(z_1)z_2 + z_2\big(- k(z_1) - d(z_1)z_2\big) = -z_2^2 d(z_1) \leq 0 \ \ \forall z \in U.$$

By Theorem 5.2 (with $G = \mathbb{R}^2$), it follows that 0 is a stable equilibrium.

Finally, set $d = 0$ and let k be the identity map. In this case, the Liénard system reduces to the harmonic oscillator $\ddot{y} + y = 0$. Hypotheses (i) and (ii) clearly hold and so the equilibrium 0 is stable but is not asymptotically stable since (maximal) solutions of the harmonic oscillator have the property that $\|(y(t), \dot{y}(t))\| = \|(y(0), \dot{y}(0))\|$ for all $t \in \mathbb{R}$.

Exercise 5.13

Let U and V be as in Corollary 5.17. Stability of the equilibrium 0 is an immediate consequence of Theorem 5.2. The remaining issue is to establish attractivity. Let $\varepsilon > 0$ be such that $\overline{\mathbb{B}}(0, \varepsilon) \subset U$. By stability, there exists $\delta > 0$ such that, if $\xi \in \overline{\mathbb{B}}(0, \delta)$, then $x(t) \in \overline{\mathbb{B}}(0, \varepsilon)$ for all $t \in \mathbb{R}_+$ and for every maximal solution x with $x(0) = \xi$. Let x be any such solution. By boundedness of x and continuity of f, we may infer boundedness of $\dot{x}$ and so x is uniformly continuous. Since $V_f(x(t)) \leq 0$ for all $t \in \mathbb{R}_+$, it follows that $V \circ x$ is bounded ($0 \leq V(x(t)) \leq V(\xi)$ for all $t \in \mathbb{R}_+$) and non-increasing. Hence, $V \circ x$ converges, in particular, there exists $c \in [0, V(\xi)]$ such that $V(x(t)) \to c$ as $t \to \infty$. Therefore,

$$\lim_{t \to \infty} \int_0^t V_f(x(s))\mathrm{d}s = \lim_{t \to \infty} V(x(t)) - V(\xi) = c - V(\xi).$$

Furthermore, by continuity of V_f, together with uniform continuity and boundedness of x, $V_f \circ x$ is uniformly continuous. By Barbălat's lemma (Lemma 5.9), we may conclude that $V_f(x(t)) \to 0$ as $t \to \infty$. Seeking a contradiction, suppose that $x(t) \not\to 0$ as $t \to \infty$. Then there exist $\theta \in (0, \varepsilon)$ and a sequence (t_n) in $\mathbb{R}_+$ with $t_n \to \infty$ as $n \to \infty$ and $\|x(t_n)\| \geq \theta$. By continuity and negativity of V_f on the annulus $A := \{z \in U : \theta \leq \|z\| \leq \varepsilon\}$, there exists $\mu > 0$ such that $V_f(z) \leq -\mu$ for all $z \in A$. Therefore, $V_f(x(t_n)) \leq -\mu$ for all $n \in \mathbb{N}$, which contradicts the fact that $V_f(x(t)) \to 0$ as $t \to \infty$.

Exercise 5.14

By attractivity of the equilibrium, there exists $\varepsilon > 0$ such that

$$\lim_{t \to \infty} \varphi(t, \zeta) = 0 \ \ \forall \zeta \in \overline{\mathbb{B}}(0, 2\varepsilon). \tag{$*$}$$

Let $\xi \in \mathcal{A}$ be arbitrary. It suffices to show that ξ has a neighbourhood U contained in $\mathcal{A}$. Since $\xi \in \mathcal{A}$, there exists $T \geq 0$ such that $(T, \eta) \in \mathrm{dom}(\varphi)$ and $\|\varphi(T, \xi)\| \leq \varepsilon$. By openness of $\mathrm{dom}(\varphi)$ and continuity of φ (see Theorem 4.34), there exists $\delta > 0$ such that

$$\|\varphi(T, \eta) - \varphi(T, \xi)\| \leq \varepsilon \ \ \forall \eta \in U := \overline{\mathbb{B}}(\xi, \delta).$$

Therefore,

$$\|\varphi(T, \eta)\| \leq \|\varphi(T, \eta) - \varphi(T, \xi)\| + \|\varphi(T, \xi)\| \leq 2\varepsilon \ \ \forall \eta \in U.$$

By $(*)$, it follows that

$$\lim_{t \to \infty} \varphi(t + T, \eta) = \lim_{t \to \infty} \varphi(t, \varphi(T, \eta)) = 0 \ \ \forall \eta \in U.$$

Therefore, the neighbourhood U of ξ is contained in $\mathcal{A}$. Since $\xi \in \mathcal{A}$ is arbitrary, it follows that $\mathcal{A}$ is an open set.

Exercise 5.16

The Lorenz system is of the form $\dot{x} = f(x)$, with continuously differentiable $f: \mathbb{R}^3 \to \mathbb{R}^3$ given by

$$f(z) = f(z_1, z_2, z_3) := \left(\sigma(z_2 - z_1), rz_1 - z_2 - z_1 z_3, z_1 z_2 - bz_3\right)$$

with $\sigma > 0$, $b > 0$ and $0 < r < 1$. Consider the function $V: \mathbb{R}^3 \to \mathbb{R}$ given by $V(z) = V(z_1, z_2, z_3) := rz_1^2 + \sigma z_2^2 + \sigma z_3^2$. Clearly, $V(0) = 0$, $V(z) > 0$ for all $z \in \mathbb{R}^3 \backslash \{0\}$ and V is radially unbounded. Moreover,

$$V_f(z_1, z_2, z_3) = 2r\sigma z_1(z_2 - z_1) + 2\sigma z_2(rz_1 - z_2 - z_1 z_3) + 2\sigma z_3(z_1 z_2 - bz_3)$$

$$= -2\sigma(rz_1^2 - 2rz_1 z_2 + z_2^2) - 2b\sigma z_3^2 \quad \forall (z_1, z_2, z_3) \in \mathbb{R}^3,$$

Since $0 < r < 1$, we may choose ρ such that $0 < r < \rho < 1$. Write $\mu := \min\{r(1 - \rho), (1 - r/\rho)\}$. Then, $\mu > 0$ and, since $2z_1 z_2 \leq \rho z_1^2 + z_2^2/\rho$, we have

$$rz_1^2 - 2rz_1 z_2 + z_2^2 \geq r(1 - \rho)z_1^2 + (1 - r/\rho)z_2^2 \geq \mu\left(z_1^2 + z_2^2\right).$$

Therefore,

$$V_f(z) = V_f(z_1, z_2, z_3) \leq -2\sigma\left(\mu z_1^2 + \mu z_2^2 + bz_3^2\right) \leq 0 \; \forall z \in \mathbb{R}^3$$

Moreover, $V_f^{-1}(0) = \{0\}$. Hence, by Theorem 5.22, the equilibrium 0 is globally asymptotically stable.

Exercise 5.17

(a) A routine calculation gives $(\nabla V)(z_1, z_2) = 2\left(z_1, \, z_2(1 + z_2^2)^{-2}\right)$ for all $(z_1, z_2) \in \mathbb{R}^2$.

If $z_1^2 z_2^2 \geq 1$, then

$$V_f(z_1, z_2) = 2\left(-z_1^2 + z_2^2(1 + z_2^2)^{-2}\right) = 2(1 + z_2^2)^{-2}\left((1 - z_1^2 z_2^2)z_2^2 - z_1^2 - 2z_1^2 z_2^2\right) < 0.$$

If $z_1^2 z_2^2 < 1$ and $(z_1, z_2) \neq 0$, then

$$V_f(z_1, z_2) = 2\left(-z_1^2 + (2z_1^2 z_2^4 - z_2^2)(1 + z_2^2)^{-2}\right)$$

$$= 2(1 + z_2^2)^{-2}\left((z_1^2 z_2^2 - 1)z_2^2 - z_1^2 - 2z_1^2 z_2^2\right) < 0.$$

Clearly, $V(0) = 0$ and $V(z) > 0$ for all $z \in \mathbb{R}^2 \backslash \{0\}$. By Corollary 5.17, it follows that the equilibrium 0 is asymptotically stable.

(b) Let $\xi = (\xi_1, \xi_2) \in \mathbb{R}^2$ be such that $\xi_1^2 \xi_2^2 \geq 1$. Then $x: \mathbb{R} \to \mathbb{R}^2$ given by

$$x(t) = (x_1(t), x_2(t)) = (e^{-t}\xi_1, e^t\xi_2)$$

solves the initial-value problem $\dot{x} = f(x)$, $x(0) = \xi$. Indeed, $x(0) = \xi$. Also, $\dot{x}_1 = -x_1$, $\dot{x}_2 = x_2$ and $x_1^2(t)x_2^2(t) = \xi_1^2 \xi_2^2 \geq 1$, showing that $\dot{x} = f(x)$. Since $|x_2(t)| \to \infty$ as $t \to \infty$, we may conclude that 0 is not globally asymptotically stable.

(c) Setting $z_n = (0, n)$, it follows that $\|z_n\| = n \to \infty$ and $V(z_n) = n^2/(1 + n^2) \to 1$ as $n \to \infty$. Hence, V is not radially unbounded.

Exercise 5.18

Write $M = (M_{ij})$, where M_{ij} denotes the entry in row i and column j of M. Then, for $k = 1, \ldots, N$,

$$q(z) = \sum_{i=1}^{N} \sum_{j=1}^{N} M_{ij} z_i z_j = \sum_{i \neq k} \sum_{j \neq k} M_{ij} z_i z_j + \sum_{j \neq k} M_{kj} z_k z_j + \sum_{i \neq k} M_{ik} z_i z_k + M_{kk} z_k^2$$

and so

$$(\partial_k q)(z) = \sum_{j \neq k} M_{kj} z_j + \sum_{i \neq k} M_{ik} z_i + 2 M_{kk} z_k = \sum_{j=1}^{N} M_{kj} z_j + \sum_{i=1}^{N} M_{ik} z_i$$

$$= k\text{-th component of } (M + M^*) z.$$

Therefore, $(\nabla q)(z) = (M + M^*) z$ for all $z \in \mathbb{R}^N$.

Exercise 5.22
Define $f \colon \mathbb{R}^3 \to \mathbb{R}^3$ by

$$f(z) = \big(f_1(z_1, z_2, z_3), f_2(z_1, z_2, z_3), f_3(z_1, z_2, z_3) \big)$$

$$= \big(-2z_1 + z_1^2 |z_3| + z_2 , \; z_1 \sin z_3 - z_2 + 4z_3 , \; z_1 z_2 - z_2 z_3 - z_3 \big).$$

A straightforward calculation reveals that all (nine) first partial derivatives $\partial_i f_j(0)$ exist. However, f is not differentiable at points $(z_1, z_2, 0)$ with $z_1 \neq 0$ and so the hypotheses of Corollary 5.29 fail to hold.

Exercise 5.23
The Lorenz system is of the form $\dot{x} = f(x)$, with continuously differentiable $f \colon \mathbb{R}^3 \to \mathbb{R}^3$ given by

$$f(z) = f(z_1, z_2, z_3) := \big(\sigma(z_2 - z_1), r z_1 - z_2 - z_1 z_3, z_1 z_2 - b z_3 \big).$$

Therefore,

$$A := (Df)(0) = \begin{pmatrix} -\sigma & \sigma & 0 \\ r & -1 & 0 \\ 0 & 0 & -b \end{pmatrix}$$

with characteristic polynomial given by $(\lambda + b)(\lambda^2 + (\sigma + 1)\lambda + \sigma(1 - r))$. Given that $\sigma > 0$ and $r > 1$, it immediately follows that A has a positive eigenvalue. Therefore, by Theorem 5.31, 0 is an unstable equilibrium of the Lorenz system.

Exercise 5.24
As in Exercise 5.16, define $V \colon \mathbb{R}^3 \to \mathbb{R}$ by $V(0z) = V(z_1, z_2, z_3) := r z_1^2 + \sigma z_2^2 + \sigma z_3^2$. Writing $a_1 := \min\{r, \sigma\} > 0$ and $a_2 := \max\{r, \sigma\}$, we have $a_1 \|z\|^2 \leq V(z) \leq a_2 \|z\|^2$ for all $z \in \mathbb{R}^3$. Furthermore, by the calculation in the solution to Exercise 5.16, we have $V_f(z) \leq -a_3 \|z\|^2$ for all $z \in \mathbb{R}^3$, where $a_3 := 2\sigma \min\{\mu, b\} > 0$. Therefore, by Theorem 5.35, 0 is an exponentially stable equilibrium.

Exercise 5.26
The claim follows from a straightforward application of Proposition 4.20.

Exercise 5.29
(a) & (b) Since $\psi(\cdot, 0) = 0$ and $g(s)s \leq 0$ for all $s \in \mathbb{R}_+$, it follows that $0 \leq \psi(t, \xi) \leq \xi$ for all $\xi \in \mathbb{R}_+$ and all $t \in I_\xi \cap \mathbb{R}_+$, where I_ξ denotes the maximal interval of existence of the solution of the initial-value problem $\dot{x} = g(x)$, $x(0) = \xi$. Therefore, $\mathbb{R}_+ \subset I_\xi$ for all $\xi \in \mathbb{R}_+$ (by Theorem 4.11) and so $\mathbb{R}_+ \times \mathbb{R}_+ \subset \operatorname{dom}(\psi)$.

(c) Since $g(s)s < 0$ for all $s > 0$, we may infer that, for every, $\xi > 0$, $\psi(\cdot, \xi)$ is decreasing and $\psi(t, \xi) \to 0$ as $t \to \infty$. Moreover, if $0 \leq \xi_1 < \xi_2$, then $\psi(t, \xi_1) < \psi(t, \xi_2)$ for all $t \in \mathbb{R}_+$ (by Corollary 4.36). Define $\theta \colon \mathbb{R}_+ \times \mathbb{R}_= \to \mathbb{R}_+$ by $\theta(r, t) := \psi(t, r)$. Then, for each $r > 0$, $\theta(r, \cdot)$ is decreasing and $\theta(r, t) \to 0$ as $t \to \infty$. Moreover, for each $t \in \mathbb{R}_+$, $\theta(0, t) = 0$ and, if $0 \leq r_1 < r_2$, then $\theta(r_1, t) < \theta(r_2, t)$. Furthermore, by continuity of ψ, $\theta(\cdot, t)$ is continuous for each $t \in \mathbb{R}_+$. Therefore, for each $t \in \mathbb{R}_+$,

$\theta(\cdot, t)$ is a $\mathcal{K}$ function. We may now conclude that θ is of class $\mathcal{KL}$.

Exercise 5.30
Define $c \colon \mathbb{R}_+ \to \mathbb{R}_+$ by

$$c(s) := \sup\{|V_f(z, w)| \colon \|z\| \le b_1(s),\ \|w\| \le s\} \quad \forall\, s \in \mathbb{R}_+$$

and observe that c is non-decreasing, with $c(0) = 0$. Moreover, $b_3(s) = c(s) + b_2(b_1(s))$ for all $s \in \mathbb{R}_+$. Since $b_1, b_2 \in \mathcal{K}_\infty$, it follows that $b_2 \circ b_1$ is in $\mathcal{K}_\infty$. Therefore, to conclude that b_3 is in $\mathcal{K}_\infty$ it suffices to show that the function c is continuous. Continuity at $s = 0$ is clear. Let $s > 0$ be arbitrary. We will show that c is continuous at s. Let (s_n) be a sequence in $\mathbb{R}_+$ with $s_n \to s$ as $n \to \infty$. We may assume that $s_n > 0$ for all $n \in \mathbb{N}$. For each $n \in \mathbb{N}$, define $\rho_n := \min\{s_n, s\}$ and $\sigma_n := \max\{s_n, s\} \ge \rho_n > 0$. Then $|c(s) - c(s_n)| = c(\sigma_n) - c(\rho_n)$ for all $n \in \mathbb{N}$ and so, to conclude that c is continuous at s, it is sufficient to show that $\lim_{n \to \infty} \big(c(\sigma_n) - c(\rho_n)\big) = 0$. For each $n \in \mathbb{N}$, the set

$$K_n := \{(z, w) \colon \|z\| \le b_1(\sigma_n),\ \|w\| \le \sigma_n\}$$

is compact which, together with continuity of $(z, w) \mapsto |V_f(z, w)|$, ensures the existence of $(y_n, v_n) \in K_n$ such that $c(\sigma_n) = |V_f(y_n, v_n)|$. Define sequences (z_n) and (w_n) by

$$z_n := \frac{b_1(\rho_n)}{b_1(\sigma_n)}\, y_n, \quad w_n := \frac{\rho_n}{\sigma_n}\, v_n$$

and observe that

$$\|z_n\| \le b_1(\rho_n), \quad \|w_n\| \le \rho_n \quad \forall\, n \in \mathbb{N}.$$

Therefore,

$$|V_f(z_n, w_n)| \le c(\rho_n) \le c(\sigma_n) = |V_f(y_n, v_n)| \quad \forall\, n \in \mathbb{N}. \tag{$*$}$$

By boundedness of the sequence (σ_n), there exists $\sigma > 0$ such that $\sigma_n \le \sigma$ for all $n \in \mathbb{N}$. Define $K := \{(z, w) \colon \|z\| \le b_1(\sigma),\ \|w\| \le \sigma\}$. Then K is compact and is such that $(y_n, v_n), (z_n, w_n) \in K$ for all $n \in \mathbb{N}$. Let $\varepsilon > 0$ be arbitrary. Since V_f is uniformly continuous on K, there exists $\delta > 0$ such that, for all $(z, w), (y, v) \in K$,

$$\|z - y\| + \|w - v\| \le \delta \implies \big| |V_f(z, w)| - |V_f(y, v)| \big| \le \varepsilon. \tag{$**$}$$

Since $\lim_{n \to \infty} \rho_n = \lim_{n \to \infty} \sigma_n = s > 0$, we may infer that, as $n \to \infty$,

$$\|z_n - y_n\| = \left(1 - \frac{b_1(\rho_n)}{b_1(\sigma_n)}\right) \|y_n\| \to 0 \quad \text{and} \quad \|w_n - v_n\| = \left(1 - \frac{\rho_n}{\sigma_n}\right) \|v_n\| \to 0,$$

and so there exists $N \in \mathbb{N}$ such that $\|z_n - y_n\| + \|w_n - v_n\| \le \delta$ for all $n \ge N$. The conjunction of $(*)$ and $(**)$ now gives $0 \le c(\sigma_n) - c(\rho_n) \le \varepsilon$ for all $n \ge N$ and so $\lim_{n \to \infty}(c(\sigma_n) - c(\rho_n)) = 0$, completing the proof.

Exercise 5.31
It is clear that $a_1(0) = 0 = a_2(0)$ and that the functions a_1 and a_2 are non-decreasing and are continuous at 0. Let $s > 0$ be arbitrary. Let (s_n) be a sequence in $\mathbb{R}_+$ with $s_n \to s$ as $n \to \infty$. Since $s > 0$, we may assume that $s_n > 0$ for all $n \in \mathbb{N}$. Let $\varepsilon > 0$ be arbitrary. We will establish continuity at s of both a_1 and a_2 by showing that, for $i = 1, 2$, there exists $N \in \mathbb{N}$ such that

$$|a_i(s) - a_i(s_n)| \le \varepsilon \quad \forall\, n \ge N. \tag{$\dagger$}$$

For each $n \in \mathbb{N}$, define $\rho_n := \min\{s, s_n\} > 0$ and $\sigma_n := \max\{s, s_n\} \ge \rho_n$. Clearly, $\lim_{n \to \infty} \rho_n = \lim_{n \to \infty} \sigma_n = s > 0$ and so there exist $\rho > 0$ and $\sigma > 0$ such that $\rho \le \rho_n \le \sigma_n \le \sigma$ for all $n \in \mathbb{N}$. Observe that $\sigma_n / \rho_n \le \sigma / \rho$ for all $n \in \mathbb{N}$ and

$\sigma_n/\rho_n \to 1$ as $n \to \infty$. Since W is radially unbounded, there exists $r \geq \rho$ such that $W(y) > a_1(\sigma)$ for all y with $\|y\| > r$. Write $R := r\sigma/\rho \geq \sigma$ and set $K := \overline{\mathbb{B}}(0, R)$. Since W is uniformly continuous on K, there exists $\delta > 0$ such that, for all $y, z \in K$,

$$\|y - z\| \leq \delta \implies |W(y) - W(z)| \leq \varepsilon. \tag{$\dagger\dagger$}$$

(a) First, we prove continuity of a_2 at s. By continuity of W, for each $n \in \mathbb{N}$, there exists y_n, with $\|y_n\| \leq \sigma_n$, such that $a_2(\sigma_n) = \sup\{W(y)\colon \|y\| \leq \sigma_n\} = W(y_n)$. For each $n \in \mathbb{N}$, set $z_n := (\rho_n/\sigma_n)y_n$. Then $\|z_n\| \leq \rho_n$ and

$$0 \leq a_2(\sigma_n) - a_2(\rho_n) \leq W(y_n) - W(z_n) \ \ \forall\, n \in \mathbb{N}. \tag{$*$}$$

Observe that the sequences (y_n) and (z_n) are in K and, since $\rho_n/\sigma_n \to 1$ as $n \to \infty$, we have

$$\|y_n - z_n\| = \left(1 - \frac{\rho_n}{\sigma_n}\right)\|y_n\| \to 0 \ \text{ as } n \to \infty.$$

In particular, there exists $N \in \mathbb{N}$ so that $\|y_n - z_n\| \leq \delta$ for all $n \geq N$ which, in conjunction with ($*$) and ($\dagger\dagger$), gives

$$|a_2(s) - a_2(s_n)| = a_2(\sigma_n) - a_2(\rho_n) \leq \varepsilon \ \ \forall\, n \geq N.$$

Therefore, ($\dagger$) holds for $i = 2$ and so a_2 is continuous at s.
(b) Next, we prove that a_1 is continuous at s. Recall that, for all y with $\|y\| > r$, we have $W(y) > a_1(\sigma) \geq a_1(\rho_n)$ for all $n \in \mathbb{N}$. Therefore,

$$a_1(\rho_n) = \inf\{W(y)\colon \rho_n \leq \|y\|\} = \inf\{W(y)\colon \rho_n \leq \|y\| \leq r\} \ \ \forall\, n \in \mathbb{N}$$

and so there exists a sequence (y_n), such that $\rho_n \leq \|y_n\| \leq r$ and $a_1(\rho_n) = W(y_n)$ for all $n \in \mathbb{N}$. Define the sequence (z_n) by $z_n := (\sigma_n/\rho_n)y_n$. Then

$$\sigma_n \leq \frac{\sigma_n}{\rho_n}\|y_n\| = \|z_n\| \leq \frac{r\sigma}{\rho} = R \ \ \forall\, n \in \mathbb{N}.$$

Therefore,

$$0 \leq a_1(\sigma_n) - a_1(\rho_n) \leq W(z_n) - W(y_n) \ \ \forall\, n \in \mathbb{N}. \tag{$**$}$$

Observe that the sequences (y_n) and (z_n) are in K and, since $\sigma_n/\rho_n \to 1$ as $n \to \infty$, we have

$$\|y_n - z_n\| = \left(\frac{\sigma_n}{\rho_n} - 1\right)\|y_n\| \to 0 \ \text{ as } n \to \infty.$$

In particular, there exists $N \in \mathbb{N}$ so that $\|y_n - z_n\| \leq \delta$ for all $n \geq N$ which, in conjunction with ($**$) and ($\dagger\dagger$), gives

$$|a_1(s) - a_1(s_n)| = a_1(\sigma_n) - a_1(\rho_n) \leq \varepsilon \ \ \forall\, n \geq N.$$

Therefore, ($\dagger$) holds for $i = 1$ and so a_1 is continuous at s.

Exercise 5.32
(a) Define $V\colon \mathbb{R} \to \mathbb{R}$ by $V(z) = z^2/2$. Then

$$V_f(z, v) = -z^2(1 + 2z^2) + z(1 + z^2)v^2 = -z^4 + (1 + z^2)(zv^2 - z^2) \ \forall\, (z, v) \in \mathbb{R} \times \mathbb{R}.$$

Therefore, for $|z| \geq v^2$, we have $V_f(z, v) \leq -z^4$ and so, an application of Corollary 5.44 (with b_1 and b_2 given by $b_1(s) = s^2$ and $b_2(s) = s^4$) shows that the system is ISS.

(b) Define $V : \mathbb{R}^2 \to \mathbb{R}$ by $V(z) = V(z_1, z_2) := z_1^2/2 + z_2^4/4$. By Lemma 5.46, there exist $a_1, a_2 \in \mathcal{K}_\infty$ such that

$$a_1(\|z\|) \leq V(z) \leq a_2(\|z\|) \; \forall\, z \in \mathbb{R}^2.$$

Moreover,

$$V_f(z, v) = -z_1^2 - z_2^4 + z_1 z_2^2 + z_2^3 v \leq -z_1^2/2 - z_2^4/2 + z_2^3 v \; \forall\, (z, v) \in \mathbb{R}^2 \times \mathbb{R}.$$

Let $\mu > 0$. By Young's inequality[1]

$$z_2^3 v = (\mu z_2^3)(v/\mu) \leq (\mu z_2^3)^{4/3}/(4/3) + (v/\mu)^4/4 \; \forall\, (z_2, v) \in \mathbb{R} \times \mathbb{R},$$

and, setting $\mu = 3^{-3/4}$, we have $z_2^3 v \leq z_2^4/4 + 27v^4/4$. Therefore,

$$V_f(z, v) \leq -V(z) + 27v^4/4 \; \forall\, (z, v) \in \mathbb{R}^2 \times \mathbb{R}.$$

An application of Theorem 5.41 (with $a_3 = a_1$ and a_4 given by $a_4(s) = 27s^4/4$) shows that the system is ISS.

(c) Define $V : \mathbb{R}^2 \to \mathbb{R}$ by $V(z) = V(z_1, z_2) := \|z\|^2/2$. Then

$$V_f(z, v) = V_f(z_1, z_2, v_1, v_2) = -z_1^2 - z_2^4 + z_1 v_1 + z_2 v_2 \; \forall\, (z, v) \in \mathbb{R}^2 \times \mathbb{R}^2.$$

For all $(z_1, v_1), (z_2, v_2) \in \mathbb{R} \times \mathbb{R}$, $z_1 v_1 \leq (z_1^2 + v_1^2)/2$ and $z_2 v_2 \leq z_2^4/4 + 3v_2^{4/3}/4$ (by Young's inequality). Therefore, defining $W_1, W_2 : \mathbb{R}^2 \to \mathbb{R}_+$ by $W_1(z) := z_1^2/2 + 3z_2^4/4$ and $W_2(v) = W_2(v_1, v_2) := v_1^2/2 + 3v_2^{4/3}/4$, we have

$$V_f(z, v) \leq -W_1(z) + W_2(v) \; \forall\, (z, v) \in \mathbb{R}^2 \times \mathbb{R}^2.$$

By Lemma 5.46, there exist $a_3, a_4 \in \mathcal{K}_\infty$ such that

$$a_3(\|z\|) \leq W_1(z) \; \forall\, z \in \mathbb{R}^2 \;\; \text{and} \;\; W_2(v) \leq a_4(\|v\|) \; \forall\, v \in \mathbb{R}^2.$$

Therefore,

$$V_f(z, v) \leq -a_3(\|z\|) + a_4(\|v\|) \; \forall\, (z, v) \in \mathbb{R}^2 \times \mathbb{R}^2,$$

and so, by Theorem 5.41, it follows that the system is ISS.

Chapter 6

Exercise 6.1
By direct calculation

$$\mathcal{C}(A_c, b_c) = (b_c, A_c b_c, \ldots, A_c^{n-1} b_c) = \begin{pmatrix} 0 & 0 & \ldots & 0 & 1 \\ 0 & 0 & \ldots & 1 & * \\ \vdots & \vdots & \ddots & \vdots & \vdots \\ 0 & 1 & \ldots & * & * \\ 1 & * & \ldots & * & * \end{pmatrix},$$

and so $\operatorname{rk}\mathcal{C}(A_c, b_c) = N$. Hence, (A_c, b_c) is controllable.

---

[1] William Henry Young (1863-1942), English. Young's inequality says that if $a, b \geq 0$ and $p, q > 0$ are such that $1/p + 1/q = 1$, then $ab \leq a^p/p + b^q/q$.

Exercise 6.4

We first show that, for all $k \in \mathbb{N}$,

$$\hat{A}^k = A^k + \sum_{i=0}^{k-1} A^{k-1-i} b f^* \hat{A}^i \,. \tag{*}$$

For $k = 1$, formula $(*)$ reduces to $\hat{A} = A + bf^*$, which is trivially true (by the definition of $\hat{A}$). Assume now that formula $(*)$ is true for $k = m$. Then,

$$\hat{A}^{m+1} = \hat{A}\hat{A}^m = A\Big(A^m + \sum_{i=0}^{m-1} A^{m-1-i} bf^* \hat{A}^i\Big) + bf^* \hat{A}^m$$

$$= A^{m+1} + \sum_{i=0}^{m-1} A^{m-i} bf^* \hat{A}^i + bf^* \hat{A}^m$$

$$= A^{m+1} + \sum_{i=0}^{m} A^{m-i} bf^* \hat{A}^i \,,$$

which is $(*)$ for $k = m + 1$. We conclude that formula $(*)$ is true for all $k \in \mathbb{N}$.
Write $P(z) = \sum_{n=0}^{N} a_n z^n$, with $a_n \in \mathbb{R}$, $n = 1, \dots, N$ and $a_N = 1$. Using $(*)$, we obtain

$$a_n \hat{A}^n = a_n A^n + a_n \sum_{i=0}^{n-1} A^{n-1-i} bf^* \hat{A}^i, \quad n = 1, \dots, N.$$

Therefore, there exist $g_n \in \mathbb{R}^N$, $n = 0, \dots, N-1$, such that

$$P(\hat{A}) = \sum_{n=0}^{N} a_n \hat{A}^n = P(A) + bg_0^* + Abg_1^* + \cdots + A^{N-1} bf^*$$

By the Cayley-Hamilton theorem, $P(\hat{A}) = 0$, and thus,

$$P(A) = -\big(bg_0^* + Abg_1^* + \cdots + A^{n-2} bg_{n-2}^* + A^{N-1} bf^*\big)\,.$$

Writing $G := \big(g_0, g_1, \cdots, g_{N-1}, f\big) \in \mathbb{R}^{N \times N}$, the above formula for $P(A)$ can be written in the form $P(A) = -\mathcal{C}(A, b)G^*$ and so $G^* = -\mathcal{C}(A, b)^{-1} P(A)$, where $\mathcal{C}(A, b)^{-1}$ exists by controllability. Since the last row of G^* coincides with f^*, it follows that

$$f^* = -(0, \dots, 0, 1)\mathcal{C}(A, b)^{-1} P(A).$$

Exercise 6.6

(a) Let $S \subset \mathbb{R}^L$ be a proper algebraic set. Then there exists a real polynomial Γ in L variables, not equal to the zero polynomial, such that $S = \{z \in \mathbb{R}^L : \Gamma(z) = 0\}$. Set $S^c := \mathbb{R}^L \backslash S$. If $w \in S^c$, then $\Gamma(w) \neq 0$ and by continuity of Γ there exists a neighbourhood $W \subset \mathbb{R}^L$ of w such that $W \subset S^c$. Consequently, S^c is open. Next we show that S^c is dense in $\mathbb{R}^L$. Seeking a contradiction, suppose that S^c is not dense in $\mathbb{R}^L$. Then there exists $z \in S$ and an open neighbourhood $Z \subset \mathbb{R}^L$ of z such that $Z \subset S$. The polynomial Γ_0 defined by $\Gamma_0(s) := \Gamma(s + z)$ for all $s \in \mathbb{R}^L$ has the property that $\Gamma_0(s) = 0$ for all $s \in Z_0$, where $Z_0 := \{s - z : s \in Z\}$. Obviously, Z_0 is an open neighbourhood of 0 and it follows from repeated partial differentiation that all coefficients of Γ_0 are zero. Thus, Γ_0 is the zero polynomial and so is Γ, yielding the desired contradiction.

(b) We prove the claim by induction over L. Trivially, the claim is true for $L = 1$. Let S be a proper algebraic set in $\mathbb{R}^{L+1}$. Then there exists a non-zero polynomial Γ in $L+1$ variables such that $S = \{z \in \mathbb{R}^{L+1} : \Gamma(z) = 0\}$. Write Γ in the form

$$\Gamma(s_1, \ldots, s_{L+1}) = \sum_{i=0}^{k} \Delta_i(s_1, \ldots, s_L) s_{L+1}^i, \qquad (*)$$

where the Δ_i, $0 \le i \le k$, are polynomials in L variables. Set

$$Z := \bigcap_{i=1}^{k} Z_i, \quad \text{where } Z_i := \{z \in \mathbb{R}^L : \Delta_i(z) = 0\}, \quad 1 \le i \le k,$$

and let λ_L denote Lebesgue measure in $\mathbb{R}^L$. Since Γ is not the zero polynomial, there exists $j \in \{1, \ldots, k\}$ such that Δ_j is not the zero polynomial, and so, Z_j is a proper algebraic set in $\mathbb{R}^L$. By induction hypothesis, $\lambda_L(Z_j) = 0$, and consequently, $\lambda_L(Z) = 0$. Let $\sigma : \mathbb{R}^{L+1} \to \{0, 1\}$ be the characteristic function of S. Defining $\rho : \mathbb{R}^L \to \mathbb{R}$ by

$$\rho(s_1, \ldots, s_L) := \int_{-\infty}^{\infty} \sigma(s_1, \ldots, s_L, s_{L+1}) ds_{L+1},$$

it follows from Fubini's theorem[2] that

$$\lambda_{L+1}(S) = \int_{\mathbb{R}^{L+1}} \sigma(s_1, \ldots, s_{L+1}) ds_1 \ldots ds_{L+1} = \int_{\mathbb{R}^L} \rho(s_1, \ldots, s_L) ds_1 \ldots ds_L. \quad (**)$$

Note that if $(s_1, \ldots, s_L) \in \mathbb{R}^L \backslash Z$, then, invoking $(*)$, we conclude that there are at most finitely many (not more than k) numbers $z \in \mathbb{R}$ such that $(s_1, \ldots, s_L, z) \in S$. Therefore, $\rho(s_1, \ldots, s_L) = 0$ for all $(s_1, \ldots, s_L) \in \mathbb{R}^L \backslash Z$ and, since $\lambda_L(Z) = 0$, it now follows from $(**)$ that $\lambda_{L+1}(S) = 0$.

Exercise 6.7

The monic polynomial P is given by $P(s) = (s+1)(s+2)(s+5) = s^3 + 8s^2 + 17s + 10$. Set

$$v = \begin{pmatrix} 1 \\ 0 \end{pmatrix}, \quad b = Bv = \begin{pmatrix} 1 \\ 0 \\ 0 \end{pmatrix}, \quad E = \begin{pmatrix} 0 & 0 & 0 \\ 1 & 0 & 0 \end{pmatrix},$$

in which case we have

$$A + BE = \begin{pmatrix} 0 & 0 & 2 \\ 1 & 2 & 0 \\ 2 & 0 & 1 \end{pmatrix}, \quad \mathcal{C}(A + BE, b) = \begin{pmatrix} 1 & 0 & 4 \\ 0 & 1 & 2 \\ 0 & 2 & 2 \end{pmatrix}.$$

The matrix $\mathcal{C}(A + BE, b)$ has full rank and so $(A + BE, b)$ is controllable. Moreover,

$$\mathcal{C}(A + BE, b)^{-1} = \begin{pmatrix} * & * & * \\ * & * & * \\ 0 & 1 & -1/2 \end{pmatrix}, \quad P(A + BE) = \begin{pmatrix} 46 & 0 & 60 \\ 41 & 84 & 22 \\ 60 & 0 & 76 \end{pmatrix}.$$

Therefore,

$$f^* = -(0, 0, 1)\mathcal{C}(A + BE, b)^{-1} P(A + BE) = (-11, -84, 16)$$

---

[2] Guido Fubini (1897-1943), Italian.

and

$$F = E + vf^* = \begin{pmatrix} -11 & -84 & 16 \\ 1 & 0 & 0 \end{pmatrix}.$$

Exercise 6.10

Let $\lambda \in \sigma(A)$. If λ is uncontrollable, then an argument identical to that used in the proof of the necessity part of the eigenvalue-assignment theorem (Theorem 6.3) shows that $\lambda \in \sigma(A + BF)$ for all $F \in \mathbb{R}^{M \times N}$.

Conversely, assume that $\lambda \in \sigma(A + BF)$ for all $F \in \mathbb{R}^{M \times N}$. Then any monic real polynomial P of degree N such that $P(\lambda) \neq 0$ cannot be assigned to (A, B) and therefore, by the eigenvalue-assignment theorem, (A, B) is not controllable. If $B = 0$, then, trivially, $\mathrm{rk}(\lambda I - A, 0) = \mathrm{rk}(\lambda I - A) < N$, showing that λ is uncontrollable. Let $B \neq 0$. Then, without loss of generality, we may assume that A and B take the form (Kalman controllability decomposition, Lemma 3.10):

$$A = \begin{pmatrix} A_1 & A_2 \\ 0 & A_3 \end{pmatrix} \quad \text{and} \quad B = \begin{pmatrix} B_1 \\ 0 \end{pmatrix},$$

where the pair (A_1, B_1) is controllable. For every $F = (F_1, F_2) \in \mathbb{R}^{M \times N}$, we have

$$A + BF = \begin{pmatrix} A_1 + B_1 F_1 & A_2 + B_1 F_2 \\ 0 & A_3 \end{pmatrix}, \quad \sigma(A + BF) = \sigma(A_1 + B_1 F_1) \cup \sigma(A_3),$$

where the second identity follows form Theorem A.7. Since (A_1, B_1) is controllable, Theorem 6.3 ensures that we can choose F_1 such that $\lambda \notin \sigma(A_1 + B_1 F_1)$. Consequently, $\lambda \in \sigma(A_3)$ and thus, $\mathrm{rk}(\lambda I - A, B) < N$, showing that λ is uncontrollable.

Exercise 6.11

Since, for all $z \in \mathbb{R}^N$, $(\nabla V)(z) = Pz$ and $\langle Pz, Az \rangle = \langle PAz, z \rangle = \langle A^* Pz, z \rangle$, we have

$$\langle (\nabla V)(z), Az \rangle = \langle Pz, Az \rangle = \langle (PA + A^* P)z, z \rangle / 2. \tag{$*$}$$

It is now immediate that, if $PA + A^* P = 0$, then $\langle (\nabla V)(z), Az \rangle = 0$ for all $z \in \mathbb{R}^N$. Conversely, assume that $\langle (\nabla V)(z), Az \rangle = 0$ for all $z \in \mathbb{R}^N$. Then, by $(*)$, the matrix $Q := PA + A^* P$ satisfies $\langle Qz, z \rangle = 0$ for all $z \in \mathbb{R}^N$. Let $y, z \in \mathbb{R}^N$ be arbitrary. Exploiting the symmetry of Q, we have $\langle Qy, z \rangle = \langle Qz, y \rangle$. Therefore

$$0 = \langle Q(y + z), y + z \rangle = \langle Qy, y \rangle + \langle Qz, z \rangle + 2\langle Qy, z \rangle = 2\langle Qy, z \rangle.$$

and, since y and z are arbitrary, it follows that $Q = 0$.

Exercise 6.12

Note that

$$\mathrm{span}\{Az, Bz, z\} = \mathrm{span}\left\{ \begin{pmatrix} z_2 \\ -z_1 \end{pmatrix}, \begin{pmatrix} 0 \\ z_1 \end{pmatrix}, \begin{pmatrix} z_1 \\ z_2 \end{pmatrix} \right\} = \mathbb{R}^2 \ \forall z \in \mathbb{R}^2 \backslash \{0\}.$$

Since $\mathrm{ad}^1(A, B) = I$, it follows that

$$\mathrm{span}\{Az, Bz, \mathrm{ad}^1(A, B)z, \mathrm{ad}^2(A, B)z, \ldots\} = \mathrm{span}\{Az, Bz, z\} = \mathbb{R}^2 \ \forall z \in \mathbb{R}^2 \backslash \{0\}.$$

Noting that $A + A^* = 0$, it follows from Corollary 6.15 that the feedback law $u(t) = -\langle x(t), Bx(t) \rangle = -x_1(t)x_2(t)$ is globally asymptotically stabilizing.

Exercise 6.13

(a) Let $N = 1$, $A = 1$ and $S = \{1\} \subset \mathbb{R}$. Then S is A-invariant, but S is not positively

$\exp(At)$-invariant because $\exp(At) = e^t \neq 1$ for all $t > 0$.

Let $N = 1$, $A = -1$ and $S = (0, \infty) \subset \mathbb{R}$. For each $\xi \in S$, we have $\exp(At)\xi = e^{-t}\xi \in S$ for all $t \in \mathbb{R}$ and so S is $\exp(At)$-invariant. However, S is not A-invariant because, for each $\xi \in S$, $A\xi = -\xi \notin S$.

(b) Let $S \subset \mathbb{R}^N$ be a subspace; since S is finite dimensional, it is closed. Assume that S is A-invariant. Set $E_n(t) := \sum_{k=0}^n (1/k!)(At)^k$ for all $n \in \mathbb{N}$ and let $\xi \in S$. Since S is an A-invariant subspace, we have $E_n(t)\xi \in S$ for all $n \in \mathbb{N}$ and all $t \in \mathbb{R}_+$. By closedness of S, it follows that $\lim_{n\to\infty} E_n(t)\xi = \exp(At)\xi$ is in S for all $t \in \mathbb{R}_+$. Now assume that the subspace S is positively $\exp(At)$-invariant. Let $\xi \in S$ be arbitrary. Then, for each $n \in \mathbb{N}$, $\zeta_n := n(\exp(An^{-1}) - I)\xi$ is in S and so, by closedness of S, $A\xi = \lim_{n\to\infty} \zeta_n \in S$. Therefore, S is A-invariant.

(c) Let $N = 1$, $A = 1$ and $S = [1, \infty) \subset \mathbb{R}$. For each $\xi \in S$, we have $\exp(At)\xi = e^t\xi \in S$ for all $t \in \mathbb{R}_+$ and so S is positively $\exp(At)$-invariant. However, S is not $\exp(At)$-invariant because, for each $\xi \in S$, $\exp(At)\xi = e^t\xi \to 0$ as $t \to -\infty$.

(d) Let $S \subset \mathbb{R}^N$ be a subspace. As a finite-dimensional subspace S is closed. By part (b), if S is positively $\exp(At)$-invariant, then S is A-invariant, and thus, by the closedness and subspace property of S, we conclude that $\exp(At)\xi = \sum_{k=0}^{\infty}(1/k!)(At)^k\xi$ is in S for all $\xi \in S$ and all $t \in \mathbb{R}$.

Exercise 6.14

(a) Writing $x_1(t) = y(t)$, $x_2(t) = \dot{y}(t)$ and $x_3(t) = z(t)$, we have $\dot{x}(t) = Ax(t) + u(t)Bx(t)$ with $A, B \in \mathbb{R}^{3\times3}$ as given.

(b) By direct calculation, we have

$$\mathrm{ad}^1(A, B) = [A, B] = AB - BA = \begin{pmatrix} 0 & -1 & 0 \\ -1 & 0 & 0 \\ 0 & 0 & 0 \end{pmatrix},$$

$$\mathrm{ad}^2(A, B) = [A, \mathrm{ad}^1(A, B)] = \begin{pmatrix} -2 & 0 & 0 \\ 0 & 2 & 0 \\ 0 & 0 & 0 \end{pmatrix}.$$

By induction, we find that, for all $k \in \mathbb{N}$,

$$\mathrm{ad}^{2k-1}(A, B) = (-4)^{k-1}\mathrm{ad}^1(A, B), \quad \mathrm{ad}^{2k}(A, B) = (-4)^{k-1}\mathrm{ad}^2(A, B).$$

Therefore,

$$\mathrm{span}\{Az, Bz, \mathrm{ad}^1(A, B)z, \ldots\} = \mathrm{span}\{Az, Bz, \mathrm{ad}^1(A, B)z, \mathrm{ad}^2(A, B)z\}$$

$$= \mathrm{span}\left\{ \begin{pmatrix} z_2 \\ -z_1 \\ 0 \end{pmatrix}, \begin{pmatrix} 0 \\ -z_2 \\ z_3 \end{pmatrix}, \begin{pmatrix} -z_2 \\ -z_1 \\ 0 \end{pmatrix}, \begin{pmatrix} -z_1 \\ z_2 \\ 0 \end{pmatrix} \right\}$$

which is not equal to $\mathbb{R}^3$ for all $z \in \mathbb{R}^3$ of the form $z = (z_1, z_2, 0)$. Therefore, the hypotheses of Corollary 6.15 fail to hold.

(c) Set $\Omega := \{(z_1, z_2, z_3) = z \in \mathbb{R}^3 : z_3(z_1^2 + z_2^2) \neq 0\}$. Observe that $A + A^* = 0$ and

$$\mathrm{span}\{Az, Bz, \mathrm{ad}^1(A, B)z, \mathrm{ad}^2(A, B)z\} = \mathbb{R}^3 \quad \forall z \in \Omega.$$

Setting $\Gamma := \{z \in \mathbb{R}^3 : \langle z, Bz \rangle = 0\} = \{(z_1, z_2, z_3) \in \mathbb{R}^3 : z_2^2 - z_3^2 = 0\}$, we see that $(\mathbb{R}^3 \backslash \Omega) \cap \Gamma = \{(z_1, z_2, z_3) \in \mathbb{R}^3 : z_2 = 0 = z_3\}$ and the only positively $\exp(At)$-invariant subset thereof is $\{0\}$. Therefore, by Theorem 6.15, we may conclude that the feedback $u(t) = -\langle x(t), Bx(t) \rangle = x_2^2(t) - x_3^2(t) = \dot{y}^2(t) - z^2(t)$ is globally asymptotically stabilizing.

Exercise 6.16
Noting that, for all $s \in \mathbb{C}$ and all $\alpha \in \mathbb{R}$, $\mathrm{rk}\,(sI - A, b) = \mathrm{rk}\,(sI - (A - \alpha bc^*), b)$ and $\mathrm{rk}\,(sI - A^*, c) = \mathrm{rk}\,(sI - (A - \alpha bc^*)^*, c)$, the requisite results follow by the Hautus criteria for controllability and observability (Theorems 3.11 and 3.21).

Exercise 6.18
Setting $A_\alpha := A - \alpha bc^*$, then, as in the proof of Theorem 6.17, (A_α, b, c^*) is a minimal realization of the strictly-proper rational function $\hat{G}_\alpha := \hat{G}/(1 + \alpha\hat{G})$. By assumption $\hat{G}_\alpha$ is positive real and so, by the positive real lemma (Lemma 6.18), there exist a symmetric positive-definite matrix $P \in \mathbb{R}^{N \times N}$ and a vector $l \in \mathbb{R}^N$ such that $PA_\alpha + A_\alpha^* P = -ll^*$ and $Pb = c$. Let k_α, f, φ and V be as in the proof of Theorem 6.17. Then

$$V_f(z) = \langle (\nabla V)(z), A_\alpha z - bk_\alpha(c^* z) \rangle = -(l^* z)^2 - 2(c^* z)k_\alpha(c^* z)$$
$$\leq -2(c^* z)k_\alpha(c^* z) \ \ \forall z \in \mathbb{R}^N.$$

(a) Assume $k \in S[\alpha, \infty)$. Then $wk_\alpha(w) = wk(w) - \alpha w^2 \geq 0$ for all $w \in \mathbb{R}$ and so $V_f(z) \leq 0$ for all $z \in \mathbb{R}^N$. By the same argument as that used in the proof of Theorem 6.17, it follows that (6.39) holds.
(b) Now assume that $k \in S(\alpha, \infty)$. Then $wk_\alpha(w) > 0$ for all $w \in \mathbb{R}\backslash\{0\}$. Therefore, $V_f(z) < 0$ for all $z \notin \ker c^*$ and so $V_f^{-1}(0) \subset \ker c^*$. The same argument (based on LaSalle's invariance principle) as that used in the proof of Theorem 6.17 now applies to conclude that the equilibrium is globally asymptotically stable.

Exercise 6.19
In this case,

$$A = \begin{pmatrix} 0 & 1 \\ -\mu & 0 \end{pmatrix}, \quad b = c = \begin{pmatrix} 0 \\ 1 \end{pmatrix}$$

and so $\hat{G}$ is given by $\hat{G}(s) = s/(s^2 + \mu)$, which has simple poles at $\pm i\sqrt{\mu}$, each with residue $1/2$. Moreover, $\mathrm{Re}\,\hat{G}(i\omega) = 0$ for all $w \neq \pm\sqrt{\mu}$. By Lemma 6.16, $\hat{G}$ is positive real and the requisite results follow from Theorem 6.19 (with $\alpha = 0$).

Exercise 6.20
(a) Let $\tilde{A}$, $\tilde{b}$ and $\tilde{c}$ be as in the proof of Theorem 6.21. Furthermore, let $f: \mathbb{R}^{N+1} \to \mathbb{R}^{N+1}$ be the locally Lipschitz function given by

$$f(z) := \tilde{A}z - \tilde{b}k(\tilde{c}^* z) + \tilde{d}, \quad \text{where } \tilde{d} := \begin{pmatrix} 0 \\ \gamma\rho \end{pmatrix} \in \mathbb{R}^{N+1}.$$

Then the initial-value problem (6.49) may be expressed in the form

$$\dot{\eta}(t) = f(\eta(t)), \quad \eta(0) = \begin{pmatrix} \xi \\ \zeta \end{pmatrix}, \quad \text{where } \eta(t) := \begin{pmatrix} x(t) \\ u(t) \end{pmatrix}.$$

By the global Lipschitz property of k, there exists $\lambda > 0$ such that $|k(\tilde{c}^* z) - k(0)| \leq \lambda|\tilde{c}^* z| \leq \lambda\|\tilde{c}\|\|z\|$ for all $z \in \mathbb{R}^{N+1}$. Therefore,

$$\|f(z)\| \leq \|\tilde{A}\|\|z\| + \lambda\|\tilde{b}\|\|\tilde{c}\|\|z\| + \|\tilde{b}\||k(0)| + \|\tilde{d}\| \ \ \forall z \in \mathbb{R}^{N+1}.$$

Writing $L := \max\{\|\tilde{A}\| + \lambda\|\tilde{b}\|\|\tilde{c}\|, \|\tilde{b}\||k(0)| + \|\tilde{d}\|\}$, we have

$$\|f(z)\| \leq L(1 + \|z\|) \ \ \forall z \in \mathbb{R}^{N+1}.$$

By Proposition 4.12, it now follows that the maximal solution of the initial-value problem (6.49) has interval of existence $\mathbb{R}$.

(b) For all $t \in \mathbb{R}_+$, we have

$$\dot{z}(t) = \dot{x}(t) = Ax(t) + bk(u(t)) = Az(t) + b\big(k(u(t)) - k(u^\rho)\big)$$
$$= Az(t) + b\big(k(v(t) + u^\rho) - k(u^\rho)\big) = Az(t) + b\tilde{k}(v(t))$$

and

$$\dot{v}(t) = \dot{u}(t) = \gamma\big(\rho - c^* x(t)\big) = \gamma\big(\rho - c^* z(t) + c^* A^{-1} bk(u^\rho)\big)$$
$$= \gamma\big(\rho - c^* z(t) - \hat{G}(0)k(u^\rho)\big) = \gamma\big(\rho - c^* z(t) - \rho\big) = -\gamma c^* z(t).$$

(c) Let $s \in \mathbb{C}$ and $z \in \mathbb{C}^N$ be arbitrary and assume that $z^*(sI - \tilde{A}, \tilde{b}) = 0$, where $\tilde{A}$ and $\tilde{b}$ are given by (6.53). By the Hautus criterion for controllability, it is sufficient to show that $z = 0$. Writing $z^* = (w^*, \bar{v})$, where $w \in \mathbb{C}^N$ and $v \in \mathbb{C}$, we obtain

$$z^*(sI - \tilde{A}, \tilde{b}) = (w^*(sI - A) + \bar{v}\gamma c^*, \ s\bar{v}, \ -w^* b) = 0.$$

Assume that $s \neq 0$. Then $v = 0$, and thus $w^*(sI - A, b) = 0$. Since (A, b) is controllable, the Hautus criterion for controllability implies that $w = 0$, and hence, $z = 0$. Now assume that $s = 0$. Then

$$-w^* A + \bar{v}\gamma c^* = 0, \quad w^* b = 0, \tag{$*$}$$

and consequently,

$$\bar{v}\gamma\hat{G}(0) = -\bar{v}\gamma c^* A^{-1} b = 0.$$

Since $\gamma\hat{G}(0) > 0$, we now conclude that $v = 0$. By ($*$), $w^*(-A, b) = 0$, and so controllability of (A, b) together with the Hautus criterion yields that $w = 0$, and hence, $z = 0$.

(d) Observability of $(\tilde{c}^*, \tilde{A})$ follows from an argument similar to that employed in the solution of part (c).

Bibliography

[1] R.P. Agarwal and D. O'Regan, *An Introduction to Ordinary Differential Equations*, Springer-Verlag, New York, 2008.

[2] H. Amann. *Ordinary Differential Equations*, Walter de Gruyter, Berlin, 1990.

[3] P.J. Antsaklis and A.N. Michel. *A Linear Systems Primer*, Birkhäuser, Boston, 2007.

[4] M. Braun. *Differential Equations and Their Applications*, 4th edition, Springer-Verlag, New York, 1993.

[5] T.W. Gamelin. *Complex Analysis*, Springer-Verlag, New York, 2001.

[6] E. Hairer and G. Wanner. *Analysis by its History*, Springer-Verlag, New York, 1996.

[7] P.R. Halmos. *Naive Set Theory*, Springer-Verlag, New York, 1994.

[8] D. Hinrichsen and A.J. Pritchard. *Mathematical Systems Theory I*, Springer-Verlag, Berlin, 2005.

[9] I. Kaplansky. *Set Theory and Metric Spaces*, 2nd edition, Chelsea Publishing Company, New York, 1977.

[10] W.G. Kelley and A.C. Peterson, *The Theory of Differential Equations: Classical and Qualitative*, 2nd Edition, Springer-Verlag, New York, 2010.

[11] S. Lang. *Linear Algebra*, 3rd edition, Springer-Verlag, New York, 2010.

[12] P.D. Lax. *Linear Algebra and Its Applications*, 2nd edition, John Wiley & Sons, Hoboken, NJ, 2007.

H. Logemann and E. P. Ryan, *Ordinary Differential Equations*,
Springer Undergraduate Mathematics Series,
DOI: 10.1007/978-1-4471-6398-5, © Springer-Verlag London 2014

[13] C.D. Meyer. *Matrix Analysis and Applied Linear Algebra*, SIAM, Philadelphia, 2000.

[14] R.K. Miller and A.N. Michel. *Ordinary Differential Equations*, Dover Publications, Mineola, NY, 2007.

[15] J.W. Polderman and J.C. Willems. *Introduction to Mathematical Systems Theory*, Springer-Verlag, New York, 1998.

[16] H.A. Priestley. *An Introduction to Complex Analysis*, 2nd edition, Oxford University Press, Oxford, 2003.

[17] M.H. Protter and C.B. Morrey. *A First Course in Real Analysis*, 2nd Edition, Springer-Verlag, New York, 1991.

[18] T.C. Sideris, *Ordinary Differential Equations and Dynamical Systems*, Atlantis Press, Paris, 2013.

[19] E.D. Sontag. *Mathematical Control Theory*, 2nd edition, Springer-Verlag, New York, 1998.

[20] W.J. Terrell. *Stability and Stabilization*, Princeton University Press, Princeton, NJ, 2009.

[21] W. Walter. *Ordinary Differential Equations*, Springer-Verlag, New York, 1998.

Index

H. Logemann and E. P. Ryan, *Ordinary Differential Equations*,
Springer Undergraduate Mathematics Series,
DOI: 10.1007/978-1-4471-6398-5, © Springer-Verlag London 2014